Compendium of Plant Genomes

More information about this series at http://www.springer.com/series/11805

Takeshi Nishio · Hiroyasu Kitashiba
Editors

The Radish Genome

Springer

Editors
Takeshi Nishio
Graduate School of Agricultural Science
Tohoku University
Sendai, Miyagi
Japan

Hiroyasu Kitashiba
Graduate School of Agricultural Science
Tohoku University
Sendai, Miyagi
Japan

ISSN 2199-4781 ISSN 2199-479X (electronic)
Compendium of Plant Genomes
ISBN 978-3-319-59252-7 ISBN 978-3-319-59253-4 (eBook)
DOI 10.1007/978-3-319-59253-4

Library of Congress Control Number: 2017948205

Printed on acid-free paper

This Springer imprint is published by Springer Nature
The registered company is Springer International Publishing AG
The registered company address is: Gewerbestrasse 11, 6330 Cham, Switzerland

This book series is dedicated to
my wife Phullara, and our children Sourav,
and Devleena

Chittaranjan Kole

Preface to the Series

Genome sequencing has emerged as the leading discipline in the plant sciences coinciding with the start of the new century. For much of the twentieth century, plant geneticists were only successful in delineating putative chromosomal location, function, and changes in genes indirectly through the use of a number of 'markers' physically linked to them. These included visible or morphological, cytological, protein, and molecular or DNA markers. Among them, the first DNA marker, the RFLPs, introduced a revolutionary change in plant genetics and breeding in the mid-1980s, mainly because of their infinite number and thus potential to cover maximum chromosomal regions, phenotypic neutrality, absence of epistasis, and codominant nature. An array of other hybridization-based markers PCR-based markers, and markers based on both facilitated construction of genetic linkage maps, mapping of genes controlling simply inherited traits and even gene clusters (QTLs) controlling polygenic traits in a large number of model and crop plants. During this period a number of new mapping populations beyond F2 were utilized and a number of computer programs were developed for map construction, mapping of genes, and for mapping of polygenic clusters or QTLs. Molecular markers were also used in studies of evolution and phylogenetic relationship, genetic diversity, DNA-fingerprinting and map-based cloning. Markers tightly linked to the genes were used in crop improvement employing the so-called marker-assisted selection. These strategies of molecular genetic mapping and molecular breeding made a spectacular impact during the last one and a half decades of the twentieth century. But still they remained 'indirect' approaches for elucidation and utilization of plant genomes since much of the chromosomes remained unknown and the complete chemical depiction of them was yet to be unraveled.

Physical mapping of genomes was the obvious consequence that facilitated development of the 'genomic resources' including BAC and YAC libraries to develop physical maps in some plant genomes. Subsequently, integrated genetic-physical maps were also developed in many plants. This led to the concept of structural genomics. Later on, emphasis was laid on EST and transcriptome analysis to decipher the function of the active gene sequences leading to another concept defined as functional genomics. The advent of techniques of bacteriophage gene and DNA sequencing in the 1970s was extended to facilitate sequencing of these genomic resources in the last decade of the twentieth century.

As expected, sequencing of chromosomal regions would have led to too much data to store, characterize, and utilize with the-then available computer software could handle. But development of information technology made the life of biologists easier by leading to a swift and sweet marriage of biology and informatics and a new subject was born—bioinformatics.

Thus, evolution of the concepts, strategies and tools of sequencing and bioinformatics reinforced the subject of genomics—structural and functional. Today, genome sequencing has traveled much beyond biology and involves biophysics, biochemistry and bioinformatics!

Thanks to the efforts of both public and private agencies, genome sequencing strategies are evolving very fast, leading to cheaper, quicker and automated techniques right from clone-by-clone and whole-genome shotgun approaches to a succession of second generation sequencing methods. Development of software of different generations facilitated this genome sequencing. At the same time, newer concepts and strategies were emerging to handle sequencing of the complex genomes, particularly the polyploids.

It became a reality to chemically—and so directly—define plant genomes, popularly called whole-genome sequencing or simply genome sequencing.

The history of plant genome sequencing will always cite the sequencing of the genome of the model plant *Arabidopsis thaliana* in 2000 that was followed by sequencing the genome of the crop and model plant rice in 2002. Since then, the number of sequenced genomes of higher plants has been increasing exponentially, mainly due to the development of cheaper and quicker genomic techniques and, most importantly, development of collaborative platforms such as national and international consortia involving partners from public and/or private agencies.

As I write this preface for the first volume of the new series "Compendium of Plant Genomes", a net search tells me that complete or nearly-complete whole-genome sequencing of 45 crop plants, eight crop and model plants, eight model plants, 15 crop progenitors and relatives, and three basal plants are accomplished, the majority of which are in the public domain. This means that we nowadays know many of our model and crop plants chemically, i.e., directly, and we may depict them and utilize them precisely better than ever. Genome sequencing has covered all groups of crop plants. Hence, information on the precise depiction of plant genomes and the scope of their utilization is growing rapidly every day. However, the information is scattered in research articles and review papers in journals and dedicated web pages of the consortia and databases. There is no compilation of plant genomes and the opportunity of using the information in sequence-assisted breeding or further genomic studies. This is the underlying rationale for starting this book series, with each volume dedicated to a particular plant.

Plant genome science has emerged as an important subject in academia, and the present compendium of plant genomes will be highly useful both to students and teaching faculties. Most importantly, research scientists involved in genomics research will have access to systematic deliberations on the plant genomes of their interest. Elucidation of plant genomes is not onlyof interest for the geneticists and breeders, but also for practitioners of an array of plant science disciplines, such as taxonomy, evolution, cytology, physiology, pathology, entomology, nematology, crop production, bio-chemistry, and obviously bioinformatics. It must be mentioned that information regarding each plant genome is ever-growing. The contents of the volumes of this compendium are therefore focusing on the basic aspects of the genomes and their utility. They include information on the academic and/ or economic importance of the plants, description of their genomes from a molecular genetic and cytogenetic point of view, and the genomic resources developed. Detailed deliberations focus on the background history of the national and international genome initiatives, public and private partners involved, strategies and genomic resources and tools utilized, enumeration on the sequences and their assembly, repetitive sequences, gene annotation, and genome duplication. In addition, synteny with other sequences, comparison of gene families, and, most importantly, potential of the genome sequence information for gene pool characterization through genotyping by sequencing (GBS) and genetic improvement of crop plants have been described. As expected, there is a lot of variation of these topics in the volumes based on the information available on the crop, model, or reference plants.

I must confess that as the series editor it has been a daunting task for me to work on such a huge and broad knowledge base that spans so many diverse plant species. However, pioneering scientists with life-time experience and expertise on the particular crops did excellent jobs editing the respective volumes. I myself have been a small science worker on plant genomes since the mid-1980s and that provided me the opportunity to personally know several stalwarts of plant genomics from all over the globe. Most, if not all, of the volume editors are my longtime friends and colleagues. It has been highly comfortable and enriching for me to work with them on this book series. To be honest, while working on this series I have been and will remain a student first, a science worker second, and a series editor last. And I must express my gratitude to the volume editors and the chapter authors for providing me the opportunity to work with them on this compendium.

I also wish to mention here my thanks and gratitude to the Springer staff, Dr. Christina Eckey and Dr. Jutta Lindenborn in particular, for all their constant and cordial support right from the inception of the idea.

I always had to set aside additional hours to edit books besides my professional and personal commitments—hours I could and should have given to my wife, Phullara, and our kids, Sourav, and Devleena. I must mention that they not only allowed me the freedom to take away those hours from them but also offered their support in the editing job itself. I am really not sure whether my dedication of this compendium to them will suffice to do justice to their sacrifices for the interest of science and the science community.

Kalyani, India Chittaranjan Kole

Preface

Radish (*Raphanus sativus* L.) is an important vegetable in East Asia, e.g., Japan, China, and Korea, but not so much in Europe, America, Africa, or Oceania. Therefore, studies on radish are predominantly performed by Asian researchers, and not by Europeans or Americans. In the East Asia, radish roots are thick, long, and mostly white, whereas those are tiny and mostly red in Europe. The East Asian type is called East Asian big long radish, white radish, or daikon (in Japanese), and the European radish type is called European small radish or European garden radish. In this e-book, we use terms "Asian big radish" and "European small radish" for them. In Asian big radish, there is a great variation in root size and shape, especially in Japanese radish. In Southeast Asia, pods are consumed as a vegetable, and seeds are used as oil seeds. On the other hand, *Raphanus raphanistrum*, a species in the same genus as radish, is a serious weed in America and Australia, but not so much in Asia. The recent globalization will change the food culture, and Asian big radish may become an important vegetable in the western countries.

Radish is closely related to the genus *Brassica*, which contains many important crops such as rapeseed, cabbage, broccoli, Chinese cabbage, turnip, and mustard. Genetic and genomic studies of *Brassica* have been intensively carried out, but these studies of radish have been left behind those of *Brassica*. Phylogenetic relationship of radish with *Brassica* species is not established. Although radish is crossable with *R. raphanistrum* and hybrids between radish and *R. raphanistrum* are fully fertile, they have different species names. Asian wild radish growing at seashores is *R. sativus* var. *raphanistroides*. Relationships between the wild radish and cultivated radish are controversial. Besides the importance of radish as a crop itself, radish and wild radish are important genetic resources for breeding of *Brassica* crops.

Development of the next-generation sequencers accelerated whole-genome sequencing of many crop species. Two groups in Japan and one Korean group carried out the whole-genome sequencing of radish independently. Although the group of Tohoku University and Kazusa DNA Research Institute, where high-density linkage maps of DNA markers were constructed to form a standard linkage map, published a draft genome sequence of radish first, the other Japanese group and the Korean group published radish genome sequences with more data soon after that. These genome sequence data will contribute to identification of genes important for breeding of radish and

Brassica crops. Since these three groups joined in publishing this e-book, we hope this cooperative book publication accelerates collaboration of them for integration of three radish genome sequences to make a standard reference genome sequence of radish.

Dr. Chittaranjan Kole recommended us to edit a book of "The Radish Genome". For publishing this book, we have been supported by Dr. Christina Eckey of Springer. We are much grateful to them for their helpful suggestions and encouragements.

Miyagi, Japan

Takeshi Nishio
Hiroyasu Kitashiba

Contents

Contributors

Sang Woo Bang Faculty of Agriculture, Utsunomiya University, Utsunomiya, Japan

Su Ryun Choi Molecular Genetics and Genomics Laboratory, Department of Horticulture, Chungnam National University, Daejeon, Korea

Katsunori Hatakeyama Faculty of Agriculture, Iwate University, Morioka, Japan

Hideki Hirakawa Kazusa DNA Research Institute, Kisarazu, Chiba, Japan

Masahiko Ishida Institute of Vegetable and Floriculture Science, NARO, Tsukuba, Japan

Young-Min Jeong Department of Life Science, The Catholic University of Korea, Bucheon, Korea

Tomohiro Kakizaki Institute of Vegetable and Floriculture Science, NARO, Tsu, Mie, Japan

Hiroyasu Kitashiba Graduate School of Agricultural Science, Tohoku University, Sendai, Japan

Xixiang Li Institute of Vegetables and Flowers, Chinese Academy of Agricultural Sciences, Beijing, China

Yong Pyo Lim Molecular Genetics and Genomics Laboratory, Department of Horticulture, Chungnam National University, Daejeon, Korea

Yuki Mitsui Tokyo University of Agriculture, Tokyo, Japan

Jeong-Hwan Mun Department of Bioscience and Bioinformatics, Myongji University, Yongin, Korea

Satoshi Niikura Tohoku Seed Co., Utsunomiya, Tochigi, Japan

Takeshi Nishio Graduate School of Agricultural Science, Tohoku University, Aoba-ku, Sendai, Japan

Koji Sakamoto Takii Plant Breeding and Experiment Station, Konan, Shiga, Japan

Kenta Shirasawa Kazusa DNA Research Institute, Kisarazu, Japan

Yuyan Sun Institute of Vegetables and Flowers, Chinese Academy of Agricultural Sciences, Beijing, China

Yoshihito Takahata Faculty of Agriculture, Iwate University, Morioka, Japan

Toru Terachi Faculty of Life Sciences, Kyoto Sangyo University, Kyoto, Japan

Qingbiao Wang National Engineering Research Center for Vegetables, Haidian, Beijing, People's Republic of China; Beijing Vegetable Research Center, Beijing Academy of Agriculture and Forestry Sciences, Beijing, People's Republic of China

Hiroshi Yamagishi Faculty of Life Sciences, Kyoto Sangyo University, Kita-ku, Kyoto, Japan

Shuji Yokoi Graduate School of Life and Environmental Sciences, Osaka Prefecture University, Naka-ku, Sakai, Japan

Hee-Ju Yu Department of Life Science, The Catholic University of Korea, Bucheon, Korea

Xiaona Yu Molecular Genetics and Genomics Laboratory, Department of Horticulture, Chungnam National University, Daejeon, Korea

Li Zhang National Engineering Research Center for Vegetables, Haidian, Beijing, People's Republic of China; Beijing Vegetable Research Center, Beijing Academy of Agriculture and Forestry Sciences, Beijing, People's Republic of China

1 Economic and Academic Importance of Radish

Takeshi Nishio

Abstract

Radish is important as a root vegetable, a leafy vegetable, a fruit vegetable, an oil crop, and also as a cover plant. The economic importance and characteristics of radish differ between the East and the West of the world. In the East, there are radish cultivars having large roots with various shapes called "Asian big radish" and those grown for production of immature pods or oil seeds, whereas radish is a small vegetable grown within one month in the West. Asian big radish is expected to eventually become popular in the West. Radish belongs to the genus *Raphanus*, but is similar to the *Brassica* species except for the shape of pods and seeds. Despite their similarities, the order of genes in chromosomes is quite different between *Raphanus* and *Brassica*. Radish genome sequences have been published from three groups using similar cultivars, and therefore, collaboration for combining sequence data is considered to be effective for determination of more reliable genome sequences. Some radish lines have high salt tolerance and disease resistance different from *Brassica* crops. Radish also has a characteristic glucosinolate composition. Since radish can be crossed with *Brassica* species, it is also important as a genetic resource for *Brassica* crop breeding.

1.1 Introduction

Radish is a Brassicaceae crop, mainly used as a root vegetable. Roots of radish have large variations in size and shape. European small radish has a small round root ca. 2–3 cm in diameter, whereas "Sakurajima-daikon" in Japan has a large round root of more than 30 cm. Most widely grown radish cultivars in Asia have roots of cylindrical shape ca. 10 cm in diameter and ca.

T. Nishio (✉)
Graduate School of Agricultural Science, Tohoku University, 468-1, Aza-Aoba, Aramaki, Aoba-ku, Sendai 980-0845, Japan
e-mail: nishio@tohoku.ac.jp

T. Nishio and H. Kitashiba (eds.), *The Radish Genome*,
Compendium of Plant Genomes, DOI 10.1007/978-3-319-59253-4_1

40 cm in length, whereas "Moriguchi-daikon" in Japan has a long cylindrical root of more than 2 m with a diameter of ca. 3 cm. The color of radish roots is also various. The major Asian radish has a white thick root and is called "East Asian big long radish," "white radish," or "daikon" in Japanese ("Asian big radish" hereafter), whereas European small radish, which is commonly called "radish," and the root surface of some landraces in Asia are red, although the red part develops from a hypocotyl. There are also purple, green, and black colors on the root surface in other landraces. Some cultivars in China are red or green inside the roots.

Young plants removed by thinning are used as a leafy vegetable. "Kosena-daikon," a Japanese radish landrace, which has a small root, is a cultivar specialized as a leafy vegetable. Immature pods of rat's tail radish, which is cultivated in Southeast Asia, e.g., Thailand and India, are used as a vegetable. The root of rat's tail radish is not thick. A radish sprout called "Kaiware-daikon" is popular as a vegetable in Japan, and the recent vogue of Japanese cuisine has also made it popular in Western countries. Since radish seeds contain a high amount of oil, ca. 40%, radish is also produced as an oil crop (Ahuja et al. 1987). Furthermore, cultivation of radish as a cover crop to avoid soil erosion and to suppress weeds has become popular in the USA and Canada (Weil and Kremen 2007).

The species name of radish is *Raphanus sativus*, its genus being different from that of the major Brassicaceae oil crops and vegetables, i.e., *Brassica*. European small radish is similar to red turnip landraces in *Brassica rapa*. "Shogoin-daikon," a Japanese radish landrace having a large round root, is very similar to "Shogoin-kabu," a Japanese turnip landrace. It is not easy to distinguish radish from turnip at a vegetable market. The greatest morphological difference between radish and turnip is the shape of siliques (Fig. 1.1). Siliques of Brassicaceae plants have two parts, i.e., a beak formed from the sub-stylar region and a part having valves, called the valvar portion (Gomez-Campo 1980). Although *Brassica* and most Brassicaceae species have seeds in the valvar portion without seeds in the beak, radish has seeds in the beak and the valvar portion of radish is absent. Seeds of radish are arranged in a line in a silique and are much larger than those of *Brassica* and most other Brassicaceae species, which are arranged in two lines in a silique. Radish siliques do not dehisce and can float on water.

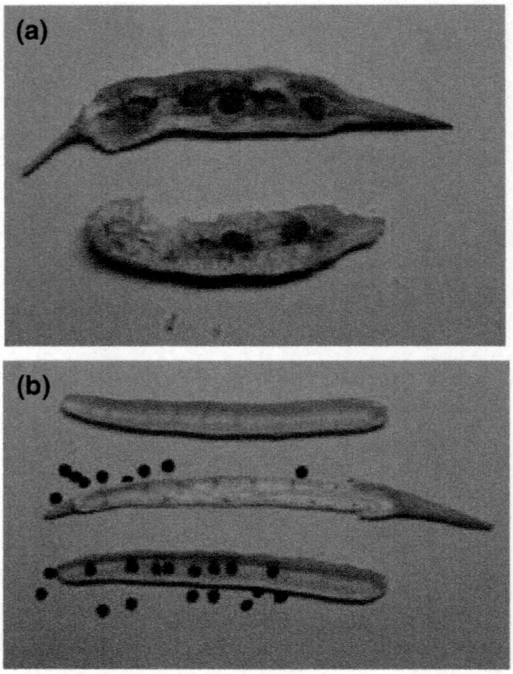

Fig. 1.1 Seed pods (siliques) of radish (**a**) and turnip (**b**)

Raphanus sativus is closely related to *Raphanus raphanistrum*, which is a wild radish distributed as a weed in farmlands, roadsides, and coastal dunes in Europe, North America, and Australia. *R. sativus* has white or pale purple flowers, whereas *R. raphanistrum* has yellow flowers, although some lines of *R. raphanistrum* have white flowers. Siliques of *R. raphanistrum* are longer than those of *R. sativus* and can be fragmented into pieces containing a single seed. *R. raphanistrum* can be crossed with *R. sativus*, and hybrid progeny have normal seed fertility, indicating that these two species have the same genome. In Asian countries, *R. raphanistrum* is not common as a farmland weed, whereas *R. sativus* var. *raphanistroides* is widely distributed in coastal dunes. *R. sativus* var. *raphanistroides* can grow near the seaside but is rarely found in farmlands or riversides, suggesting that it has high salt tolerance with low competitive ability as a weed in farmlands or riversides.

Although radish and *Brassica* species are distantly related, intergeneric hybrids between them can be obtained and radish is a useful genetic resource for breeding of *Brassica* crops especially in breeding for disease resistance and stress tolerance. In this chapter, biological and economical characteristics of radish are outlined.

1.2 Classification of Radish

Raphanus sativus and *R. raphanistrum* belong to the subtribe *Raphaninae*, which is different from the subtribe containing *Brassica, Sinapis eruca*, and *Diplotaxis*, i.e., *Brassisineae*. The morphology of silique is an important taxonomic criterion to separate *Raphanus* from *Brassica*. However, recent molecular taxonomic studies using DNA polymorphism data of chloroplast DNA (cpDNA) and nuclear DNA do not support the grouping of *Brassica, Sinapis eruca*, and *Diplotaxis* excluding *Raphanus*, but suggest that *Raphanus* is closely related to *Brassica rapa* and *Brassica oleracea* (see Chap. 2).

The tribe *Brassiceae* contains seven subtribes: *Brassicinae, Cakilinae, Moricandiinae, Raphaninae, Savignyinae, Vellinae*, and *Zillinae*. Molecular phylogeny based on cpDNA restriction site polymorphism has indicated that the tribe *Brassiceae* is separated into seven lineages, i.e., Rapa/Oleracea lineage, Nigra lineage, Crambe lineage, Cakile lineage, Vella lineage, Savingya lineage, and Zilla lineage. Cakile lineage, Vella lineage, Savingya lineage, and Zilla lineage correspond to *Cakilinae, Vellinae, Savignyinae*, and *Zillinae*, respectively, but species in *Brassicinae, Raphaninae*, and *Moricandiinae* are intermingled and reclassified into Rapa/Oleracea lineage containing *B. rapa, B. oleracea, Eruca vesicaria, Erucastrum abyssinicum, Diplotaxis erucoides, Moricandia arvensis*, etc., Nigra lineage containing *Brassica nigra, Brassica fluticulosa, Sinapis arvensis, Sinapis alba, Erucastrum varium*, etc., and Crambe lineage (Warwick and Black 1997; Warwick and Hall 2009). *R. sativus* and *R. raphanistrum* belong to the Rapa/Oleracea lineage. Phylogenetic studies using nucleotide sequence data of the nuclear *S*-locus related 1 (*SLR1*) gene (Inaba and Nishio 2002) and the nuclear ribosomal internal transcribed spacer (ITS) and chloroplast *trn*L intron sequences (Warwick and Sauder 2005) have shown a slightly different classification of species, but *R. sativus* has always been found to belong to a group of *B. rapa* and *B. oleracea*. These findings suggest that morphological change of the silique from the *Brassica* type to the *Raphanus* type in the Rapa/Oleracea lineage occurred independently from that in the Crambe lineage, the Cakile lineage, and the Zilla lineage, which also have siliques of the *Raphanus* type.

1.3 Genome of Radish

Radish is a monogenomic species having a chromosome number of gametes (*n*) of nine. Although it is monogenomic, most genes have three similar copies in a genome as in the *Brassica* monogenomic species, suggesting that genome triplication has occurred after divergence from an ancestor of *Arabidopsis thaliana*. The chromosome number is the same as that of *B. oleracea*, but the genome structure of radish is quite different from those of *B. oleracea* and closely related *B. rapa*, as has been suggested by a classical genome study analyzing meiotic chromosome pairing (Harberd and McArthur 1980; Mizushima 1980). Genome syntenies of radish with *B. rapa* or *B. oleracea* can be observed in short regions, but there are few overall syntenies (Li et al. 2011; Shirasawa et al. 2011), suggesting that genome rearrangements have occurred many times in the speciation processes of these species (Mun et al. 2015; see Chap. 5). This is in contrast to the relationships between rice and other Poaceae species (Bennetzen and Freeling 1997) and between tomato and other *Solanum* species (Peters et al. 2012).

Genome size of radish has been estimated to be from 526 (Arumuganathan and Earle 1991) to 573 Mb (Johnston et al. 2005) by flow cytometry. Unigene sequences of radish have been published in the RadishBase (http://bioinfo.bti.cornell.edu/cgi-bin/radish/index.cgi) (Shen et al. 2013). The sequences of the whole radish genome were first reported by Kitashiba et al. (2014) and are available in "*Raphanus sativus* Genome Data Base" (http://radish.kazusa.or.jp). Using short-read genomic sequences of 191.1 Gb (246.5 times the radish genome size estimated by them) obtained by next-generation sequencers and both end sequences of 20,736 BAC clones, 76,592 scaffolds of 402 Mb spanning 75.9% of estimated genomic size containing 61,572 predicted genes were obtained. To a high-density linkage map of 2553 DNA markers, 1345 scaffolds were assigned. Using next-generation sequencing data 121.8 times the radish genome size, 40,123 scaffolds spanning 393.3 Mb have been constructed (Mitsui et al. 2015). Jeong et al. (2016) have also read the radish genome sequences and assembled them into 10,674 scaffolds spanning 426 Mb. In these studies, similar cultivars of Asian big radish were used for determination of the genome sequence. Reassembling of sequences or overlapping of scaffolds obtained in these studies may enable construction of much longer scaffolds covering large parts of the radish genome. Although these studies have so far been performed with the support of different organizations, collaboration of these teams is expected (see Chaps. 3 and 4).

1.4 Characteristic Traits of Radish

Raphanus sativus is an allogamous species having self-incompatibility. Therefore, populations of radish landraces and wild radish have high genetic polymorphism, which is reflected by DNA polymorphism in a cultivar (see Chap. 6). Self-incompatibility can be overcome by bud pollination or high-concentration CO_2 treatment (Niikura and Matsuura 2000), and selfed progeny can be obtained. However, they show inbreeding depression, making it difficult to obtain inbred lines.

Like self-incompatibility of *Brassica* species, *S*-receptor kinease (SRK) is the recognition molecule of the stigma, and SP11, also called SCR, is the recognition molecule of pollen (Okamoto et al. 2004). The genes of these recognition molecules, i.e., *SRK* and *SP11/SCR*, have multiple alleles and are inherited as one set, which is called *S* haplotype, by progeny. There are many *S* haplotypes in *R. sativus* (Sakamoto et al. 1998; see Chap. 13), and nucleotide sequences of *SRK* and *SP11/SCR* of some *S* haplotypes in *R. sativus* are similar to those of *B. rapa*, suggesting that *S* haplotypes possessed by an ancient species were inherited by species in both *Raphanus* and *Brassica* without great alteration of nucleotide sequences (Okamoto et al. 2004).

Most radish cultivars are vegetables having thick roots. Size, shape, and color of roots are important traits of radish. Quantitative trait loci (QTLs) controlling root shape and color have been reported (Tsuro et al. 2008; Hashida et al. 2013). However, root thickening is influenced by other traits, such as flowering. In our QTL analysis of root thickness using a progeny obtained by crossing between "Aokubi" having a white thick root with green color on the top and a cultivar of rat's tail radish, a QTL having the highest LOD score corresponded to a QTL for bolting time (unpublished). This may be natural because early flowering is considered to have an adverse effect on root thickening. Transcriptome analysis of growing roots has been performed, and genes related to root thickening were identified (see Chaps. 8 and 9).

The color of root surface in radish is derived from anthocyanins. Pigments of red color and purple color of radish cultivars are pelargonidin and cyanidin, respectively. A hybrid between a red root line and a white root line was found to have purple roots, suggesting that the red and the white have knockout mutations in different genes responsible for cyanidin synthesis and the functional alleles in the red and the white acted as complementary genes. Alleles of the gene for flavonoid 3′-hydroxylase (F3′H) in red root cultivars have been revealed to have insertions of a Ty3/gypsy transposon or a helitron (Ozeki 2010). A gene for dihydroflavonol reductase (*RsDFR*) and a gene for anthocyanidin synthase (*RsANS*) are expressed in the epidermal tissues of roots of a red-skinned cultivar, whereas these genes are not expressed in those of a white-skinned cultivar (Park et al. 2011).

The large seed size is a remarkable trait of radish among Brassicaceae species. Seed size of radish is about five times in weight that of *B. rapa*. Due to the large seed size, cotyledons and hypocotyls of seedlings are larger than those of *Brassica*. The larger size of seedlings facilitates cultivation by direct sowing of seeds in the field and produces bigger sprouts than *Brassica*. The large seed trait is related to the shape of siliques. Amphidiploid plants of intergeneric hybrids between *R. sativus* and *B. rapa* have siliques of an intermediate type, which has both the beak and the valvar portion with a few seeds (Fig. 1.2). The amphidiploid plants are partially sterile, but a small number of seeds can be obtained. The size of the seeds is also intermediate between *Raphanus* and *Brassica*.

Fig. 1.2 Flowers (**a**) and a silique (**b**) of an amphidiploid intergeneric hybrid between turnip and radish. Diploid hybrid plants between them are sterile, whereas amphidiploid (tetraploid) hybrid plants are fertile and set seeds. (Provided by Dr. K. Tonosaki)

The pungency of radish is due to isothiocyanates, which are hydrolyzed products of glucosinolates. The content of glucosinolates has a great influence on the taste of grated fresh radish and radish salad. The major glucosinolate in radish roots is glucoraphasatin (also called 4-methylthio-3-butenyl glucosinolate, dehydroerucin), and variation of the glucosinolate composition is limited. However, there is a great variation in the contents of glucoraphasatin in Japanese radish cultivars (Ishida et al. 2015). QTLs controlling glucosinolate content in the root have been reported (Zou et al. 2013), and candidate genes responsible for the content have been inferred. Recently, a mutant having a high amount of glucoerucin without glucoraphasatin has been selected (Ishida et al. 2015), and the gene responsible for this mutation has been identified (Kakizaki et al. 2017).

Although radish generally has high salt tolerance, such tolerance of *R. sativus* var. *raphanistroides* is especially high (Nasu et al. 2012). Although genetic analysis of salt tolerance of *R. sativus* var. *raphanistroides* has not yet advanced, the genes for salt tolerance of radish will be useful for breeding of *Brassica* crops. In the production of radish, high-temperature stress has become a serious problem. The core of a radish root turns reddish brown by the high-temperature stress, resulting in unmarketable products. However, since there is a variation in sensitivity to the high-temperature stress, it should be possible to develop a tolerant cultivar.

After bolting, radish roots become fibrous and unsuitable for markets. Therefore, a late bolting trait is preferred. One the other hand, rat's tail radish or cultivars for oil production are requested to flower even in tropical regions. For floral induction, vernalization is required, but rat's tail radish can flower without vernalization. There is a large variation in vernalization requirement in radish as in many other winter crops. In a QTL analysis of bolting time using a progeny obtained by a cross between "Aokubi" and rat's tail radish, a QTL having a significant LOD score was detected in a region containing an *FLC* gene (see Chap. 11).

Although clubroot caused by *Plasmodiophora brassicae* is a serious problem in the production of *Brassica* vegetables, Japanese and Korean cultivars of radish are generally tolerant to clubroot. Radish can be used as a source of resistance genes for the breeding of *Brassica* vegetables (Akaba et al. 2009). A QTL controlling clubroot resistance has been mapped on LG1 (Kamei et al. 2010), corresponding to Rs5 (Kitashiba et al. 2014, Chap. 3). In radish cultivation, *Fusarium* yellow caused by *Fusarium oxysporum* is one of the most serious diseases. Leaves wilt and the vascular tissue in a root is browned by this disease. Some cultivars and landraces have relatively high resistance to *Fusarium* yellow (see Chap. 12).

Radish is recalcitrant in tissue and cell cultures because of its low plant regeneration ability. There have been few reports on successful protoplast culture, anther or isolated microspore culture (Takahata et al. 1996), and plant transformation (Park et al. 2005; Cho et al. 2008). The difficulty of plant transformation makes it impossible to demonstrate the function of isolated genes in radish. Development of an efficient plant transformation technique is indispensable for basic and applied studies in radish. Since there must be genetic variation in plant regeneration ability, identification of cultivars or lines having high regeneration ability would be the first step for the development of in vitro culture techniques.

1.5 Use of Radish as a Vegetable, a Feed, and Materials

Although radish was once the most cultivated vegetable in Japan, its production has been decreasing because consumption of radish pickles has recently decreased. Exploring novel ways of utilization is required for increasing the economic value of radish. European small radish is used for salad after slicing or rough cutting. Asian big radish has not been used for salad, but shredded white radish, which is a garnish of sliced raw fish, is increasingly used in a salad. Since crispiness is important for the shredded radish, white radish cannot be replaced by turnip roots. The white radish root can be stored and is inexpensive, and therefore, it is a good material for use in salad.

Tempura and shabu-shabu, popular Japanese dishes, are garnished with grated radish, which contains amylase and other enzymes to help digestion. Grated radish is also served with grilled fish or dried young sardines. Grated radish is pungent because of its high content of isothiocyanates. By disruption of cells by grating, glucosinolates are hydrolyzed by the enzyme myrosinase, which is stored in the vacuoles of myrosin cells. Isothiocyanates have anti-bacterial and anti-fungal activity, and some isothiocyanates have been revealed to have anti-carcinogenic activity (Ishida et al. 2015). The isothiocyanate produced from the major glucosinolate in roots, i.e., 4MTB-GSL, has also been reported to be able to reduce cell proliferation and induce apoptosis in cancer cell lines with very limited toxicity toward normal human T-lymphocytes (Papi et al. 2008). Due to the increasing popularity of Japanese cuisine, consumption of grated radish will most probably increase.

A popular boiled dish containing white radish is "oden," which is served even in convenience stores in Japan. In Chinese cuisine, Asian big radish is mainly served in boiled dishes. Although digestive enzymes and isothiocyanates in radish roots are lost, pieces of radish roots absorb soup, resulting in good taste. There are various types of radish pickles in Japan, e.g., takuanzuke, bettarazuke, asazuke, and fukujinzuke, but consumption of radish pickles has decreased. In Korea, Asian big radish is used as a material of kimuchi.

Radish sprouts, called "kaiware-daikon," are consumed with sushi. Consumption of kaiware-daikon has decreased in Japan after the incident of bacteria O157 tainting of kaiware-daikon in 1996. As fresh green vegetables, only kaiware-daikon, cucumber, and labiate leaves are served with sushi. The international popularity of sushi may result in increased consumption of kaiware-daikon. As in radish roots, kaiware-daikon contains 4MTB-GSL, a major glucosinolate (Barillari et al. 2005; Papi et al. 2008).

Like *Brassica* seed oil, radish seed oil contains a high amount of erucic acid, about 30–40% (Ahuja et al. 1987), which is considered to be harmful as a food. Erucic acid is known to be suitable for industrial use, such as for use as a lubricant. It is promoted to make hair velvety, and radish seed oil is marketed as cosmetics. Residue of oil extraction can be used for extraction of glucosinolates. The major glucosinolate in radish seeds has been reported to be glucoraphenin, which is hydrolyzed into sulforaphene (Barillari et al. 2005). Sulforaphene has been indicated to have 1.3–1.5 times higher activity to inhibit mutagenesis induced by cooked food mutagens than sulforaphane (Shishu 2009).

Availability of radish as a winter cover crop to avoid soil erosion and loss of soil nitrogen has been reported (Weil and Kremen 2007). Radish can alleviate soil compaction by drilling with thick roots, which are killed by cold temperature, resulting in holes. Radish can suppress weeds more effectively than other cover crops, e.g., rye and rape. Furthermore, cover crops of *Brassica* can reduce soil nematodes (Weil and Kremen 2007), and radish is also considered to be effective in control of nematodes.

1.6 Breeding of Radish and Use of Radish as a Breeding Material

Most modern cultivars of radish are hybrid cultivars. Old cultivars and landraces are mass selected cultivars having some genetic variations within a cultivar. Since inbreeding depression is remarkable, efforts to enhance uniformity of a cultivar result in decrease of plant vigor. As a root vegetable cultivar, uniformity of root shape and color is necessary, but some variations in leaf shape and other traits can be allowed. To develop a cultivar having high uniformity, hybrid breeding has been used (see Chap. 15). The parents of a hybrid cultivar are inbred lines. Although repeated selfings to develop inbred lines make plants less vigorous, some lines showing less inbreeding depression can be selected from a large number of inbred lines. These inbred lines are maintained and propagated by bud pollination by hand. Since the number of seeds obtained by one time of the hand pollination is small, double cross or three-way cross is preferred to single cross, which can provide more uniform progeny than double cross.

In the production of hybrid seeds, self-incompatibility or male sterility is used for avoiding self-fertilization. Self-incompatibility is sometimes unstable under some stress conditions, e.g., high temperature, and contamination of hybrid seeds by selfed seeds may increase. Since such contamination spoils the quality of hybrid seeds, examination of contaminated seeds using isozyme or DNA markers is indispensable. To reduce the percentage of selfed seeds, selection of parental lines having strong and stable self-incompatibility is required. The strength of self-incompatibility depends on *S* haplotypes. Therefore, a simple technique for *S* haplotype identification is required. The methods developed for identification of *S* haplotypes in *B. rapa* and *B. oleracea* can also be used in radish (Nishio et al. 1996; Sakamoto et al. 1998; Niikura and Matsuura 1998; Lim et al. 2002).

Although it is difficult to use male sterility in hybrid seed production by double cross, risk of contamination of selfed seeds in single cross or three-way cross is lower in the use of male sterility than that in the use of self-incompatibility. Furthermore, restoration of pollen fertility in male-sterile lines under some growing conditions, e.g., high temperature, can be easily noticed by visual inspection of plants. The source of the well-known cytoplasmic male sterility used in rapeseed breeding is radish, i.e., Ogura cytoplasm (see Chap. 7). The same cytoplasm has been found in landraces in radish, *R. sativus* var. *raphanistroides*, and *R. raphanistrum* (Yamagishi and Terachi 1996). Although a restorer gene is not necessary in hybrid breeding of root vegetables, a gene useful in the breeding of rapeseed, possibly also of rat's tail radish, has been identified (Koizuka et al. 2003).

Mutation breeding of radish has not been successfully performed because it is difficult to distinguish induced mutations and variations present in radish cultivars. Furthermore, most mutated traits are controlled by recessive alleles and are not suitable for hybrid breeding, which combines dominant alleles of parents. However, all the landraces having characteristic traits have been developed by spontaneous mutations. Recently, a cultivar having a new composition of glucosinolate has been developed by selecting a spontaneous mutation in a landrace (Ishida et al. 2015). Techniques for reverse-genetic selection of mutants have been developed. Although TILLING (McCallum et al. 2000), which has been successfully used for reverse-genetic selection of mutants in many plant species, is not suitable for use in radish cultivars, which has many gene variations in a cultivar, mutant selection using a next-generation sequencer, e.g., KeyPoint ® of Keygene, may be available for use in mutation breeding of radish. Another cost-effective technique for reverse-genetic selection is required.

Because of the difficulty of plant regeneration, genetic engineering has not been used in radish breeding. However, many GMO cultivars have been developed in soybean, which is also recalcitrant in tissue and cell culture. The reason why such efforts have not been made in radish may be lower economic value of radish than that of soybean. Another problem of using genetic engineering in radish breeding may be wide distribution of wild radish in the world. Stress tolerance genes and insect resistance genes can be transmitted to wild plants, providing higher competitive ability. Since flowering is not necessary in root vegetable production, the use of a sterility gene, e.g., pollen sterility gene used in the seed production technology (DuPont-Pioneer), may be useful.

Intergeneric hybrids between *R. sativus* (as a female parent) and *B. oleracea*, named *Raphanobrassica* (Karpechenko 1924), and between *B. rapa* (as a female parent) and *R. sativus*, named *Brassicoraphanus* (Terasawa 1932), have long been studied (see Chap. 14). Efficiency of hybrid production without embryo culture between *B. rapa* and *R. sativus* depends on cultivars of *B. rapa*, and three QTLs controlling hybrid formation ability have been identified (Tonosaki et al. 2013). Crossability of *R. sativus* with *Brassica* suggests that *Brassica* genetic resources may be available for use in radish breeding. Since *R. sativus* has valuable traits for *Brassica* breeding, e.g., disease resistance (Akaba et al. 2009), nematode resistance (Lelivelt et al. 1993), stress tolerance, and a different composition of glucosinolates, the economic value of radish as a breeding material of several *Brassica* crop species can be emphasized.

References

Ahuja KL, Singh H, Raheja RK, Labana KS (1987) The oil content and fatty acid composition of various genotypes of cauliflower, turnip and radish. Plant Foods Hum Nutr 37:33–40

Akaba M, Kaneko Y, Hatakeyama K, Ishida M, Bang SW, Matsuzawa Y (2009) Identification and evaluation of clubroot resistance of radish chromosome using a *Brassica napus–Raphanus sativus* monosomic addition line. Breed Sci 59:203–206

Arumuganathan K, Earle ED (1991) Nuclear DNA content of some important species. Plant Mol Biol Rep 9:208–218

Barillari J, Cervellati R, Paolini M, Tatibouet A, Rollin P, Iori R (2005) Isolation of 4-methylthio-3-butenyl glucosinolate from *Raphanus sativus* sprouts (kaiware daikon) and its redox properties. J Agric Food Chem 53:9890–9896

Bennetzen J, Freeling M (1997) The unified grass genome: synergy in synteny. Genome Res 7:301–306

Cho MA, Min SR, Ko SM, Liu JR, Choi PS (2008) *Agrobacterium*-mediated genetic transformation of radish (*Raphanus sativus* L.). Plant Biotechnol 25:205–208

Gomez-Campo C (1980) Morphology and morphotaxonomy of the tribe *Brassiceae*. In: Tsunoda S et al (eds) *Brassca* crops and wild allies. Japan Scientific Societies Press, Tokyo, pp 3–31

Harberd DJ, McArthur ED (1980) Meiotic analysis of some species and genus hybrids in the *Brassiceae*. In: Tsunoda S et al (eds) *Brassca* crops and wild allies. Japan Scientific Societies Press, Tokyo, pp 65–87

Hashida T, Nakatsuji R, Budahn H, Schrader O, Peterka H, Fujimura T, Kubo N, Hirai M (2013) Construction of a chromosome-assigned, sequence-tagged linkage map for the radish, *Raphanus sativus* L. and QTL analysis of morphological traits. Breed Sci 63:218–226

Inaba R, Nishio T (2002) Phylogenetic analysis of *Brassiceae* based on the nucleotide sequences of the *S*-locus related gene, *SLR1*. Theor Appl Genet 105:1159–1165

Ishida M, Kakizaki T, Morimitsu Y, Ohara T, Hatakeyama K, Yoshiaki H, Kohori J, Nishio T (2015) Novel glucosinolate composition lacking 4-methylthio-3-butenyl glucosinolate in Japanese white radish (*Raphanus sativus* L.). Thoer Appl Genet 128:2037–2046

Jeong Y-M, Kim N, Ahn B, Oh M, Chung W-H, Chung H, Jeong S, Lim K-B, Hwang Y-J, Kim G-B, Baek S, Choi S-B, Hyung D-J, Lee S-W, Sohn S-H, Kwon S-J, Jin M, Seol Y-J, Chae W, Choi K, Park B-S, Yu H-J, Mun J-H (2016) Elucidating the triplicated ancestral genome structure of radish based on chromosome-level comparison with the *Brassica* genomes. Theor Appl Genet 129:1357–1372

Johnston JS, Pepper AE, Hall AE, Chen ZJ, Hodnett G, Drabek J, Lopez R, Price HJ (2005) Evolution of genome size in Brassicaceae. Ann Bot 95:229–235

Kakizaki T, Kitashiba H, Zou Z, Li F, Fukino N, Ohara T, Nishio T, Ishida M (2017) A 2-oxoglutarate-dependent dioxygenase mediates the biosynthesis of glucoraphasatin in radish. Plant Physiol 173: 1583–1593

Karpechenko GD (1924) Hybrids of *Raphanus sativus* L. x *Brassica oleracea* L. J Genet 14:375–396

Kamei A, Tsuro M, Kubo N, Hayashi T, Wang N, Fujimura T, Hirai M (2010) QTL mapping of clubroot resistance in radish (*Raphanus sativus* L.). Theor Appl Genet 120:1021–1027

Kitashiba H, Li F, Hirakawa H, Kawanabe T, Zou Z, Hasegawa Y, Tonosaki K, Shirasawa S, Fukushima A, Yokoi S, Takahata Y, Kakizaki T, Ishida M, Okamoto S, Sakamoto K, Shirasawa K, Tabata S, Nishio T (2014) Draft sequences of the radish (*Raphanus sativus* L.) genome. DNA Res 21:481–490

Koizuka N, Imai R, Fujimoto H, Hayakawa T, Kimura Y, Kohno-Murase J, Sakai T, Kawasaki S, Imamura J (2003) Genetic characterization of a pentatricopeptide repeat protein gene, *orf587*, that restores fertility in the cytoplasmic male-sterile Kosena radish. Plant J 34:407–415

Lelivelt CLC, Lang W, Dolstra O (1993) Intergeneric crosses for the transfer of resistance to the beet cyst nematode from *Raphanus sativus* to *Brassica napus*. Euphytica 58:111–120

Li F, Hasegawa Y, Saito M, Shirasawa S, Fukushima A, Ito T, Fujii H, Kishitani S, Kitashiba H, Nishio T (2011) Extensive chromosome homoeology among Brassiceae species were revealed by comparative genetic mapping with high-density EST-based SNP markers in radish (*Raphanus sativus* L.). DNA Res 18:401–411

Lim SH, Cho HJ, Lee SJ, Cho YH, Kim BD (2002) Identification and classification of *S* haplotypes in *Raphanus sativus* by PCR-RFLP of the *S* locus glycoprotein (*SLG*) gene and the *S* locus receptor kinase (*SRK*) gene. Theor Appl Genet 104:1253–1262

McCallum CM, Comai L, Greene EA, Henikoff S (2000) Targeted screening for induced mutations. Nat Biotechnol 18:455–457

Mitsui Y, Shimomura M, Komatsu K, Namiki N, Shibata-Hatta M, Imai M, Katayose Y, Mukai Y, Kanamori H, Kurita K, Kagami T, Wakatsuki A, Ohyanagi H, Ikawa H, Minaka N, Nakagawa K, Shiwa Y, Sasaki T (2015) The radish genome and comprehensive gene expression profile of tuberous root formation and development. Sci Rep 5:10835

Mizushima U (1980) Genome analysis in *Brasica* and allied genera. In: Tsunoda S et al (eds) *Brassca* crops and wild allies. Japan Scientific Societies Press, Tokyo, pp 89–106

Mun J-H, Chung H, Chung W-H, Oh M, Jeong Y-M, Kim N, Ahn BO, Park B-S, Park S, Lim K-B, Hwang Y-J, Yu H-J (2015) Construction of a reference genetic map of Raphanus sativus based on genotyping by whole-genome resequencing. Theor Appl Genet 128:259–272

Nasu S, Kitashiba H, Nishio T (2012) "Na-no-hana Project" for recovery from the Tsunami disaster by producing salinity-tolerant oilseed rape lines: Selection of salinity-tolerant lines of *Brassica* crops. J Integr Field Sci 9:33–37

Niikura S, Matsuura S (1998) Identification of self-incompatibility alleles (*S*) by PCR-RFLP in radish (*Raphanus sativus* L.). Euphytica 102:379–384

Niikura S, Matsuura S (2000) Genetic analysis of the reaction level of self-incompatibility to a 4% CO_2 gas treatment in the radish (*Raphanus sativus* L.). Theor Appl Genet 101:1189–1193

Nishio T, Kusaba M, Watanabe M, Hinata K (1996) Registration of *S* alleles in *Brassica campestris* L by the restriction fragment sizes of *SLG*s. Theor Appl Genet 92:388–394

Okamoto S, Sato Y, Sakamoto K, Nishio T (2004) Distribution of similar self-incompatibility (*S*) haplotypes in different genera, *Raphanus* and *Brassica*. Sex Plant Reprod 17:33–39

Ozeki Y (2010) Study on the relationship between the anthocyanin molecular species and color phenotype of roots in the inbred lines of red radish (*Raphanus sativus* L.). Japan Food Chem Res Found Res Rep 16:33–39 (in Japanese with English summary)

Papi A, Orlandi M, Bartolini G, Barillari J, Iori R, Paolini M, Ferroni F, Fumo MG, Pedulli GF, Velgimigli L (2008) Cytotoxic and antioxidant activity of 4-methylthio-3-butenyl isothiocyanate from *Raphanus sativus* L. (kaiware daikon) sprouts. J Agric Food Chem 56:875–883

Park B-J, Liu Z, Kanno A, Kameya T (2005) Transformation of radish (*Raphanus sativus* L.) via sonication and vacuum infiltration of germinated seeds with *Agrobacterium* harboring a group 3 LEA gene from *B. napus*. Plant Cell Rep 24:494–500

Park NI, Xu H, Li X, Jang IH, Park S, Ahn GH, Lim YP, Kim SJ, Park SU (2011) Anthocyanin accumulation and expression of anthocyanin biosynthetic genes in radish (*Raphanus sativus*). J Agric Food Chem 59:6034–6039

Peters SA, Bargsten JW, Szinay D, van de Belt J, Visser RGF, Bai Y, de Jong H (2012) Structural homology in the Solanaceae: analysis of genomic regions in support of synteny studies in tomato, potato and pepper. Plant J 71:602–614

Sakamoto K, Kusaba M, Nishio T (1998) Polymorphism of the *S*-locus glycoprotein gene (*SLG*) and the *S*-locus related gene (*SLR1*) in *Raphanus sativus* L. and self-incompatible ornamental plants in the Brassicaceae. Mol Gen Genet 258:397–403

Shen D, Shu H, Huang M, Zheng Y, Li X, Fei Z (2013) RadishBase: a database for genomics and genetics of radish. Plant Cell Physiol 54(1–6):e3

Shirasawa K, Oyama M, Hirakawa H, Sato S, Tabata S, Fujioka T, Kimizuka-Takagi C, Sasamoto S, Watanabe A, Kato M, Kishida Y, Kohara M, Takahashi C, Tsuruoka H, Wada T, Sakai T, Isobe S (2011) An EST-SSR linkage map of *Raphanus sativus* and comparative genomics of the Brassicaceae. DNA Res 18:221–232

Shishu KIP (2009) Inhibition of cooked food-induced mutagenesis by dietary constituents: comparison of two natural isothiocyanates. Food Chem 112:977–981

Takahata Y, Komatsu H, Kaizuma N (1996) Microspore culture of radish (*Raphanus sativus L.*): influence of genotype and culture conditions on embryogenesis. Plant Cell Rep 16:163–166

Terasawa Y (1932) Tetraploid Bastarde von *Brassica chinensis* L. x *Raphanus sativus* L. Jpn J Genet 7:183–185 (In Japanese)

Tonosaki K, Michiba K, Bang SW, Kitashiba H, Kaneko Y, Nishio T (2013) Genetic analysis of hybrid seed formation ability of *Brassica rapa* in intergeneric crossings with *Raphanus sativus*. Theor Appl Genet 126:837–846

Tsuro M, Suwabe K, Kubo N, Matsumoto S, Hirai M (2008) Mapping of QTLs controlling root shape and red pigmentation in radish, *Raphanus sativus* L. Breed Sci 58:55–61

Warwick SI, Black LD (1997) Phylogenetic implications of chloroplast DNA restriction site variation in subtribes Raphaninae and Cakilinae (Brassicaceae, tribe Brassiceae). Can J Bot 75:960–973

Warwick SI, Hall JC (2009) Phylogeny of *Brassica* and wild relatives. In: Gupta SK (ed) Biology and breeding of crucifers. CRC Press, London, pp 19–36

Warwick SI, Sauder CA (2005) Phylogeny of tribe Brassiceae (Brassicaceae) based on chloroplast restriction site polymorphisms and nuclear ribosomal internal transcribed spacer and chloroplast *trn*L intron sequences. Can J Bot 83:467–483

Weil RR, Kremen A (2007) Thinking across and beyond disciplines to make cover crops pay. J Sci Food Agric 87:551–557

Yamagishi H, Terachi T (1996) Molecular and biological studies on male-sterile cytoplasm in the Cruciferae III. Distributtion of Ogura-type cytoplasm among Japanese wild radishes and Asian radish cultivars. Theor Appl Genet 93:325–332

Zou Z, Ishida M, Li F, Kakizaki T, Suzuki S, Kitashiba H, Nishio T (2013) QTL analysis using SNP markers developed by next-generation sequencing for identification of candidate genes controlling 4-methylthio-3-butenyl glucosinolate contents in roots of radish, *Raphanus sativus* L. PLoS ONE 8:e53541

2 Speciation and Diversification of Radish

Hiroshi Yamagishi

Abstract

Even though the main cultivation of radish is restricted to East Asia (China, Japan, and Korea), the crop is used worldwide for various purposes. In addition to cultivated radish, wild radish is distributed worldwide as a weed. The species relationship of radish to *Brassica* plants, the origin of cultivated radish and the intraspecific classification of radish have all been controversial. However, recent developments in genome analysis of both nuclear and organelle genomes as well as the exploitation of transgenic *Brassica napus* and assessment of dispersal of transgenes has provided new information for phylogenetic studies of *Raphanus*. In this chapter, recent ideas about inter- and intraspecific divergence based on this information are presented. The morphological diversity of cultivated radish worldwide is also presented.

2.1 Introduction

The genus *Raphanus* belongs to the tribe *Brassiceae*. In the tribe, the genus *Brassica* is also involved. While *Brassica* contains six cultivated species and many wild species, *Raphanus* has only one cultivated species (*R. sativus*) and a few wild species. Both the nuclear genes and cytoplasm of cultivated radish have long been used as breeding materials for *Brassica* crops by intergeneric hybridization and back-crossing. However, the relationships between *Brassica* and *Raphanus*, and between wild and cultivated *Raphanus* species have been controversial.

Recent developments in genome analyses using molecular techniques for both the nuclear and organelle genomes initiated insights into speciation in the *Brassiceae* after whole genome triplication (WGT) and the origin of the genus *Raphanus* and cultivated radish within it. On the other hand, given the wide distribution of transgenic rapeseed (*B. napus*), the probability of spontaneous hybridization between *Brassica* and *Raphanus*, especially the wild species *R. raphanistrum*, which grows worldwide as a

H. Yamagishi (✉)
Faculty of Life Sciences, Kyoto Sangyo University, Motoyama, Kamigamo, Kita-ku, Kyoto 603-8555, Japan
e-mail: hiyamagi@cc.kyoto-su.ac.jp

T. Nishio and H. Kitashiba (eds.), *The Radish Genome*, Compendium of Plant Genomes, DOI 10.1007/978-3-319-59253-4_2

weed, has attracted interest from the viewpoint of the assessment of escape of transgenes.

Based on research in recent years, the relationship between *Brassica* and *Raphanus*, speciation in *Raphanus*, and the origin of cultivated radish are reviewed in this chapter. The diversification of cultivated radish is also demonstrated with many images as examples. It is expected that this chapter will serve for understanding of the current development of phylogenetic research on *Raphanus*.

2.2 *Raphanus* in the Family Brassicaceae

The genus *Raphanus* belongs to the tribe *Brassiceae* as *Brassica*. In classical taxonomy, *Brassica* is included in the subtribe *Brassicinae* (Gómez-Campo 1980; Schulz 1936), while *Raphanus* is contained in *Raphaninae*. However, based on analysis of the chloroplast genome of *Brassiceae* plants conducted by Warwick and Black (1991), two lineages of cytoplasm are present in *Brassiceae*, namely the *B. rapa/B. oleracea* lineage (hereafter, rapa lineage) and the *B. nigra* lineage (hereafter, nigra lineage). *Raphanus* belongs to the rapa lineage, and clearly belongs to *Brassicinae*. Based on molecular studies, Al-Shehbaz et al. (2006) postulated that some of the most commonly known genera of *Brassiceae* including *Raphanus* need to be abandoned, and that nomenclatural changes are needed. Their postulation means that *Raphanus* needs to be included in the same genus as *Brassica*. Although Pradhan et al. (1992) found that *Raphanus* species belong to the rapa lineage based on RFLP data of chloroplast DNA, they are included in the same clade with *B. nigra* based on RFLP data of mitochondrial DNA. The relationship of the variations between the two organelle genomes in *Raphanus* remains to be studied in more detail.

Yang et al. (2002) ascertained that *Raphanus* is closely related to the rapa lineage by sequencing of the chloroplast noncoding region between *trn*T and *trn*F and between *trn*D and *trn*T. However, they also observed based on sequences of the internal transcribed spacer region of 18S–25S nuclear ribosomal DNA that *Raphanus* is more closely related to *B. nigra* than to the *B. oleracea/B. rapa* clade (Yang et al. 1999), a finding that conflict with the results from chloroplast DNA. This discrepancy between the results based on the chloroplast and nuclear genomes was also observed by Warwick and Sauder (2005), who also found that *R. sativus* and *R. raphanistrum* form a monophyletic group for the nuclear internal transcribed spacer region as well as RFLP and sequence data of the chloroplast genome. From these results, Yang et al. (2002) hypothesized that *Raphanus* was derived from a hybridization between the nigra lineage and rapa lineage, with the latter as the female parent. They also estimated the date of this hybridization event to be 60% of the divergence time between the two *Brassica* lineages. Lysak et al. (2005) estimated the time of the split between nigra and rapa lineages as 7.9–14.6 million years ago (Mya). Combining the two estimations by Yang et al. (2002) and Lysak et al. (2005), the hybridization between the two lineages and the production of *Raphanus* are inferred to be about 4.7–8.8 Mya.

On the other hand, recent results of whole genome analysis of *Raphanus* indicate that the divergence between *Raphanus* and *Brassica* was much older than the estimation above. Moghe et al. (2014) determined the whole genome of *R. raphanistrum* and inferred that the WGT in Brassicaceae species was 24–29 Mya and that *Raphanus* diverged from *B. rapa* 13–19 Mya, while Mitsui et al. (2015) estimated that the divergence between *R. sativus* and *B. rapa* was about 16.7 Mya. Interestingly, the two estimations on the original date of *Raphanus* coincide well with each other, although the estimations of the WGT event are slightly different [24–29 Mya by Moghe et al. (2014) and 15.6–27.4 Mya by Mitsui et al. (2015)]. The relationship between the WGT and divergence of *Raphanus* from *Brassica* remains to be studied further.

2.3 Intergeneric Hybridization Between *Brassica* and *Raphanus*

Artificial intergeneric hybridization between *Brassica* and *Raphanus* has been conducted for long periods, mainly for breeding crop *Brassica* species. Hybridization and breeding uses are shown in Chap. 14. Along with the development and distribution of transgenic *Brassica*, mainly *B. napus*, data on the possibility of hybridization between *B. napus* and weedy *R. raphanistrum* have accumulated from the viewpoint of risk assessment of escape of transgenes in rapeseed cultivars. These data would also help in understanding the phylogenetic relationships between *B. napus* and *Raphanus*.

Eber et al. (1994) observed spontaneous hybridization between male-sterile oilseed rape and weedy *Brassica adpressa* or *R. raphanistrum*. They found that hybrids were produced much more easily between *B. napus* and *R. raphanistrum* than between *B. napus* and *B. adpressa*. Further, the hybrids showed high chromosome pairing and the presence of multivalents, which indicated that recombination between the chromosomes of *B. napus* and *R. raphanistrum* is possible. Also, Baranger et al. (1995) observed that more than half of the seeds produced by male-sterile rapeseed were interspecific hybrids with *R. raphanistrum* having the genome constitution of ACRr ($2n = 28$).

The ability to form interspecific hybrids with *R. raphanistrum* under natural conditions differs significantly between rapeseed lines. Differences in both the ability to accept wild radish pollen grains that germinate on the stigma and the rate of fertilization of ovules were observed between oilseed rape cultivars (Guéritaine et al. 2003). From the interspecific hybrids produced, Chèvre et al. (1998) obtained BC_1 and BC_2 progeny with *R. raphanistrum* as pollen parent. While interspecific F_1 plants produced very low numbers of BC_1 seeds (on average 0.12 seeds/100 flowers), the BC_1 and BC_2 generations showed much higher seed fertility (BC_1 = 1.4 seeds/100 flowers, BC_2 = 7.9 seeds/100 flowers, respectively). These observations indicate that interspecific barriers are largely overcome between these two species and that transgenes flow from rapeseed to weedy *R. raphanistrum*.

In comparison with interspecific hybridization with *B. napus* as the seed parent, hybridization with *R. raphanistrum* as the seed parent is much more difficult. Darmency et al. (1998) observed that as many as three interspecific hybrids per 100 plants of wild radish as the seed parent were produced, whereas male-sterile oilseed rape plants surrounded by wild radish could produce up to 37 hybrids per plant. Crossing ability between reciprocal combinations differed by a factor of 1000. However, there was a difference in interspecific hybridization capability within a population of wild radish (Guéritaine and Darmancy 2001). Guéritaine and Darmancy (2001) observed that the ratio of fertilized ovules to the total number of ovules in ovaries where pollen tubes arrived ranged from 0.02 to 0.51 among the wild radish plants, and they estimated that pollen germination and ovule fertilization represent minor barriers to interspecific hybridization between oilseed rape and wild radish. The possibility of spontaneous hybridization between *B. napus* and *R. raphanistrum* would be another criterion to consider in the classification of *Raphanus* at the genus level, in addition to the molecular analyses mentioned previously.

2.4 Wild and Cultivated Species of *Raphanus*

Based on classical taxonomy, *Raphanus* includes six wild species and one cultivated species (Kitamura 1958). The wild species are *R. aucheri*, *R. rostratus*, *R. raphanistrum*, *R. microcarpus*, *R. maritimus*, and *R. landra*, while the cultivated species is *R. sativus*. However, Crisp (1995) stated that the taxonomy of radish is confused, and in the past, it was often based on inadequate knowledge of the variation in the species. He presented an interpretation that includes all forms and their wild relatives (*R. raphanistrum*, *R. landra*, and *R. maritimus*) in *R. sativus*, because all of these taxa have $2n = 2x = 18$ and are cross-fertile. The recent nomenclature trend is to categorize the wild species as *R. raphanistrum* and place three subspecies within it, namely ssp. *raphanistrum*, *landra*, and *maritimus* (Sahli et al. 2008; Warwick and Francis 2005). Similarly, Cheam and Code (1995) described that *R. raphanistrum* is sometimes treated as an aggregate, which includes *R. maritimus* and other taxa as subspecies.

Cheam and Code (1995) and Warwick and Francis (2005) respectively demonstrated the morphological characteristics of *R. raphanistrum* growing as a weed in Australia and Canada. Cheam and Code (1995) described that various petal colors such as white, purple, pink, and pale yellow, often with light or dark distinct veins are observed in wild species growing in Australia. Yellow or white petals are the most common, and pink petals are rare. They also stated that the fruit is a pod, constricted between seeds, with a slender, seedless beak. On the other hand, Warwick and Francis (2005) mentioned that Canadian weedy radish *R. raphanistrum* can usually be distinguished from the cultivated radish, *R. sativus*, by its slender, non-fleshy taproot, yellow flowers (occasionally white), more prominent petal venation if white-flowered, and its woodier, more slender, multi-ribbed and deeply segmented fruits. The comparisons of petal color and fruit morphology between *R. raphanistrum* and *R. sativus* are shown in Figs. 2.1 and 2.2. Warwick and Francis also conveyed the notion by Rollins (1993) that frequent hybridization between *R. raphanistrum* and naturalized weedy escaped *R. sativus* in North America had increased the morphological complexity of both species.

In fact, there have been various experiments showing that hybridization between wild and cultivated species of *Raphanus* is very easy. We conducted artificial hybridization between a radish cultivar and several lines of *R. raphanistrum* (Yamagishi and Terachi 1997). The number of seeds per pollinated flower ranged from 2.22 to 4.10 when cultivated radish was used as the seed parent, while the number varied from 0.06

Fig. 2.1 Flowers of wild and cultivated radishes. *Left* cultivated radish 'MS-Gensuke', *right Raphanus raphanistrum* (RS–15), *center* F_1 hybrid between 'MS-Gensuke' and *R. raphanistrum*

Fig. 2.2 Pods of wild and cultivated radishes. *Top R. raphanistrum* (RS-15), *bottom* 'MS-Gensuke'

to 0.24 in the reciprocal cross, showing a difference depending on the direction of hybridization. However, Eber et al. (1998) observed numerous hybrids of cultivated and wild radish in field experiments. In their experiments, *R. sativus* as female parent was pollinated spontaneously with *R. raphanistrum* growing <200 m away. On the other hand, Snow et al. (2001) found that interspecific F_1 hybrids between *R. raphanistrum* as the female parent and cultivated radish (*R. sativus*) as the pollen parent had lower fitness than wild plants. But they also observed that after three years, plants with white flowers, the gene for which is derived from crop radish, were present at frequencies of 8–22% of wild plants, demonstrating the persistence of crop genes in wild radish populations. Further, Bett and Lydiate (2003) produced a hybrid between *R. sativus* and *R. raphanistrum*. They observed meiosis of the F_1 hybrid and segregation of RFLP markers for all chromosomes. Their findings indicated that the chromosomes of the *Raphanus* hybrid were completely homologous and collinear. They interpreted the results as suggesting that *R. sativus* and *R. raphanistrum* could be more usefully considered as different subspecies rather than different species.

Another interesting observation on wild radish was made by Hegde et al. (2006), who collected wild radish growing in California and compared variation of its morphological and allozyme characteristics with *R. raphanistrum* from Europe and cultivated radishes. California wild radishes had a specific combination of traits different from the two *Raphanus* species. The results suggested the local extinction of two *Raphanus* species in California via replacement by their hybrid-derived descendants. Based on these experiments on the hybridization between *R. sativus* and *R. raphanistrum*, we can conclude that the two species should be considered as a single species, as Warwick and Francis (2005) described.

To determine the origin of California wild radish described above, Ridley et al. (2008)

compared chloroplast DNA among European cultivars (4 cultivars), *R. raphanistrum* (6 populations) and California wild radish (11 populations). The chloroplast DNA was classified into 8 haplotypes, but cultivated radish and *R. raphanistrum* shared no chloroplast DNA haplotypes. On the other hand, both *R. sativus* and *R. raphanistrum* shared haplotypes with California wild radish, evidencing bidirectional hybridization between the progenitor species in the creation of the California lineage. They also found that multiple cultivars and multiple European source populations of *R. raphanistrum* contributed to the diversity of chloroplast DNA.

2.5 Origin of Cultivated Radish

As mentioned above for the case of California wild radish, analyses of organelle genomes (chloroplast and mitochondria) provide evidence for origin of cultivated or wild plants. From this viewpoint, we have been comparing chloroplast or mitochondrial genomes among wild and cultivated radishes. Based on PCR of mitochondrial *cox1* and *orfB* genes, five haplotypes were found, and all of them were commonly present in *R. raphanistrum* and *R. sativus*, although contrasting features were found among cultivated radishes depending on the region of cultivation (Yamagishi and Terachi 2003). The results demonstrated that cultivated radishes have multiple origins from wild radishes. Similarly, we determined the DNA sequence of the chloroplast genome in the region between *trn*L and *psb*G for 59 lines or cultivars of wild and cultivated radish (Yamagishi et al. 2009). The sequences were divided into 11 types. Though seven of them were unique to wild radish, and one was specific to cultivated radish, the other three haplotypes were shared between wild and cultivated radishes, another demonstration of the multiple origins of cultivated radish. This observation on the chloroplast genome was ascertained by Lü et al. (2008). They analyzed nucleotide sequence variations of the *trn*K/*mat*K region in cultivated

Fig. 2.3 Habitat of Japanese wild radish along the seacoast

and wild radish species. The radish sequences were classified into three groups, and both wild and cultivated radishes belonged to all three clusters. The three kinds of cultivated radish, European small radish, Asian big radish, and black radish, each belonged to a different cluster of the phylogenetic tree, suggesting independent multiple origins of these varieties.

We also analyzed the DNA sequence of the mitochondrial causal gene, *orf138*, of Ogura cytoplasmic male sterility (Yamagishi and Terachi 2001). We presumed the origin of Japanese wild radish from the distribution of haplotypes of *orf138*, in addition to the distribution of the nuclear fertility restoration gene that suppresses the expression of *orf138* in wild and cultivated radishes (Yamagishi 1998). Japanese wild radish grows along the seaside in almost all parts of Japan (Fig. 2.3) and it does not develop a swollen root (Fig. 2.4). We inferred that Japanese wild radish is a descendant of *R. raphanistrum*, and in Japanese wild radish, several independent mutations occurred in *orf138*. We also inferred that thereafter a few types of Japanese wild radish were picked up into cultivation. Examples of the domestication of Japanese wild radish are local cultivars such as 'Sabaka' and 'Kozena' (or 'Kosena'), shown in Fig. 2.5 (Yamagishi and Yamashita 2009). 'Kozena' is a unique local cultivar whose root is not swollen; the leaves are preserved for dietary use. Although these are considered to be old local events, even recently, intentional breeding was conducted using Japanese wild radish as breeding materials, as in the case of 'Izumo-orochi daikon' in Shimane Prefecture (Fig. 2.6).

2.6 Diversity of Cultivated Radish

Radish is cultivated almost worldwide, and there are various ways to use it, as described by Crisp (1995) and Kaneko and Matsuzawa (1993). The main parts used are the enlarged root and

Fig. 2.4 Roots of Japanese wild radish (five plants on the *left*) in comparison with cultivated radishes (four plants on the *right*)

Fig. 2.5 Japanese local cultivars domesticated from wild radish. **a** 'Sabaka' and **b** 'Kozena'

Fig. 2.6 A cultivar bred from Japanese wild radish in Shimane Prefecture, 'Izumo-orochi daikon'

hypocotyl, and based on their size, there are two broad types. One is a small-rooted, short-season European type, and the other is a large-rooted type grown mostly in Asia. In addition to these two, there are cultivars for special use. One cultivar is a rat's tail radish grown in Southeast Asia, whose young and extremely long pods are eaten, while another is cultivated for oilseed in South Asia.

Including these types, cultivated radish has been classified into five varieties. The varieties classified by two research groups and their correspondence are shown in Table 2.1. Among the five, the names of only two varieties coincide between the two groups; namely, *niger* (black radish) and *caudatus* (rat's tail radish). Thus, not only the species classification of *Raphanus*, but also the intraspecific classification and nomenclature of *R. sativus* are still in confusion.

Additionally, I observed differences in characteristics of the cultivated radish, following the naming conventions of Lü et al. (2008), in seed size, length of pods, and plant morphology such as roots, leaves and flowers. Figure 2.7 shows seeds of European small radish ('Icycle') and Asian big radish ('Shinshu Jidaikon'). The latter has larger seeds. The longest seed diameter was 3.9 mm on average in 'Shinshu Jidaikon',

Table 2.1 Variety classifications of cultivated radish by two research groups and their correspondence

Kaneko and Matsuzawa (1993)	Lü et al. (2008)
Radicula	*Sativus* (European small radish)
Raphanistroides	*Hortensis* (East Asian big long radish)
Niger	*Niger* (black Spanish radish)
Oleifer	*Chinensis* (Chinese oil radish)
Caudatus	*Caudatus* (tail-podded radish)

Fig. 2.7 Comparison of seed size between 'Shinshu Jidaikon' (var. *hortensis*: *left*) and 'Icycle' (var. *sativus*: *right*)

Table 2.2 Variation in seed size among cultivars and wild radishes

Variety[a]	Cultivar	Seed diameter (mm)[b] Average (range[c])
sativus	Icycle	2.9 (2.5–3.3)
niger	Kurodaikon A	2.9 (2.4–3.4)
hortensis	Shinshu-Jidaikon	3.9 (3.6–4.6)
chinensis	Unnan-Aburadaikon	3.8 (3.3–4.4)
caudatus	Saya-Daikon	3.8 (3.3–4.1)
California wild radish		3.5 (3.0–4.3)
Japanese wild radish		3.8 (3.4–4.3)

[a]Nomenclature by Lü et al. (2008)
[b]The longest diameter in a seed was measured
[c]The minimum and maximum values of ten seeds are shown

but ‘Icycle’ seeds were smaller by about 1 mm (Table 2.2). Black radish has similar small seeds to the European small radish, whereas the seeds of Chinese oil radish and rat’s tail radish were as large as Asian big radish (Table 2.2).

Although Crisp (1995) described the pods of rat’s tail radish growing up to 80 cm in length, the pods I observed were about 20 cm (Fig. 2.8). However, they are truly much longer than the pods of East Asian big long radish and *R. raphanistrum*, the latter two being less than 5 cm (Fig. 2.9).

Typical morphology of various cultivars of radish is shown in Figs. 2.10, 2.11, 2.12, 2.13, 2.14, 2.15, 2.16, 2.17, 2.18, 2.19, 2.20, 2.21, 2.22 and 2.23. Comparison of the plant with a ruler or Vernier caliper in each figure helps interpret the size of the cultivar. Both the ruler

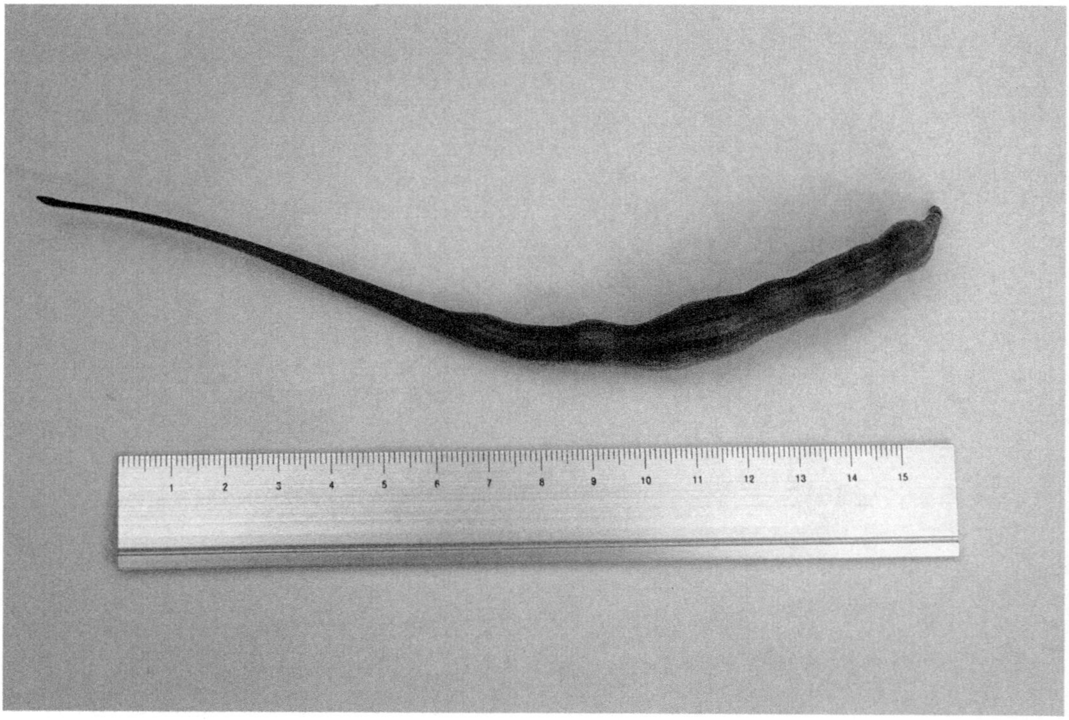

Fig. 2.8 Pod of rat’s tail radish ‘Saya-daikon’ (var. *caudatus*)

Fig. 2.9 Comparison of pod size among rat's tail radish (*top*), 'Kuki-daikon' (var. *hortensis, middle*) and *R. raphanistrum* (*bottom*)

Fig. 2.10 Plant morphology of European small radish (**a** 'Comet', **b** 'Icycle')

and caliper are 15 cm long. The figures show the differences in leaf morphology as well as the root. Figure 2.10a, b shows two cultivars of European small radish. Though both a round red cultivar and a long white cultivar exist in European small radish, the root length of both is less than 15 cm. In contrast, Fig. 2.11a, b exhibit the two types of root shape in black radish. Both types have lobed leaves.

Figures 2.12, 2.13 and 2.14 show cultivars in China. The Chinese cultivars show large variations in both root shape and root color, and also in leaf morphology. In the case of 'Ten-an-Koshin', the inside of the root is pink (Fig. 2.15). Figures 2.16, 2.17, 2.18, 2.19, 2.20, 2.21, 2.22 and 2.23 show some Japanese cultivars. 'Sakurajima-daikon' and 'Moriguchi-daikon' (Figs. 2.16 and 2.17) are famous cultivars, having the world's largest

Fig. 2.11 Plant morphology of black radish (**a** 'Kurodaikon A', **b** 'Long Swart')

Fig. 2.12 Chinese cultivar 'Aonaga-daikon'

Fig. 2.13 Chinese cultivar 'Koi'

Fig. 2.14 Chinese cultivar 'Ten-an-Koshin'

(more than 20 kg) and the longest (almost 2 m long) roots, respectively. In general, Japanese cultivars have larger roots than Chinese cultivars, but there are wide variations in root size, root shape and root color, and a Japanese cultivar with roots as small as that of European small radish is present. These figures also show the diversity of leaf shape, especially the degree of lobing, ranging from the entire leaf ('Karami-daikon': Fig. 2.18) to complex, deeply lobed leaves ('Kireba-Jidaikon': Fig. 2.19). Such large variations among the cultivars cause difficulty in classification of cultivated radish.

There are also variations in flower morphology, specifically the color and shape of petals, as shown in Fig. 2.24. The petal color varies from white to pink. Even in the white-petal cultivars, sepal color differs. Figure 2.25 shows the variation in flower size compared to black radish ('Kurodaikon A').

Finally, there is also variation within one cultivar in addition to variations among radish cultivars. I purchased one local cultivar, named 'Karami-daikon', sold in a local market (Fig. 2.26). Local cultivars have been traditionally bred by open pollination, and thus, the cultivar shows differences

Fig. 2.15 Inside of the root of 'Ten-an-Koshin'

Fig. 2.16 Japanese cultivar 'Sakurajima-daikon'

in both root shape and root color among individuals. Development of F_1 hybrids in radish breeding has mainly been conducted in private seed companies, and has dramatically increased the uniformity of such radish cultivars, as described in Chap. 15. Such tendencies in radish breeding may decrease the diversity that has been kept in local radishes.

Fig. 2.17 Japanese cultivar 'Moriguchi-daikon'

Fig. 2.18 Japanese cultivar 'Karami-daikon'

Fig. 2.19 Japanese cultivar 'Kireba-Jidaikon'

Fig. 2.20 Japanese cultivar 'Kokubu-daikon'

Fig. 2.22 Japanese cultivar 'Miyashige-Oonaga'

Fig. 2.21 Japanese cultivar 'Uchiki-Gensuke' (or 'Utsugi-Gensuke')

Fig. 2.23 Japanese cultivar 'Okinawa -Shimadaikon'

Fig. 2.24 Variation in flower morphology among Japanese cultivars (**a** 'Awabansei', **b** 'Moriguchi-daikon', **c** 'Kokubu-daikon')

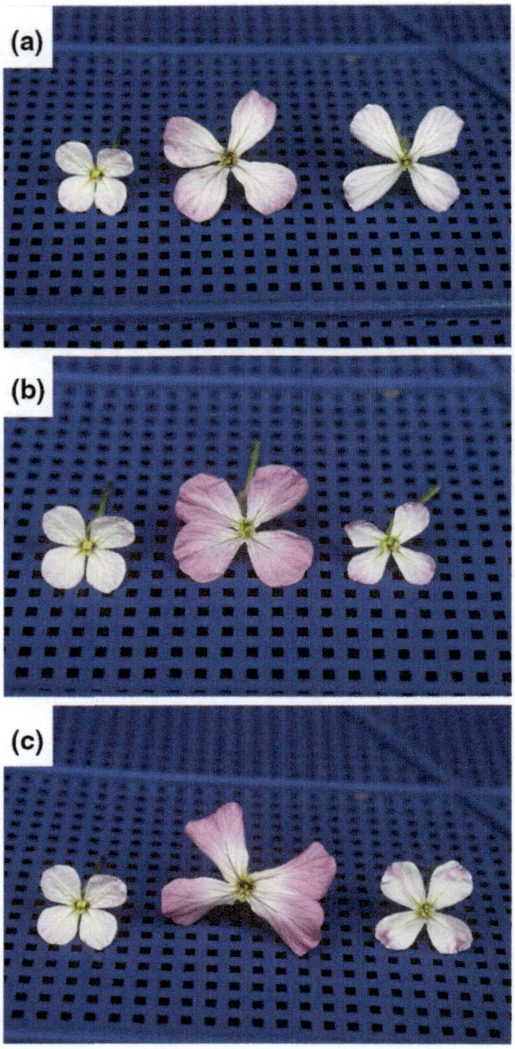

Fig. 2.25 Flower size of cultivated radishes in comparison with black radish 'Kurodaikon A' (*left* in all panels, **a**–**c**.). **a** *Center* 'Fumei' (Chinese cultivar), *right* 'Taibyo-Minowase' (Japanese cultivar). **b** *Center* 'Shinshu-Jidaikon' (Japanese cultivar), *right* 'Long Swart' (black radish). **c** *Center* 'Sakurajima-daikon' (Japanese cultivar), *right* 'Aonaga-daikon' (Chinese cultivar)

Fig. 2.26 Variation in root shape and color among individual plants within a local cultivar of Japanese radish named 'Karami-daikon'

References

Al-Shehbaz IA, Beilstein MA, Kellogg EA (2006) Systematics and phylogeny of the Brassicaceae (Cruciferae): an overview. Plant Syst Evol 259: 89–120

Baranger A, Chèvre AM, Eber F, Renard M (1995) Effect of oilseed rape genotype on the spontaneous hybridization rate with a weedy species: an assessment of transgene dispersal. Theor Appl Genet 91:956–963

Bett KE, Lydiate DJ (2003) Genetic analysis and genome mapping in *Raphanus*. Genome 46:423–430

Cheam AH, Code GR (1995) The biology of Australian weeds 24. *Raphanus raphanistrum* L. Plant Protect Q 10:2–13

Chèvre AM, Eber F, Baranger A, Hureau G, Barret P, Picault H, Renard M (1998) Characterization of backcross generations obtained under field conditions from oilseed rape—wild radish F_1 interspecific hybrids: an assessment of transgene dispersal. Theor Appl Genet 97:90–98

Crisp P (1995) Radish. In: Smartt J, Simmonds NW (eds) Evolution of crop plants. Longman Scientific and Technical, London, pp 86–89

Darmency H, Lefol E, Fleury A (1998) Spontaneous hybridizations between oilseed rape and wild radish. Mol Ecol 7:1467–1473

Eber F, Boucherie R, Broucqsault L-M, Bouchet Y, Chèvre A-M (1998) Spontaneous hybridization between vegetable crops and weeds. 1: garden radish (*Raphanus sativus* L.) and wild mustard (*Sinapis arvensis* L.). Agronomie 18:489–497

Eber F, Chèvre AM, Baranger A, Vallée P, Tanguy X, Renard H (1994) Spontaneous hybridization between a male-sterile oilseed rape and two weeds. Theor Appl Genet 88:362–368

Gómez-Campo C (1980) Morphology and morphotaxonomy of the Tribe Brassiceae. In: Tsunoda S, Hinata K, Gómez-Campo C (eds) *Brassica* crops and wild allies. Japan Scientific Societies Press, Tokyo, pp 3–31

Guéritaine G, Bonavent JF, Darmency H (2003) Variation of prezygotic barriers in the interspecific hybridization between oilseed rape and wild radish. Euphytica 130:349–353

Guéritaine G, Darmancy H (2001) Polymorphism for interspecific hybridization within a population of wild radish (*Raphanus raphanistrum*) pollinated by oilseed rape (*Brassica napus*). Sex Plant Reprod 14:169–172

Hegde SG, Nason JD, Glegg JM, Ellstrand MC (2006) The evolution of California's wild radish has resulted in the extinction of its progenitors. Evolution 60:1187–1197

Kaneko Y, Matsuzawa Y (1993) Radish. In: Kalloo G, Bergh BO (eds) Genetic improvement of vegetable crops. Pergamon Press, London, pp 487–510

Kitamura S (1958) Varieties and transitions of radish. In: Nishiyama I (ed) Japanese radish. Japanese Science Society, Tokyo, pp 1–19

Lü N, Yamane K, Ohnishi O (2008) Genetic diversity of cultivated and wild radish and phylogenetic relationships among *Raphanus* and *Brassica* species revealed by the analysis of *trn*K/*mat*K sequence. Breed Sci 58:15–22

Lysak MA, Koch MA, Pecinka A, Schubert I (2005) Chromosome triplication found across the tribe *Brassiceae*. Genome Res 15:516–525

Mitsui Y, Shimomura M, Komatsu K, Namiki N, Shibata-Hatta M, Imai M, Katayose Y, Mukai Y, Kanamori H, Kurita K, Kagami T, Wakatsuki A, Ohyanagi H, Ikawa H, Minaka N, Nakagawa K, Shiwa Y, Sasaki T (2015) The radish genome and comprehensive gene expression profile of tuberous root formation and development. Sci Rep 5:10835. doi:10.1038/srep10835

Moghe GD, Hufnagel DE, Tang H, Xiao Y, Dworkin I, Town CD, Conner JK, Shiu SH (2014) Consequences of whole-genome triplication as revealed by comparative genomic analyses of the wild radish *Raphanus raphanistrum* and three other Brassicaceae species. Plant Cell 26:1925–1937

Pradhan AK, Prakash S, Mukhopadhyay A, Pental D (1992) Phylogeny of *Brassica* and allied genera based on variation in chloroplast and mitochondrial DNA patterns: molecular and taxonomic classifications are incongruous. Theor Appl Genet 85:331–340

Ridley CE, Kim S-C, Ellstrand NC (2008) Bidirectional history of hybridization in California wild radish, *Raphanus sativus* (Brassicaceae), as revealed by chloroplast DNA. Am J Bot 95:1437–1442

Rollins RC (1993) The Cruciferae of continental North America. Stanford University Press, Stanford, CA

Sahli HF, Conner JK, Shaw FH, Howe S, Lale A (2008) Adaptive differentiation of quantitative traits in the globally distributed weed, wild radish (*Raphanus raphanistrum*). Genetics 180:945–955

Schulz OE (1936) Cruciferae. In: Engler A, Prantle K (eds) Die natürlichen Pflanzenfamilien, vol 17, 2nd edn. Engelmann, Leipzig, pp 227–658

Snow AA, Uthus KL, Culley TM (2001) Fitness of hybrids between weedy and cultivated radish: implications for weed evolution. Ecol Appl 11:934–943

Warwick SI, Black LD (1991) Molecular systematics of *Brassica* and allied genera (Subtribe Brassicinae, Brassiceae)—chloroplast genome and cytodeme congruence. Theor Appl Genet 82:81–92

Warwick SI, Francis A (2005) The biology of Canadian weeds. 132. *Raphanus raphanistrum,* L. Can J Plant Sci 85:709–733

Warwick SI, Sauder CA (2005) Phylogeny of tribe Brassiceae (Brassicaceae) based on chloroplast restriction site polymorphisms and nuclear ribosomal internal transcribed spacer and chloroplast *trn*L intron sequences. Can J Bot 83:467–483

Yamagishi H (1998) Distribution and allelism of restorer genes for Ogura cytoplasmic male sterility in wild and cultivated radishes. Genes Genet Syst 73:79–83

Yamagishi H, Terachi T (1997) Molecular and biological studies on male-sterile cytoplasm in the Cruciferae. IV. Ogura-type cytoplasm found in the wild radish, *Raphanus raphanistrum*. Plant Breed 116:323–329

Yamagishi H, Terachi T (2001) Intra- and inter-specific variations in the mitochondrial gene *orf138* of Ogura-type male-sterile cytoplasm from *Raphanus sativus* and *Raphanus raphanistrum*. Theor Appl Genet 103:725–732

Yamagishi H, Terachi T (2003) Multiple origins of cultivated radishes as evidenced by a comparison of the structural variations in mitochondrial DNA of *Raphanus*. Genome 46:89–94

Yamagishi H, Terachi T, Ozaki A, Ishibashi A (2009) Inter- and intraspecific sequence variations of the chloroplast genome in wild and cultivated *Raphanus*. Plant Breed 128:172–177

Yamagishi H, Yamashita Y (2009) Origin of local radish variety, 'Sabaka', in Kyoto Prefecture inferred by analyses of genes for cytoplasmic male-sterility and fertility restoration. Hort Res (Japan) 8:1–6

Yang Y-W, Lai K-N, Tai P-Y, Ma D-P, Li W-H (1999) Molecular phylogenetic studies of Brassica, Rorippa, Arabidopsis and allied genera based on the internal transcribed spacer region of 18S–25S rDNA. Mol Phylogenet Evol 13:455–462

Yang Y-W, Tai P-Y, Chen Y, Li W-H (2002) A study of the phylogeny of *Brassica rapa*, *B. nigra*, *Raphanus sativus*, and their related genera using noncoding regions of chloroplast DNA. Mol Phylogenet Euol 13:455–462

3 Genetic Maps and Whole Genome Sequences of Radish

Kenta Shirasawa and Hiroyasu Kitashiba

Abstract

Radish, *Raphanus sativus* L., is a member of Brassicaceae, to which *Arabidopsis thaliana*, a model plant in plant biology, belongs, as do other *Brassica* species including important crops. However, genetic and genomic studies of radish have been behind those of *Arabidopsis* and *Brassica*. In this decade, much effort has been made to develop genetic resources for radish, e.g., DNA markers, genetic maps, and whole genome sequences. Studies using the obtained information have revealed the genome structure of radish in terms of ancestral karyotype and have also prompted the identification of genes for agronomically important traits in radish through a map-based cloning strategy and quantitative trait locus analysis. In this chapter, we review the evolving development of radish genetic map in the past 15 years and the current status of genome sequencing of radish. We also introduce the latest strategy for the construction of a high-density genetic map using next-generation sequencing technology and propose a prospective direction of genetics and genomics research in radish which would be helpful for researchers and breeders in their efforts to promote radish breeding programs efficiently.

Electronic supplementary material **The online version of this chapter (doi:10.1007/978-3-319-59253-4_3) contains supplementary material, which is available to authorized users.**

K. Shirasawa
Kazusa DNA Research Institute, Kisarazu, Japan

H. Kitashiba (✉)
Graduate School of Agricultural Science, Tohoku University, 468-1, Aoba, Aramaki, Aobaku, Sendai 980-0845, Japan
e-mail: hiroyasu.kitashiba.c7@tohoku.ac.jp

3.1 Introduction

Molecular genetic resources, e.g., DNA markers, genetic maps, genome sequences, and mutants are essential tools for genetic and genomic studies. In *Arabidopsis thaliana*, which is a member of Brassicaceae and has been recognized as a model plant for molecular genetic studies, a numerous genetic resources have been developed. *Brassica rapa*, *Brassica nigra*, and *Brassica oleracea* are

T. Nishio and H. Kitashiba (eds.), *The Radish Genome*,
Compendium of Plant Genomes, DOI 10.1007/978-3-319-59253-4_3

diploid species having basic genomes of A, B, and C, respectively, and *Brassica juncea* (AABB), *Brassica napus* (AACC), and *Brassica carinata* (BBCC) are allotetraploid species having combinations of genomes of the diploid species (U 1935). Among them, three species, *B. rapa*, *B. oleracea,* and *B. napus* are major vegetable and oil crops. Molecular genetic resources for the three species have also been developed to accelerate genetic studies on agronomically significant traits. DNA markers, especially, are essential tools in the genetic study. For production of DNA markers such as simple sequence repeat (SSR) and polymerase chain reaction-restriction fragment length polymorphism (PCR-RFLP), also known as cleaved amplified polymorphic sequence (CAPS), information on expressed sequence tags (ESTs) and genome sequences have been utilized and a variety of genetic maps have been constructed. From 2011 onward, whole genome sequencing projects have been achieved in *B. rapa* (The *Brassica rapa* Genome Sequencing Project Consortium 2011), *B. oleracea* (Liu et al. 2014), and *B. napus* (Chalhoub et al. 2014). In addition, the genome sequences of *B. nigra* and *B. juncea* have also been reported (Yang et al. 2016). Referring to these sequences, a considerable number of single nucleotide polymorphisms (SNPs) among cultivars have been collected and used for rapid construction of higher-density genetic maps, enabling rapid and efficient genetic analyses, e.g., quantitative trait loci (QTL) and genome-wide association study (GWAS) analyses.

Radish (*Raphanus sativus* L.) is a member of Brassicaceae and closely related to *B. rapa* and *B. oleracea* (Inaba and Nishio 2002). It is widely used around the world as a vegetable crop, e.g., Asian big radish in East Asia, European small radish in Europe and the USA, and rat's tail radish in South Asia and Africa (see Chap. 2). However, since radish has been behind the other *Brassica* crops and *A. thaliana* in terms of genetic and genomic studies, genomic information on *Raphanus* has been limited. Actually, the genomic information on *Arabidopsis* and *Brassica* was used for construction of genetic maps of *Raphanus* (Bett and Lydiate 2003; Tsuro et al. 2005, 2008). To make up the dearth of studies on radish, great efforts have been made to develop genetic resources for *R. sativus*. In this chapter, we summarize the studies of genetics and genomics in *Raphanus*, focusing on genetic maps and whole genome sequences. In the final section, we introduce our latest attempt to construct a high-density genetic map by advanced DNA sequencing technologies together with anchoring the reported genome sequences.

3.2 Construction of Genetic Maps

3.2.1 First-Generation Maps

The first genetic map of genus *Raphanus*, to our knowledge, was constructed using interspecific F_2 and BC_1 populations derived from an interspecific hybrid between *R. sativus* and *R. raphanistrum* because DNA polymorphisms in this combination were expected to be higher than those in an intraspecific combination (Bett and Lydiate 2003). In this map, 144 of 177 RFLP probes derived from *Brassica* succeeded in detecting 236 polymorphisms, in which 84% of the tested *Brassica* RFLP probes were informative in the mapping populations (Table 3.1).

The first 'intraspecific' genetic map within *R. sativus* was constructed by an amplified fragment length polymorphism (AFLP) fingerprinting technology (Tsuro et al. 2005) (Table 3.1). Analysis of random amplified polymorphic DNA (RAPD) is another DNA fingerprinting technique, which is much less labor intensive, and costly than AFLP analysis. Because no genome sequence information is required for AFLP and RAPD analyses, both techniques were suitable for *Raphanus*, whose genome sequences had not been available. Indeed, several genetic maps with 156–545 AFLP loci have been constructed in collaboration with RAPD loci (Budahn et al. 2009; Kaneko et al. 2007) (Table 3.1).

Table 3.1 Genetic maps in *Raphanus* species

Cross combination	Population type	Number of linkage groups	Number of mapped loci	Marker types	Map length (cM)	References
'R7-21' and 'R11-4' (*R. raphanustrum*)	F_2 (n = 85), BC_1 (n = 54)	9	236	RFLP	844	Bett and Lydiate (2003)
'Huang-he hong-wan' (Kouga-benimaru) and 'Utsugi-Gensuke'	F_2 (n = 94)	11	241	AFLP, SSR and CAPS	675.8	Tsuro et al. (2005)
'GSK3-1' and 'HA2'	BC_1 (n = 129)	10	156	RFLP and RAPD	1239.5	Kaneko et al. (2007)
'Huang-he hong-wan' (Kouga-benimaru) and 'Utsugi-Gensuke'	F_2 (n = 104)	14	198	AFLP, SSR and CAPS	667.6	Tsuro et al. (2008)
'A22/2' and 'A317/1'	F_2 (n = 245)	9	545	RAPD, AFLP, and SSR	1517	Budahn et al. (2009)
'Huang-he hong-wan' (Kouga-benimaru) and 'Utsugi-Gensuke'	F_2 (n = 95)	18	313	AFLP and SSR	554	Kamei et al. (2010)
'GSK3-1' and 'HA2'	F_8–F_{10} (n = 155)	9	843	SSR, RFLP, and RAPD	1129.2	Shirasawa et al. (2011)
'Sayatori 26704' and 'Aokubi *S-h*'	F_2 (n = 189)	9	746	SNP	806.7	Li et al. (2011)
'NAU-Dysx' and 'NAU-Yh'	F_2 (n = 172)	9	523	SRAP, RAPD, SSR, ISSR, RAMP, and RGA	1678.2	Xu et al. (2012)
'TBS-2-5-3-2-(3)' and 'AZ26H-24-6-5-(3)'	F_2 (n = 189)	9	188	SNP	856	Zou et al. (2013)
'835' and 'B2'	F_2 (n = 222)	9	220	SSR and intron polymorphism	1041.6	Yu et al. (2013)
'Rat-tail radish' and 'Harufuku'	F_4–F_5 (n = 94)	9	322	AFLP, SSR, and STS	673.6	Hashida et al. (2013)
'Sayatori 26704' and 'Aokubi *S-h*'	F_2 (n = 189)	9	2553	SNP, CAPS, and SSR	1165.8	Kitashiba et al. (2014)
'WK10039' and 'WK10024'	F_2 (n = 93)	9	1000	SNP and PCR	1538	Mun et al. (2015)
'835' and 'B2'	F_2 (n = 210)	9	258	SSR and intron polymorphism	1135.4	Yu et al. (2016)

3.2.2 Second-Generation Maps

Generally, three types of DNA makers, RFLP, AFLP, and RAPD, do not require nucleotide sequence information of target plants. However, since the AFLP and RAPD makers are anonymous markers, it is difficult to compare genetic maps or to combine them. In plant species such as the genus *Brassica* and *Raphanus*, which have experienced a duplication or triplication of genome, some RFLP probes are apt to hybridize with multiple loci, resulting in construction of a genetic map with low accuracy. Therefore, SSR, CAPS, and SNP markers have been developed based on nucleotide sequence information of radish. Although the development of these markers is costly, genotyping analysis by RFLP, AFLP, and RAPD markers have been gradually replaced by that using SSR, CAPS, and SNP markers due to their high accuracy. Especially, SSR and SNP markers are commonly utilized for construction of genetic maps these days since an SSR marker can detect multiple loci and SNPs are the most abundant type of DNA polymorphism in genomes. Consequently, both of these markers are expected to be useful in the construction of a high-density genetic maps.

A large number of SSR markers have been developed from the ESTs and genomic sequence data of *B. rapa* (Suwabe et al. 2002, 2003, 2006). Because *Raphanus* is a species closely related to *B. rapa*, and the developed *B. rapa* SSRs have also been applied to the genus *Raphanus* to construct genetic maps (Kamei et al. 2010; Tsuro et al. 2005, 2008) (Table 3.1). However, the number of *Brassica* SSR markers applicable to the *Raphanus* genome was found to be limited, ~53%, because the two genomes are suggested to be relatively remote (Kamei et al. 2010). In order to solve this problem, 26,606 EST sequences were collected from seedlings, roots, leaves, and flowers of *R. sativus* cv. 'Utsugi-Gensuke' (Shirasawa et al. 2011) and 3800 SSRs were mined using the obtained information, 642 of which (16.9%) showed polymorphisms between an inbred line of 'Utsugi-Gensuke' and an inbred line of 'Tokinashi.' For construction of a genetic map, the 642 SSR markers yielding 662 polymorphic loci were subjected to mapping analysis using recombinant inbred lines derived from a cross between inbred lines of 'Utsugi-Gensuke' and 'Tokinashi.' Finally, 630 SSR loci together with 213 RAPD, RFLP, and a trait maker (843 loci in total) were mapped into nine linkage groups covering 1129.2 cM in total length (Table 3.1).

In parallel, a large number of SNP markers were developed by Li et al. (2011). For production of these markers, EST sequences published in the Radish Database of Michigan State University (http://radish.plantbiology.msu.edu/) were used to design primer pairs. Primer pairs from 4709 EST sequences were designed and subjected to PCR amplification using two radish inbred lines, 'Sayatori,' which is a rat's tail radish, and 'Aokubi.' Single PCR fragments were detected in 3576 primer pairs and sequenced. Consequently, SNPs were detected in 1465 DNA fragments between the parental lines. The SNPs were converted to dot-blot-SNP markers (Shiokai et al. 2010; Shirasawa et al. 2006). The 772 selected dot-blot-SNP markers were subjected to genotyping of 189 F_2 progenies derived from a crossing of 'Sayatori' and 'Aokubi.' A total of 746 SNP markers were assigned to nine linkage groups with a total length of 806.7 cM (Table 3.1).

The two genetic maps of *R. sativus* described above (Li et al. 2011; Shirasawa et al. 2011) were higher-density maps than any other previous genetic maps. Besides, the DNA markers used in the both genetic maps were produced based on the radish EST sequences. Therefore, comparisons of genome of *R. sativus* with *A. thaliana* and several *Brassica* species were performed by BLAST search. As a result, several interesting points were revealed: (1) 23 of the 24 ancestral genomic blocks (Schranz et al. 2006) are distributed in the radish genome and half of them are found to be triplicated, (2) the whole LG9 of Shirasawa et al. (2011) equal to LG8 of Li et al. (2011) shows collinearity to more than half of chromosome 1 of *A. thaliana* and to a chromosome of ancestral karyotype 1 (Schranz et al. 2006), (3) all the chromosomes in *R. sativus*, except for LG5, were inferred to have

Table 3.2 Correspondence of linkage group between reported maps

Proposed no. in this review	Linkage group number								
	Rs1	Rs2	Rs3	Rs4	Rs5	Rs6	Rs7	Rs8	Rs9
Kaneko et al. (2007)	LG9	LG5	LG8	LG4	LG3	LG2	LG1	LG7	LG6
Tsuro et al. (2008)	LG9, 10, 13	–	LG3	LG8	LG4, 7, 12	LG1	LG11	LG2	LG5, 6
Budahn et al. (2009)	g	i	d	c	e	f	h	b	a
Kamei et al. (2010)	N/A	N/A	N/A	N/A	LG1	N/A	N/A	N/A	N/A
Shirasawa et al. (2011)	LG9	LG5	LG8	LG4	LG3	LG2	LG1	LG7	LG6
Li et al. (2011)	LG8	LG4	LG6	LG3	LG2	LG1	LG9	LG7	LG5
Zou et al. (2013)	LG8	LG4	LG6	LG3	LG2	LG1	LG9	LG7	LG5
Yu et al. (2013)	LG9	LG5	LG8	LG4	LG3	LG2	LG1	LG7	LG6
Hashida et al. (2013)	Chr G	Chr I	Chr D	Chr C	Chr E	Chr F	Chr H	Chr B	Chr A
Kitashiba et al. (2014)	R1	R2	R3	R4	R5	R6	R7	R8	R9
Mun et al. (2015)	R1	R2	R3	R4	R5	R6	R7	R8	R9
Yu et al. (2016)	LG9	LG5	LG8	LG4	LG3	LG2	LG1	LG7	LG6

corresponding homoeologous chromosomes in the *Brassica* species (Li et al. 2011).

Based on the developed DNA markers described above, QTL study together with construction of a genetic map has been undertaken in radish. An SSR-based genetic map consisting of 220 marker loci, including some of those developed by Suwabe et al. (2006) and Shirasawa et al. (2011), has been constructed to detect QTLs for the *Fusarium* wilt resistance trait (Yu et al. 2013) (Table 3.1). Zou et al. (2013) constructed an SNP-based genetic map for QTL analysis of glucosinolate content using a part of the SNP markers developed by Li et al. (2011) (Table 3.1).

Differing from the types of markers described so far, sequence-related amplified polymorphism (SRAP) markers (Li and Quiros 2001) have been produced in radish (Xu et al. 2012). Two hundred and forty-eight SRAP markers together with SSR, RAPD, and so on were used to construct a genetic map with 1678.2 cM for a QTL study on cadmium accumulation (Xu et al. 2012) (Table 3.1).

Although many *Raphanus* genetic maps have been constructed as we described above, the nomenclature of linkage groups has differed between the respective genetic maps (Table 3.2). As this situation is confusing and misleading for future work on radish genetics and genomics, we propose use of the consensus linkage group names, Rs1–Rs9 (Table 3.2), as suggested by Li et al. (2011), which are based on homoeology analysis between the genetic map and the putative Brassicaceae ancestral karyotype suggested by Schranz et al. (2006).

3.3 Whole Genome Sequences of Cultivated Radish and Its Relatives

Since the release of the genome sequences of *A. thaliana* (The Arabidopsis Genome Initiative 2000) and *B. rapa* (The *Brassica rapa* Genome Sequencing Project Consortium 2011), three independent research groups have reported the genome sequences of *R. sativus* (Table 3.3). The first report was published by Kitashiba et al. (2014). Using Illumina short-read sequencing technology, paired-end sequence reads (insert size of 250 bp) were collected from 'Aokubi' inbred line having *S-h* haplotype to obtain genome contig sequences assembled with

Table 3.3 Genome sequences in *Raphanus sativus* and its relatives

Name of line sequenced	Generation	Number of scaffolds	Total length of scaffolds (Mb)	N50 (kb)	Genome coverage (%)	Repetitive sequences (Mb)	Number of predicted non-TE and non-pseudo genes	References
'Aokubi *S-h*'	S3	76,592	402	46.3	76	26.6	61,572	Kitashiba et al. (2014)
'Aokubi'	DH	40,123	383	138.2	67	36.7	54,357	Mitsui et al. (2015)
'WK10039'	S8	11,389	426	927.3	83	35.6	46,514	Jeong et al. (2016)
R. raphanistrum (weedy)	S5	68,331	254	10.1	49	N/A	38,174	Moghe et al. (2014)

SOAPdenovo followed by gap closing with GapCloser. In addition, sequence reads from a mate-pair library (insert size of 5 kb) and bacterial artificial chromosome (BAC)-end sequences have been employed to scaffold the contigs by SSPACE 2.0, resulting in 76,592 sequences (N50 = 46 kb; $\geq$300 bp) spanning 402 Mb of the radish genome, which covers approximately 75.6% of the estimated genome size of 530 Mb. In the sequences, a total of 61,572 genes are predicted by the Augustus program (Stake and Waack 2003). To anchor the sequences to the *Raphanus* chromosomes, a high-density genetic map consisting of 2553 markers has been developed by integration of two maps reported by Shirasawa et al. (2011) and Li et al. (2011), together with an additional 989 DNA markers (221 SNP markers and 768 PCR-RFLP markers) (Table 3.1). As a result, a total of 116 Mb sequences (1345 scaffolds) have been assigned to the *Raphanus* chromosomes.

The second genome sequence has been reported by Mitsui et al. (2015). In their study, two types of sequencing platforms were employed. One is the Illumina technology (paired-end sequencing of insert sizes of 300, 500 bp, 8, 20, and 40 kb) and the other is the Roche 454 method (single-end sequencing). Assembling of the sequence data with Newbler 2.7 followed by scaffolding with SSPACE 2.0 generated a total length of 383 Mb sequences (N50 = 138 kb; $\geq$500 bp), from which 54,357 genes are predicted by the Augustus program. With the SSR and SNP markers, a genetic map consisting of 1384 marker loci was constructed and assigned by 180 Mb sequences of the radish genome.

The latest one was reported by Jeong et al. (2016). The data by three types of NGS technologies, Illumina (insert size of 500 bp), Roche 454 (SE and PE for 3-, 8-, 20-, and 30-kb insert sizes), and PacBio long reads, and the end sequences of clones from two libraries, fosmid (40-kb insert size) and BAC, were employed to construct a chromosome-scale draft genome sequence of radish. The Roche 454 reads were assembled with Newbler 2.6, and the obtained scaffolds were assembled again with the PacBio reads by Celera Assembler 7.0. The contigs were then subjected to error correction with the Illumina reads and subsequently extended with the fosmid and BAC clone-end sequences. The resultant total size was 426 Mb (N50 = 19.6 kb; $\geq$100 bp) and then 344 Mb of the total size were aligned to the *Raphanus* chromosomes (N50 = 1.22 Mb) along a high-density genetic map constructed by genotyping using whole genome resequencing of an F_2 mapping population with NGS technology (Mun et al. 2015) (Table 3.1). A total of 46,514 genes were predicted by ab initio, sequence similarity, and transcriptome evidence-based predictions.

In addition to the cultivated radish genomes, the genome sequence of a wild relative, *R. raphanistrum*, has also been published (Moghe et al. 2014). From a fifth generation inbred line, Illumina sequence reads were obtained and assembled into 68,331 contigs (N50 = 10.1 kb) covering 254.6 Mb. The annotation analysis by a MAKER pipeline predicted a total of 38,174 genes.

3.4 Future Direction of *Raphanus* Genomics and Genetics

In many crops, genotyping of mapping populations has been conducted by the NGS-based whole genome resequencing (Kumar et al. 2012). However, even though the cost of sequencing has been decreasing year by year, genotyping by whole genome resequencing is still costly. In order to realize less costly genotyping, reduced-representation sequencing methods using reduced-representation libraries have been developed, e.g., genotyping by sequencing (GBS) and restriction-site associated DNA sequencing (RAD-Seq). For the purpose, we attempted to make the genetic maps reported by Li et al. (2011) and Kitashiba et al. (2014) denser with polymorphic markers by double digested RAD-seq (ddRAD-Seq) based on a method described by Shirasawa et al. (2016). Sequencing libraries for each of 94 F_2 individuals, F_1 plant, and the parental lines which were used in Li et al. (2011) and Kitashiba et al. (2014) were prepared with two restriction enzymes, *Pst*I and *Msp*I, and then subjected to sequencing analysis by HiSeq 2000 with a paired-end mode of 93 bases (DRA accession number: DRA005069). After trimming low-quality and adapter sequences, the remaining high-quality reads were mapped on either the genome sequences reported by Kitashiba et al. (2014), Mitsui et al. (2015), or Jeong et al. (2016) with the Bowtie2 program (version 2.2.3, sensitive preset of the end-to-end mode) (Langmead and Salzberg 2012). The mapping alignment rate on the sequences from Kitashiba et al. (2014) was 60.5% on average, whereas it was 88.0% and 57.6% when the reads were mapped on the sequences generated by Mitsui et al. (2015) and Jeong et al. (2016), respectively. Totals of 3736, 3795, and 2828 high-quality SNPs called with BCFtools (version 0.2.0) (Danecek et al. 2011) were obtained from the three mapping alignments of Kitashiba et al. (2014), Mitsui et al. (2015), and Jeong et al. (2016), respectively, after eliminating low-quality SNPs by VCFtools (version 0.1.12b; parameters: –maf 0.2 –max-alleles 2 –min-alleles 2 –minDP 5 –max-missing 0.5 –remove-indels) (Danecek et al. 2011). The high-quality SNP data were used for linkage analysis to construct a genetic map. Using the MSTmap program (Wu et al. 2008) with Kosambi map function, the data set was divided into nine groups with cut off threshold *P* values of 1E-10 to 1E-13, and then the order of SNPs was determined (parameters: no_map_dist, 10; no_map_size, 5; missing_threshold, 0.25; estimation_before_clustering, yes; detect_bad_data, yes; objective_function, ML). Finally, a highly dense genetic map consisting of nine chromosomes with 7480 SNPs covering 3684.0 cM was established (Rs-RAD map; Table 3.4; Fig. 3.1). The total length of the constructed map was approximately twice longer than the longest maps previously reported (Table 3.1). Therefore, we attempted to validate the map length with another mapping software program, Mapmaker (Lander et al. 1987), in which Kosambi map function was employed. The result indicated that the map lengths calculated with the two programs were comparable (Supplementary Fig. 3.1), suggesting the current map length is considered to be totally appropriate.

Out of the 7480 SNPs on the genetic map, 2758, 2633, and 2089 were derived from the mapping alignments using the reference sequences from Kitashiba et al. (2014), Mitsui et al. (2015), and Jeong et al. (2016), respectively (Table 3.4, Supplementary Table 3.1). Theoretically, the identical SNPs on the three reference genomes should be mapped in same map

Table 3.4 Statistics of a genetic map of *Raphanus saivus* (Rs-RAD map) constructed by ddRAD-Seq

Linkage group	Map length (cM)	Total			Kitashiba et al. (2014)		
		Number of SNPs	Number of assigned scaffolds	Total length of the scaffolds (bp)	Number of SNPs	Number of assigned scaffolds	Total length of the scaffolds (bp)
Rs1	268.6	640	121	4,24,68,897	238	68	55,93,494
Rs2	307.5	788	169	7,14,69,960	286	91	1,01,09,637
Rs3	233.7	543	144	4,56,80,235	203	78	53,12,400
Rs4	471.0	859	248	8,15,52,642	328	127	1,16,97,923
Rs5	484.6	975	221	7,56,47,199	346	125	1,10,58,800
Rs6	767.2	1424	296	9,11,76,308	531	153	1,36,81,802
Rs7	303.2	643	140	4,77,02,993	236	77	80,44,925
Rs8	316.1	773	168	5,18,57,038	284	94	88,16,829
Rs9	532.1	835	208	6,38,03,308	306	111	98,77,095
Total	**3684.0**	**7480**	**1715**	**571,358,580**	**2758**	**924**	**84,192,905**
Linkage group	Map length (cM)	Mitsui et al. (2015)			Jeong et al. (2016)		
		Number of SNPs	Number of assigned scaffolds	Total length of the scaffolds (bp)	Number of SNPs	Number of assigned scaffolds	Total length of the scaffolds (bp)
Rs1	268.6	225	46	1,02,86,756	177	7	2,65,88,647
Rs2	307.5	283	70	1,72,95,072	219	8	4,40,65,251
Rs3	233.7	193	57	1,05,49,510	147	9	2,98,18,325
Rs4	471.0	301	101	1,79,86,516	230	20	5,18,68,203
Rs5	484.6	352	92	1,85,50,954	277	4	4,60,37,445
Rs6	767.2	479	122	2,23,98,576	414	21	5,50,95,930
Rs7	303.2	228	53	1,20,11,202	179	10	2,76,46,866
Rs8	316.1	277	67	1,31,28,702	212	7	2,99,11,507
Rs9	532.1	295	85	1,53,95,195	234	12	3,85,31,018
Total	**3684.0**	**2633**	**693**	**137,602,483**	**2089**	**98**	**349,563,192**

positions, in other words, in a single genetic 'bin' in which tightly linked SNPs also belong. Indeed, the 7480 SNPs were distributed in 1458 genetic bins. Although the number of F_2 individuals used in the present study was 94, increasing the number of individuals in the mapping population would genetically dissect the scaffolds in a single bin, resulting in an increase in the number of genetic bins. Among the SNPs in the bins, 687 (47.1%) were common across the three reference genomes, while 145 (9.9%), 100 (6.9%), and 148 (10.2%) were specific to the genome sequences from Kitashiba et al. (2014), Mitsui et al. (2015), and Jeong et al. (2016), respectively. The remaining 44 (3.3%), 44 (3.3%), and 290 (19.9%) were not observed in the sequences from Kitashiba et al. (2014), Mitsui et al. (2015), and Jeong et al. (2016), respectively. This result indicated that 73.6% of the mapped SNPs were common between at least two genome sequences.

In the new genetic map, 924 contig sequences (total length of 84.2 Mb) from 76,592 sequences (N50 = 46 kb) spanning 402 Mb reported by Kitashiba et al. (2014) were uniquely anchored to the radish chromosomes (Supplementary Table 3.2) while one contig was mapped on two linkage groups. In addition, 693 respective

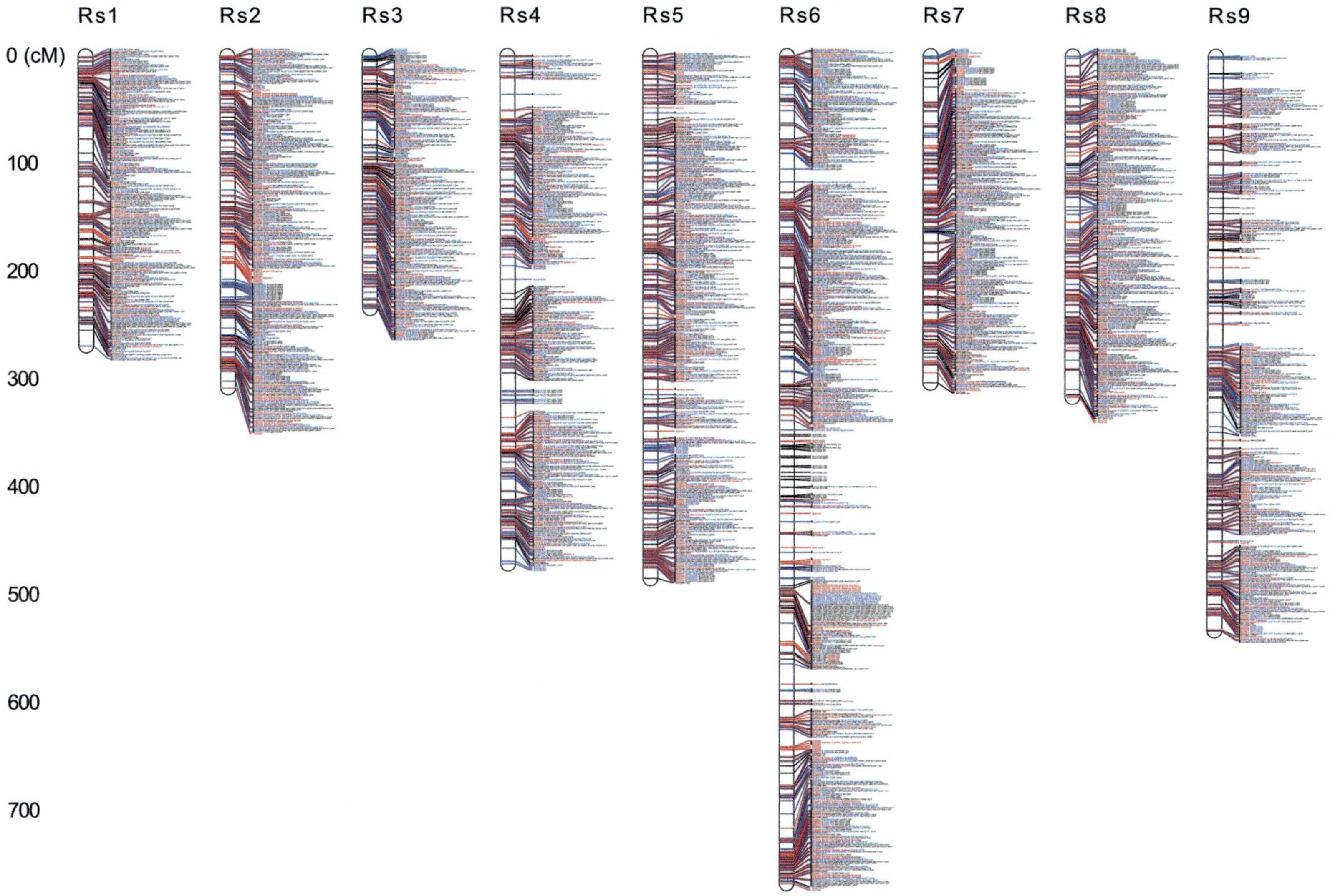

Fig. 3.1 A genetic map of *Raphanus sativus* constructed by ddRAD-seq (Rs-RAD map). SNPs detected from reference sequences reported by Kitashiba et al. (2014), Mitsui et al. (2015), and Jeong et al. (2016) are indicated in *black*, *blue*, and *red* letters, respectively

contigs (137.6 Mb) reported by Mitsui et al. (2015) and 9 pseudomolecule sequences and 89 contigs (349 Mb in total) by Jeong et al. (2016) were also anchored to the chromosomes (Supplementary Table 3.2). These results provided useful clues to clarify relationship between the sequences reported from the three research groups. It would be possible to merge the sequences into consensus genome sequences in accordance with the genetic linkages based on the current map and further sequence similarity analysis of the genome sequences as well.

In the near future, it is projected that new NGS technology will also be effective for distinguishing homological sequences derived from genome-wide triplication in the radish genome and for improving the quality of genome assemblies because NGS technology and tools for analysis of data have been advancing year by year. The long-read sequencing technology by Pacific Biosciences (Menlo Park, CA, USA) and Oxford Nanopore Technologies (Oxford, UK), the synthetic long reads by Illumina (San Diego, CA, USA), physical mapping techniques by Irys System (BioNano Genomics, San Diego, CA, USA) and Chromium System (10X Genomics, Pleasanton, CA, USA), and a new genome assembly algorithm, DenovoMAGIC (NRGene Technologies, Ramla, Israel) will be useful not only for improving the genome sequences but also for further de novo sequencing of several divergent radish lines. Furthermore, it will help us to perform pan-genome analysis with the divergent radish lines to understand the relationship between variations of phenotypes and those of individual genomes in terms of chromosomal structure and copy number as well as SNPs and indels.

Note added in proof

Another paper on the genome sequence of *R. sativus* by Zhang et al. (2015) has been published. However, we overlooked it at the time of writing this manuscript because no data had been disclosed in public databases. Therefore, the contents of this paper are not taken up in this chapter.

References

Bett KE, Lydiate DJ (2003) Genetic analysis and genome mapping in *Raphanus*. Genome 46:423–430

Budahn H, Peterka H, Mousa MA, Ding Y, Zhang S, Li J (2009) Molecular mapping in oil radish (*Raphanus sativus* L.) and QTL analysis of resistance against beet cyst nematode (*Heterodera schachtii*). Theor Appl Genet 118:775–782

Chalhoub B, Denoeud F, Liu S, Parkin IA, Tang H, Wang X, Chiquet J, Belcram H, Tong C, Samans B, Corréa M, Da Silva C, Just J, Falentin C, Koh CS, Le Clainche I, Bernard M, Bento P, Noel B, Labadie K, Alberti A, Charles M, Arnaud D, Guo H, Daviaud C, Alamery S, Jabbari K, Zhao M, Edger PP, Chelaifa H, Tack D, Lassalle G, Mestiri I, Schnel N, Le Paslier MC, Fan G, Renault V, Bayer PE, Golicz AA, Manoli S, Lee TH, Thi VH, Chalabi S, Hu Q, Fan C, Tollenaere R, Lu Y, Battail C, Shen J, Sidebottom CH, Wang X, Canaguier A, Chauveau A, Bérard A, Deniot G, Guan M, Liu Z, Sun F, Lim YP, Lyons E, Town CD, Bancroft I, Wang X, Meng J, Ma J, Pires JC, King GJ, Brunel D, Delourme R, Renard M, Aury JM, Adams KL, Batley J, Snowdon RJ, Tost J, Edwards D, Zhou Y, Hua W, Sharpe AG, Paterson AH, Guan C, Wincker P (2014) Early allopolyploid evolution in the post-Neolithic *Brassica napus* oilseed genome. Science 345:950–953

Danecek P, Auton A, Abecasis G, Albers CA, Banks E, DePristo MA, Handsaker RE, Lunter G, Marth GT, Sherry ST, McVean G, Durbin R, 1000 Genomes Project Analysis Group (2011) The variant call format and VCFtools. Bioinformatics 27:2156–2158

Hashida T, Nakatsuji R, Budahn H, Schrader O, Peterka H, Fujimura T, Kubo N, Hirai M (2013) Construction of a chromosome-assigned, sequence-tagged linkage map for the radish, *Raphanus sativus* L. and QTL analysis of morphological traits. Breed Sci 63:218–226

Inaba R, Nishio T (2002) Phylogenetic analysis of *Brassiceae* based on the nucleotide sequences of the *S*-locus related gene, *SLR1*. Theor Appl Genet 105:1159–1165

Jeong YM, Kim N, Ahn BO, Oh M, Chung WH, Chung H, Jeong S, Lim KB, Hwang YJ, Kim GB, Baek S, Choi SB, Hyung DJ, Lee SW, Sohn SH, Kwon SJ, Jin M, Seol YJ, Chae WB, Choi KJ, Park BS, Yu HJ, Mun JH (2016) Elucidating the triplicated ancestral genome structure of radish based on chromosome-level comparison with the *Brassica* genomes. Theor Appl Genet 129:1357–1372

Kamei A, Tsuro M, Kubo N, Hayashi T, Wang N, Fujimura T, Hirai M (2010) QTL mapping of clubroot resistance in radish (*Raphanus sativus* L.). Theor Appl Genet 120:1021–1027

Kaneko Y, Kimizuka-Takagi C, Bang SW, Matsuzawa Y (2007) Radish. In: Kole C (ed) Genome mapping and molecular breeding in plants, vol 5. Springer, New York, pp 141–160

Kitashiba H, Li F, Hirakawa H, Kawanabe T, Zou Z, Hasegawa Y, Tonosaki K, Shirasawa S, Fukushima A, Yokoi S, Takahata Y, Kakizaki T, Ishida M, Okamoto S, Sakamoto K, Shirasawa K, Tabata S, Nishio T (2014) Draft sequences of the radish (*Raphanus sativus* L.) genome. DNA Res 21:481–490

Kumar S, Banks TW, Cloutier S (2012) SNP discovery through next-generation sequencing and its applications. Int J Plant Genom 2012:831460

Lander ES, Green P, Abrahamson J, Barlow A, Daly MJ, Lincoln SE, Newberg LA (1987) MAPMAKER: an interactive computer package for constructing primary genetic linkage maps of experimental and natural populations. Genomics 1:174–181

Langmead B, Salzberg SL (2012) Fast gapped-read alignment with Bowtie 2. Nat Methods 9:357–359

Liu S, Liu Y, Yang X, Tong C, Edwards D, Parkin IA, Zhao M, Ma J, Yu J, Huang S, Wang X, Wang J, Lu K, Fang Z, Bancroft I, Yang TJ, Hu Q, Wang X, Yue Z, Li H, Yang L, Wu J, Zhou Q, Wang W, King GJ, Pires JC, Lu C, Wu Z, Sampath P, Wang Z, Guo H, Pan S, Yang L, Min J, Zhang D, Jin D, Li W, Belcram H, Tu J, Guan M, Qi C, Du D, Li J, Jiang L, Batley J, Sharpe AG, Park BS, Ruperao P, Cheng F, Waminal NE, Huang Y, Dong C, Wang L, Li J, Hu Z, Zhuang M, Huang Y, Huang J, Shi J, Mei D, Liu J, Lee TH, Wang J, Jin H, Li Z, Li X, Zhang J, Xiao L, Zhou Y, Liu Z, Liu X, Qin R, Tang X, Liu W, Wang Y, Zhang Y, Lee J, Kim HH, Denoeud F, Xu X, Liang X, Hua W, Wang X, Wang J, Chalhoub B, Paterson AH (2014) The *Brassica oleracea* genome reveals the asymmetrical evolution of polyploid genomes. Nat Commun 5:3930

Li F, Hasegawa Y, Saito M, Shirasawa S, Fukushima A, Ito T, Fujii H, Kishitani S, Kitashiba H, Nishio T (2011) Extensive chromosome homoeology among Brassiceae species were revealed by comparative genetic mapping with high-density EST-based SNP markers in radish (*Raphanus sativus* L.). DNA Res 18:401–411

Li G, Quiros CF (2001) Sequence-related amplified polymorphism (SRAP), a new marker system based on a simple PCR reaction: its application to mapping and gene tagging in *Brassica*. Theor Appl Genet 103:455–461

Mitsui Y, Shimomura M, Komatsu K, Namiki N, Shibata-Hatta M, Imai M, Katayose Y, Mukai Y, Kanamori H, Kurita K, Kagami T, Wakatsuki A, Ohyanagi H, Ikawa H, Minaka N, Nakagawa K, Shiwa Y, Sasaki T (2015) The radish genome and comprehensive gene expression profile of tuberous root formation and development. Sci Rep 5:10835

Moghe GD, Hufnagel DE, Tang H, Xiao Y, Dworkin I, Town CD, Conner JK, Shiu SH (2014) Consequences of whole-genome triplication as revealed by comparative genomic analyses of the wild radish *Raphanus raphanistrum* and three other Brassicaceae species. Plant Cell 26:1925–1937

Mun JH, Chung H, Chung WH, Oh M, Jeong YM, Kim N, Ahn BO, Park BS, Park S, Lim KB, Hwang YJ, Yu HJ (2015) Construction of a reference genetic map of *Raphanus sativus* based on genotyping by whole genome resequencing. Theor Appl Genet 128:259–272

Schranz ME, Lysak MA, Mitchell-Olds T (2006) The ABC's of comparative genomics in the Brassicaceae: building blocks of crucifer genomes. Trends Plant Sci 11:535–542

Shiokai S, Shirasawa K, Sato Y, Nishio T (2010) Improvement of the dot-blot-SNP technique for efficient and cost-effective genotyping. Mol Breed 25:179–185

Shirasawa K, Hirakawa H, Isobe S (2016) Analytical workflow of double-digest restriction site-associated DNA sequencing based on empirical and *in silico* optimization in tomato. DNA Res 23:145–153

Shirasawa K, Oyama M, Hirakawa H, Sato S, Tabata S, Fujioka T, Kimizuka-Takagi C, Sasamoto S, Watanabe A, Kato M, Kishida Y, Kohara M, Takahashi C, Tsuruoka H, Wada T, Sakai T, Isobe S (2011) An EST-SSR linkage map of *Raphanus sativus* and comparative genomics of the Brassicaceae. DNA Res 18:221–232

Shirasawa K, Shiokai S, Yamaguchi M, Kishitani S, Nishio T (2006) Dot-blot-SNP analysis for practical plant breeding and cultivar identification in rice. Theor Appl Genet 113:147–155

Stake M, Waack S (2003) Gene prediction with a hidden Markov model and a new intron submodel. Bioinformatics 19(Suppl 2):i215–i225

Suwabe K, Iketani H, Nunome T, Kage T, Hirai M (2002) Isolation and characterization of microsatellites in *Brassica rapa* L. Theor Appl Genet 104:1092–1098

Suwabe K, Tsukazaki H, Iketani H, Hatakeyama K, Fujimura M, Nunome T, Fukuoka H, Matsumoto S, Hirai M (2003) Identification of two loci for resistance to clubroot (*Plasmodiophora brassicae* Woronin) in *Brassica rapa* L. Theor Appl Genet 107:997–1002

Suwabe K, Tsukazaki H, Iketani H, Hatakeyama K, Kondo M, Fujimura M, Nunome T, Fukuoka H, Hirai M, Matsumoto S (2006) Simple sequence repeat-based comparative genomics between *Brassica rapa* and *Arabidopsis thaliana*: the genetic origin of clubroot resistance. Genetics 173:309–319

The Arabidopsis Genome Initiative (2000) Analysis of the genome sequence of the flowering plant *Arabidopsis thaliana*. Nature 408:796–815

The *Brassica rapa* Genome Sequencing Project Consortium (2011) The genome of the mesopolyploid crop species *Brassica rapa*. Nat Genet 43:1035–1039

Tsuro M, Suwabe K, Kubo N, Matsumoto S, Hirai M (2005) Construction of a molecular linkage map of radish (*Raphanus sativus* L.), based on AFLP and *Brassica*-SSR markers. Breed Sci 55:107–111

Tsuro M, Suwabe K, Kubo N, Matsumoto S, Hirai M (2008) Mapping of QTLs controlling root shape and red pigmentation in radish, *Raphanus sativus* L. Breed Sci 58:55–61

U N (1935) Genome-analysis in *Brassica* with special reference to the experimental formation of *B. napus* and peculiar mode of fertilization. Jpn J Bot 7:389–452

Wu Y, Bhat PR, Close TJ, Lonardi S (2008) Efficient and accurate construction of genetic linkage maps from the minimum spanning tree of a graph. PLoS Genet 4: e1000212

Xu L, Wang L, Gong Y, Dai W, Wang Y, Zhu X, Wen T, Liu L (2012) Genetic linkage map construction and QTL mapping of cadmium accumulation in radish (*Raphanus sativus* L.). Theor Appl Genet 125:659–670

Yang J, Liu D, Wang X, Ji C, Cheng F, Liu B, Hu Z, Chen S, Pental D, Ju Y, Yao P, Li X, Xie K, Zhang J, Wang J, Liu F, Ma W, Shopan J, Zheng H, Mackenzie SA, Zhang M (2016) The genome sequence of allopolyploid *Brassica juncea* and analysis of differential homoeolog gene expression influencing selection. Nat Genet 48:1225–1232

Yu X, Choi SR, Dhandapani V, Rameneni JJ, Li X, Pang W, Lee JY, Lim YP (2016) Quantitative trait loci for morphological traits and their association with functional genes in *Raphanus sativus*. Front Plant Sci 7:255

Yu X, Choi SR, Ramchiary N, Miao X, Lee SH, Sun HJ, Kim S, Ahn CH, Lim YP (2013) Comparative mapping of *Raphanus sativus* genome using *Brassica* markers and quantitative trait loci analysis for the *Fusarium* wilt resistance trait. Theor Appl Genet 126:2553–2562

Zhang X, Yue Z, Mei S, Qiu Y, Yang X, Chen X, Cheng F, Wu Z, Sun Y, Jing Y, Liu B, Shen D, Wang H, Cui N, Duan Y, Wu J, Wang J, Gan C, Wang J, Wang X, Li X (2015) A *de novo* genome of a Chinese radish cultivar. Hort Plant J 1:155–164

Zou Z, Ishida M, Li F, Kakizaki T, Suzuki S, Kitashiba H, Nishio T (2013) QTL analysis using SNP markers developed by next-generation sequencing for identification of candidate genes controlling 4-methylthio-3-butenyl glucosinolate contents in roots of radish, *Raphanus sativus* L. PLoS ONE 8:e53541

Gene Annotation and Database of the Radish

4

Hideki Hirakawa

Abstract

The genome sequences of *Raphanus sativus* var. *hortensis* were determined for the Japanese cultivars 'Aokubi S-h' (Kitashiba et al. in DNA Res 21:481–490, 2014), 'Aokubi doubled haploid (DH)' (Mitsui et al. in Sci Rep 5:10835, 2015), and the Korean cultivar 'WK10039' (Jeong et al. in Theor Appl Genet 129:1357–1372, 2016). After the genome sequencing, gene annotations, such as gene finding, domain searches, and similarity searches against several databases, were conducted. The genomic information obtained by these analyses was then released from a database. In this section, the gene annotation methods conducted in each of the three genome sequencing projects is described. In addition, several databases related to the genome sequencing, EST sequencing, and DNA markers in radish are introduced.

4.1 Introduction

In 2014, the genome sequence of the *R. sativus* var. *hortensis* Japanese cultivar 'Aokubi S-h' (Takii Seed Co., Japan) was determined (Kitashiba et al. 2014). In 2015, the genome sequence of the radish var. *hortensis* Japanese cultivar Aokubi doubled haploid (DH) line (Aokubi Soubutori Daikon DH line; Sakata Seed Co., Japan) was determined (Mitsui et al. 2015). Finally, in 2016, the genome sequence of the Korean radish cultivar 'WK10039' was determined (Jeong et al. 2016). The total lengths of the genome sequences of 'Aokubi S-h,' 'Aokubi doubled haploid,' and 'WK10039' were 402,330,269 bps (number of sequences: 76,592; N50 length: 46,262 bps; $\geq$300 bp), 383,104,752 bps (number of sequences: 40,123; N50 length: 138,635 bps; $\geq$500 bp), and 426,202,302 bps (number of sequences: 10,674; N50 length: 38,354,807 bps; $\geq$1597 bp), respectively. In WK10039, the genome sequences were determined as pseudomolecules. The gene annotation methods and databases constructed in the three genome sequencing projects, and the other databases related to radish are introduced below.

H. Hirakawa (✉)
Kazusa DNA Research Institute, Kazusa-kamatari, Kisarazu, Chiba 292-0818, Japan
e-mail: hh@kazusa.or.jp

T. Nishio and H. Kitashiba (eds.), *The Radish Genome*, Compendium of Plant Genomes, DOI 10.1007/978-3-319-59253-4_4

4.2 Gene Annotation

4.2.1 'Aokubi S-H' (Takii Seed Co., Japan)

The current version of the draft genome sequence is RSA_r1.0 (Kitashiba et al. 2014). The BioProject ID of this sequencing project is PRJDB2129. The 80,521 genes were totally predicted by Augustus 2.7 (Stanke and Waack 2003) with the *Arabidopsis thaliana* training set. Among them, 15,545 genes were similar to transposable elements (TEs) by hmmscan (Eddy 2011) searches against GyDB 2.0 (Llorens et al. 2011) with an *E*-value cutoff of 1.0, and BLASTP searches (Altschul et al. 1997) against NCBI's NR (non-redundant amino acid sequences) database (ftp://ncbi.nlm.nih.gov/blast/db/FASTA/nr.gz) with an *E*-value cutoff of 1E-10. The 64,976 genes excluding TEs were classified as 'intrinsic' (45,002 genes with start and stop codons), 'partial' (16,570 genes without start and/or stop codons), 'pseudo' (1075 genes with in-frame stop codons) or 'short' (2329 genes less than 50 amino acid residues). The product name was assigned based on the top hit of BLAST searches against NCBI's NR database. The protein domains were searched against the InterPro database (Mulder et al. 2007) by using InterProScan (Quevillon et al. 2005) with an *E*-value cutoff of 1.0. To evaluate the accuracy of the gene prediction, BLASTP searches were performed between the 85,083 genes published in the RadishDB (http://bioinfo.bti.cornell.edu/cgi-bin/radish/index.cgi) and the translated genes of RSA_r1.0 (RSA_r1.0_pep), with the result that 84,165 (98.9%) genes in RadishBase (http://bioinfo.bti.cornell.edu/cgi-bin/radish/index.cgi) were found in RSA_r1.0_pep. To characterize the functions of the predicted genes, GO slim analysis was performed based on the results of the InterProScan searches. In addition, the functional categories were assigned to the genes by BLAST searches against NCBI's KOG (eukaryotic Clusters of Orthologous Groups of proteins) database (Tatusov et al. 2003) with an *E*-value cutoff of 1E-20. The distribution of the GO categories and KOG were similar among *R. sativus*, *Brassica rapa* (v1.2; 41,019 genes), *B. oleracea* (36,862 unigenes assembled from ESTs in NCBI's dbEST; http://www.ncbi.nlm.nih.gov/dbEST/), *Raphanus raphanistrum* (22,618 dbEST unigenes assembled from ESTs in dbEST), and *A. thaliana* (TAIR10; 35,386 genes; https://www.arabidopsis.org). In the distribution of KOG IDs, the percentages of the genes in *R. sativus* were comparatively higher in the categories of 'KOG L: replication, recombination, and repair' and 'KOG D: cell cycle control, cell division, and chromosome partitioning' than those in *B. rapa* and *A. thaliana*. Clustering analysis was performed among 25,390 genes of *R. sativus*, 21,494 genes of *B. rapa*, 19,007 genes of *B. oleracea*, 15,466 genes of *R. raphanistrum*, and 18,754 genes of *A. thaliana* using the CD-HIT (Li and Godzik 2006) program, and 6110 gene clusters were found to be in common among these species. On the other hand, 8759 clusters (13,919 genes) were classified into the specific regions in *R. sativus*, 2796 clusters (3202 genes) in *B. rapa*, 8335 clusters (8925 genes) in *B. oleracea*, 2819 clusters (2997 genes) in *R. raphanistrum*, and 2654 clusters (3448 genes) in *A. thaliana*. The 1335, 1180, and 699 tRNA genes were, respectively, detected by tRNAscan-SE (Lowe and Eddy 1997) in *R. sativus* (RSA_r1.0; 61,572 genes), *B. rapa* (BRAD v1.2; http://brassicadb.org/brad/), and *A. thaliana* (TAIR10).

4.2.2 'Aokubi Doubled Haploid (DH) Line' (Sakata Seed Co., Japan)

The current version of the genome sequence is v1.0 (Mitsui et al. 2015). The BioProject ID of this sequencing project is PRJDB707. The 64,657 genes were predicted by Augustus with the *Arabidopsis thaliana* training set. The average lengths of genes and CDSs were 1724 and 1112 bps, respectively. The average number of exons per gene was 4.18. The average exon and intron lengths were 266 and 118 bps, respectively. In addition, transcript sequences (RNA-Seq) were obtained from leaves by HiSeq 2000 sequencer (Illumina, San Diego, CA, USA). The RNA-Seq reads were assembled by Trinity (Grabherr et al. 2011), and 99,096 contigs were constructed. From the contigs, 53,846 ORFs were extracted by EMBOSS (Rice et al. 2000) software. The predicted genes were subjected to BLAST searches against the NR database, and InterProScan searches against the InterPro database. To compare the genes among *R. sativus*, *A. thaliana*, *Arabidopsis lyrata*, *B. rapa*, and *Carica papaya*, CD-HIT was used to remove the paralogous genes. Then, all-against-all BLASTP searches were performed to detect orthologous genes by OrthoMCL (Li et al. 2003). Gene ontology (GO) was assigned to the genes by BLAST2GO (Conesa et al. 2005) based on BLAST searches against the NR database with an *E*-value cutoff of 1E-5. The genes were also mapped onto the Kyoto Encyclopedia of Genes and Genomes (KEGG) pathways (Ogata et al. 1999).

4.2.3 'WK10039'

The current version of the genome sequence is Rs 1.0 (Jeong et al. 2016). The BioProject ID of this project is PRJNA239785. The genome sequence has been determined to consist of pseudomolecules. RNA-Seq reads were obtained from 18 samples (seedlings, leaves, roots, and flowers) by an Illumina HiSeq sequencer (Illumina, San Diego, CA, USA). EST sequences were obtained from RadishBase (http://bioinfo.bti.cornell.edu/cgi-bin/radish/index.cgi) and NCBI's dbEST. After masking the repetitive sequences related to class I and class II by RepeatMasker (http://www.repeatmasker.org), the genes were predicted by Fgenesh+ (http://www.softberry.com), Augustus, and SNAP (Korf 2004), and the genes >300 bps were used for further analyses. In addition, GSNAP (Wu and Nacu 2010) was used for mapping transcript sequences onto the genome sequences, PASA (Haas et al. 2008) was used for evidence-based gene sets, and EVidenceModeller (Haas et al. 2008) was used to combine the *ab initio* and transcript assignments. tRNA genes were extracted by tRNAscan, and rRNA and miRNA genes were extracted using BLASTN searches with up to two mismatches. The product names of genes were assigned by BLAST searches against TrEMBL and SwissProt in UniProt (http://www.ebi.ac.uk/uniprot) with an *E*-value cutoff of 1E-5. Orthologous genes were defined among *R. sativus*, *A. thaliana*, *Capsella rubella* (pink shepherd's-purse), *B. rapa*, and *B. oleracea*. All-against-all BLASTP searches were performed, and the output results were used for OrthoMCL to conduct clustering analyses.

4.2.4 Current Status of the Gene Annotation in the Genome Sequencing Projects

As described above, the gene annotation results of the three genome sequencing projects are summarized in Table 4.1. Currently, several kinds of sequencing technology, such as the PacBio Sequel system (http://www.pacb.com/products-and-services/pacbio-systems/sequel/), Oxford Nanopore Technologies (https://www.nanoporetech.com), BioNano Genomics Irys system (http://bionanogenomics.com), 10X Genomics (http://www.10xgenomics.com), and Dovetail Genomics (http://dovetailgenomics.com), have been developed. By using these technologies, the accuracy of the genome sequence of *R. sativus* can be increased, along with the precision of the gene annotation.

4.3 Databases

4.3.1 The Raphanus Sativus Genome DataBase

The *Raphanus sativus* Genome DataBase (http://radish.kazusa.or.jp) was released from the

Table 4.1 Comparison of the assemblies in the three genome sequencing projects

		Aokubi S-h (Takii Seed Co., Japan)	Aokubi Doubled Haploid (DH) line (Sakata Seed Co., Japan)	WK10039
Sequence version		RSA_r1.0	v1.0	Rs 1.0
BioProject		PRJDB2129	PRJDB707	PRJNA239785
Gene prediction	Number of genes	80,521 (Genes: 64,976, Transposable elements: 15,545)	Genes: 64,657	Genes: 46,514
	Software	Augustus 2.7 with *Arabidopsis thaliana* training set	Augustus with *Arabidopsis thaliana* training set	Fgenesh+, Augustus, SNAP, GSNAP, PASA, EVM
De novo assembly	RNA-Seq	–	leaves	Seedlings, leaves, roots, and flowers
	Number of genes	–	99,096 contigs (53,846 ORFs)	–
	Software	–	Trinity	–
Function prediction		BLAST against NR, InterProScan against InterPro	BLAST against NR, InterProScan against InterPro	BLAST against TrEMBL and SwissProt
Functional classification		GO slim, KOG	BLAST2GO	–
Gene comparison		CD-HIT	CD-HIT, OrthoMCL	OrthoMCL
Compared species		*Brassica rapa*, *Brassica oleracea*, *Raphanus raphanistrum*, *Arabidopsis thaliana*	*Arabidopsis thaliana*, *Arabidopsis lyrata*, *Brassica rapa*, *Carica papaya*	*Arabidopsis thaliana*, *Capsella rubella*, *Brassica rapa*, *Brassica oleracea*
DB		http://radish.kazusa.or.jp	http://www.nodai-genome-d.org	http://radish-genome.org
References		*DNA Research*, **21**, 481–490 (2014)	*Scientific Reports* **5**:10835 (2015)	*Theoretical and Applied Genetics* **129**:1357–1372 (2016)

Kazusa DNA Research Institute (KDRI) (Kitashiba et al. 2014). The top page of this database is shown in Fig. 4.1. BLAST searches (BLASTN, BLASTP, and TBLASTN) against the draft genome sequence (RSA_r1.0), predicted genes, and translated genes are available at the BLAST page. SNP information among the eight cultivars, 'Aokubi', 'Sayatori', 'Taibyosobutori', 'Yumehomare', 'Sakurajima', 'AZ26H (Karamijidaikon)', 'N1–3 (Minowase)', and 'Nishimachi-Risou' are available in a tabular format at the SNP page. The sequence files of the scaffolds (RSA_r1.0), CDSs (RSA_r1.0_cds), translated genes (RSA_r1.0_pep), annotation file (RSA_r1.0_genes.gff3), and SNP genotyping table of the eight cultivars against the draft genome sequence (SNP.xls) are available.

4.3.2 NODAI Radish Genome Database

The NODAI Radish genome database (http://www.nodai-genome-d.org) was released from the Tokyo University of Agriculture (Mitsui et al. 2015). The top page of this database is shown in Fig. 4.2. BLAST searches (BLASTN, BLASTP, BLASTX, TBLASTN, and TBLASTX) against all of the scaffolds, scaffolds over 2 kb, CDSs and translated genes (version 1) are available at the BLAST page. The location of the unigenes and genes predicted by Trinity and Augustus on the scaffolds can be browsed on GBrowse (http://gmod.org/wiki/GBrowse) at the GBrowse page. The files of the scaffolds, scaffolds over 2 kb,

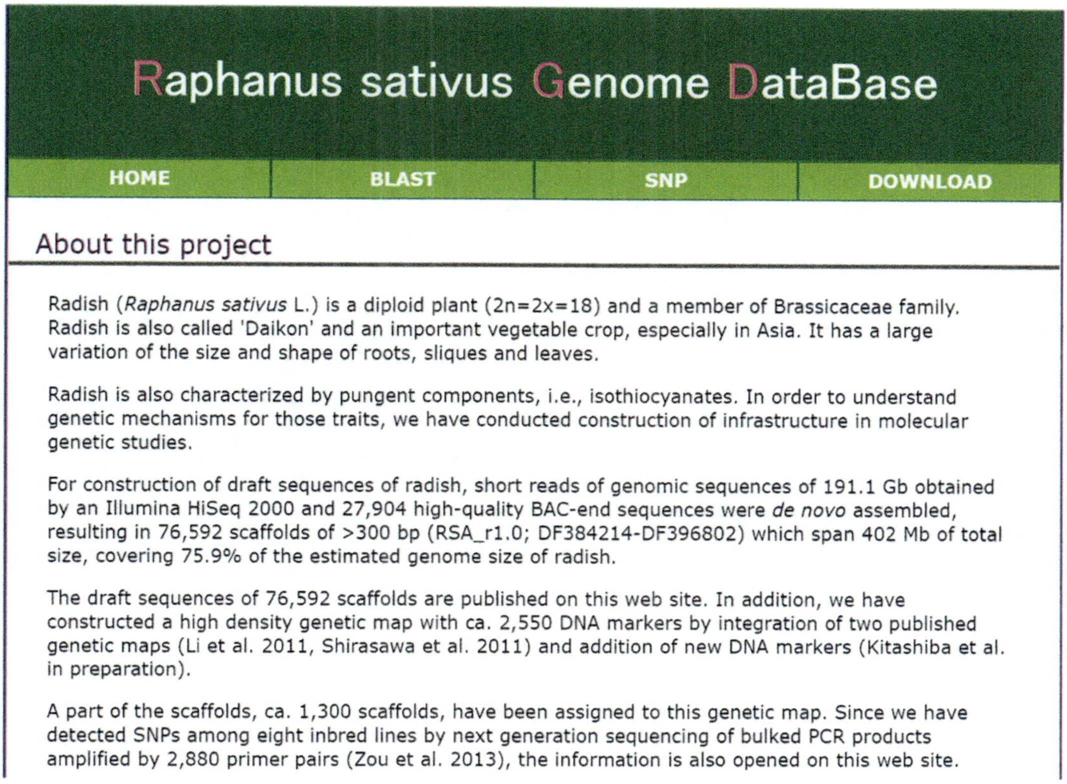

Raphanus sativus Genome DataBase

HOME | BLAST | SNP | DOWNLOAD

About this project

Radish (*Raphanus sativus* L.) is a diploid plant (2n=2x=18) and a member of Brassicaceae family. Radish is also called 'Daikon' and an important vegetable crop, especially in Asia. It has a large variation of the size and shape of roots, sliques and leaves.

Radish is also characterized by pungent components, i.e., isothiocyanates. In order to understand genetic mechanisms for those traits, we have conducted construction of infrastructure in molecular genetic studies.

For construction of draft sequences of radish, short reads of genomic sequences of 191.1 Gb obtained by an Illumina HiSeq 2000 and 27,904 high-quality BAC-end sequences were *de novo* assembled, resulting in 76,592 scaffolds of >300 bp (RSA_r1.0; DF384214-DF396802) which span 402 Mb of total size, covering 75.9% of the estimated genome size of radish.

The draft sequences of 76,592 scaffolds are published on this web site. In addition, we have constructed a high density genetic map with ca. 2,550 DNA markers by integration of two published genetic maps (Li et al. 2011, Shirasawa et al. 2011) and addition of new DNA markers (Kitashiba et al. in preparation).

A part of the scaffolds, ca. 1,300 scaffolds, have been assigned to this genetic map. Since we have detected SNPs among eight inbred lines by next generation sequencing of bulked PCR products amplified by 2,880 primer pairs (Zou et al. 2013), the information is also opened on this web site.

Fig. 4.1 Top page of *Raphanus sativus* Genome DataBase (http://radish.kazusa.or.jp)

and gene model (CDSs, peptides, and GFF) and gene expression data (RNA-Seq data with RPKM value) are also available.

4.3.3 Radish Genome Database

The Radish Genome Database (http://radish-genome.org) was generated by the Myongi University in Korea, the Catholic University of Korea, and the Korea Research Institute of Bioscience and Biotechnology in Korea (Jeong et al. 2016). The top page of this database is shown in Fig. 4.3. A keyword search against the genes in *R. sativus* Rs 1.0 is available on the Search page. With the advanced search options, plant species and chromosome numbers can be selected from *R. sativus*, *B. rapa*, *B. oleracea*, and *A. thaliana* in the keyword search. Keyword searches against ontology terms and gene loci are also available, and sequences, markers, and GO/PO annotations of genes are shown as output. In the Genome Features menu, the species, type (mRNA and CDS), name, description, and location can be browsed. Species can be selected from *R. sativus*, *B. rapa*, *B. oleracea*, and *A. thaliana*, and keyword searches can be made against the names and descriptions of these plant species. At the Gene Family page, 152 gene families of *R. sativus*, *B. rapa*, *B. oleracea*, and *A. thaliana* can be browsed. BLAST searches (BLASTN, BLASTP, BLASTX, TBLASTN, and TBLASTX) against the genome and gene sequences of *R. sativus*, *B. rapa*, *B. oleracea*, and *A. thaliana* are also available. Contig information and the number of elements, length occupied, and percentage of sequences of the repetitive sequences in version 1.2.1 can be browsed on the Repeat page. The UCSC genome browser on RadishV1.2.2 is prepared on the Genome Browser page. In the Radish ontologies, the gene ontology (Gene Ontology menu) and plant ontology (Plant

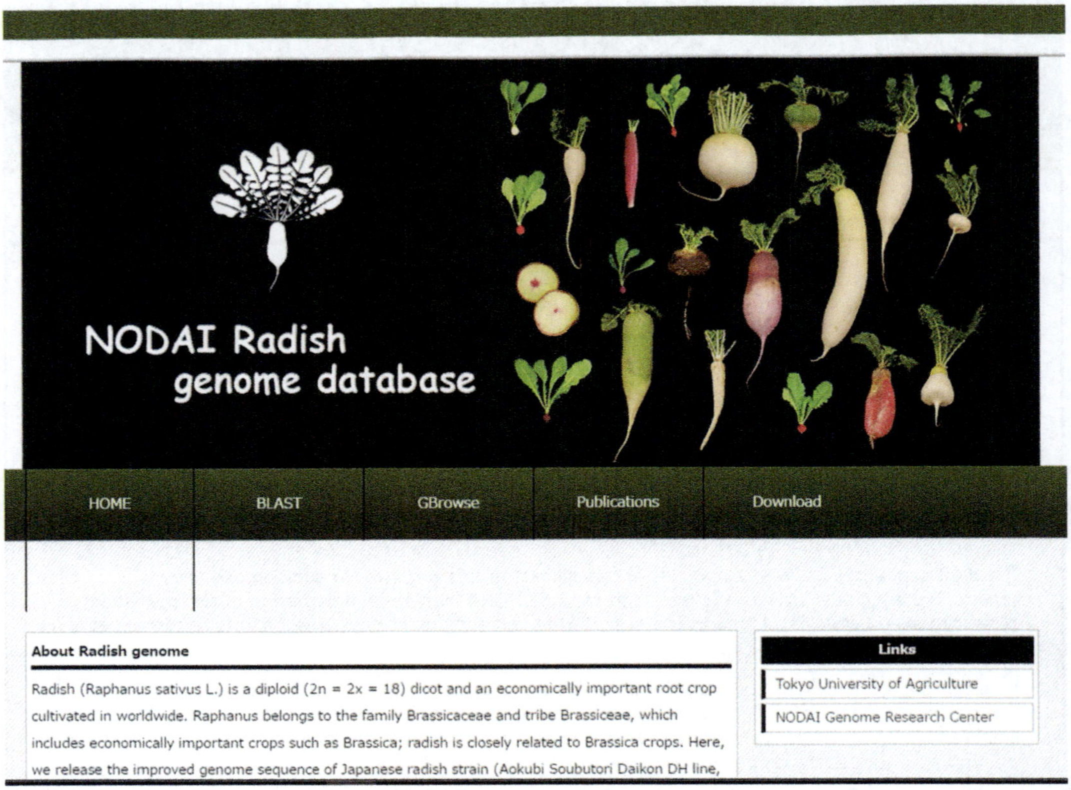

Fig. 4.2 Top page of NODAI Radish Genome Database (http://www.nodai-genome-d.org)

Ontology menu) of the genes are shown. Gene IDs and EC numbers on the metabolic pathways can be browsed at the Pathway page. The information about the F_2 mapping population and SNP and COS (Conserved Ortholog Set) maps can be browsed at the Genetic Maps page. The weighted methylome level at each organ can be browsed with a tabular format at the Methylome page. The Chromosome number, and start and end position can be inputted at the HapMap page. The genomic data related to chromosomes, genes, CDSs, and peptide sequences in Rs 1.0 are available at the Data Resources page.

4.3.4 RadishBase

RadishBase (http://bioinfo.bti.cornell.edu/cgi-bin/radish/index.cgi) (Shen et al. 2013) has been released from Cornell University. The mitochondrial genome sequences from normal cytoplasm and Ogura-type male-sterile cytoplasm are available at the Genome page. EST sequences, unigenes (version 1.0), and cDNA libraries of the radish can be browsed, respectively, at the EST and Library pages. EST can be searched under the GenBank number, unigene ID, and keyword at the Keyword page. Users can search for radish unigenes with annotation and sequence retrieval at the Batch Query page. The location of SNPs on the genome sequence, cultivar name, and alignment at the SNP position among cultivars on the unigenes can be browsed at the SNP page. Repeat sequences, number of repeats, location of SSRs, and primer sequences can be browsed at the SSR page. The radishCyc v1 (http://132.236.156.168:1555/organism-summary?object=RADISH) can be browsed for EST unigenes at the Pathway page. The references about the seven kinds of radish genetic maps can be listed on the MAP page. BLAST searches against the unigenes and EST sequences

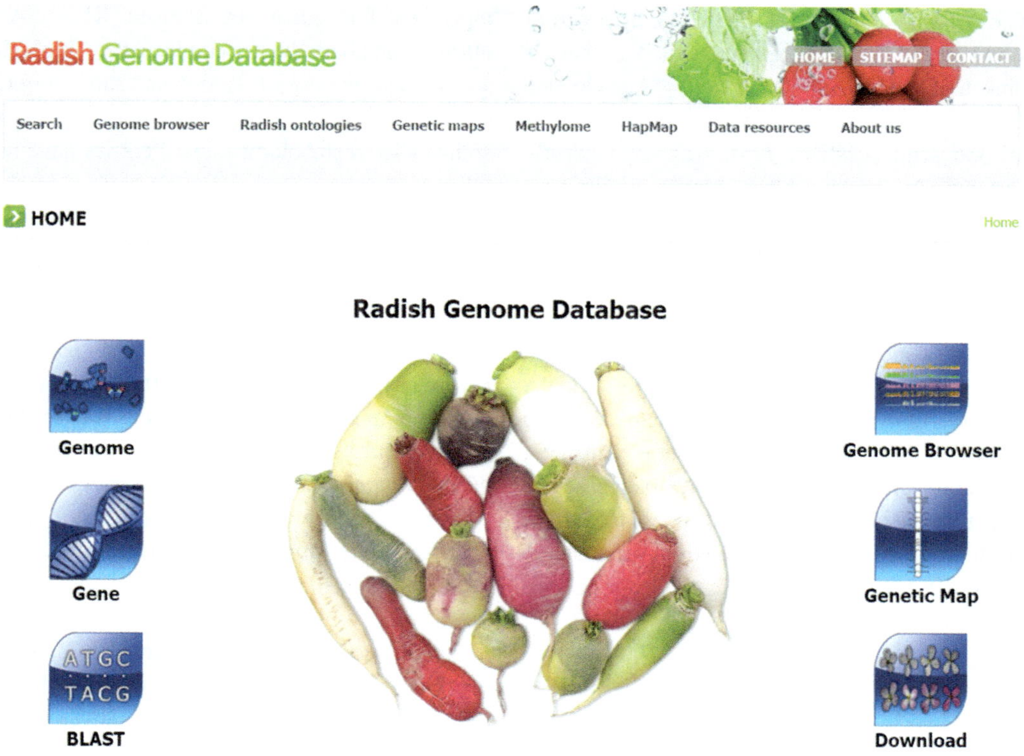

Fig. 4.3 Top page of Radish Genome Database (http://radish-genome.org)

of radish, functional classification, GO enrichment analysis of radish unigenes, and SSR identification can be performed at the Tool page. In addition, BLAST searches (BLASTN, TBLASTN, and TBLASTX) against unigenes and EST sequences and the mitochondrial genome (from normal cytoplasm and Ogura-type male-sterile cytoplasm) are available. The genes are categorized into GO terms and their description can be searched with unigene IDs. The gene ontologies, *p* value correction methods, and cutoff *p* value for significantly represented GO terms can be selected at the GO Enrichment page. The results of GO DAG (Directed Acyclic Graph) can be also browsed by a figure. Users can search SSRs by inputting the sequences, and the minimal number of repeats for di-, tri-, tetra-, penta-, and hexa-repeats. The data related to the radish sequences (EST sequences, unigene sequences, EST-unigene association, SSRs identified from EST unigenes, and EST-SNPs) and biochemical pathway are available. These files are also available at the ftp site (ftp://bioinfo.bti.cornell.edu/pub/radish).

4.3.5 RadishDB

RadishDB (http://radish.plantbiology.msu.edu/index.php/Main_Page) was initiated by the EST sequencing project collaborated by Michigan State University and the J. Craig Venter Institute. The objective of this project is to compare cDNA sequences in the radish, and the project was funded by the National Science Foundation's Plant Genome Comparative Sequencing Program. The following menus are available from the top page: About/Contact Us; The Project; cDNA Sequence/Analysis; Genome Sequence/Analysis; Outreach Activities; and Links. Information about this project (summary, background, approach, samples of radish, sequencing strategy, and sequence analysis) is provided on The Project page. The contig sequences of the two sets (set 1: 5000 clones of EST contigs; set 2: 25,000 clones from each library of EST contigs) are available. There is a link to GenBank in which 287,340 nucleotide sequences [Nucleotide (4 sequences), EST (287,336 sequences)] of *R. sativus* are currently registered. As for the linkage map, a preliminary linkage map for wild radish (*R. raphanistrum*) and a linkage map including 70 markers are available. As for SSR, two kinds of New York populations of *R. raphanistrum* are available. As for SNP, computationally identified CAPS markers from seven libraries are available from the cDNA Sequence/Analysis page. The genomic data related to genomic contigs, coding and protein sequences, genes and pseudogene annotation obtained from the genome sequencing project of *R. raphanistrum* (Moghe et al. 2014) are available from the Genome Sequence/Analysis page.

4.3.6 PGDBj

The Plant Genome DataBase Japan (PGDBj; http://pgdbj.jp) is a portal site for the integration of plant omics databases (hereafter DBs) (Asamizu et al. 2014). It has three internal DBs: (1) the DB for orthologous genes (Ortholog DB), (2) the DB for plant resources (Resource DB), and (3) the DB for DNA markers (DNA Marker DB). In the DNA marker DB, DNA markers for 80 plant species have been released, and the markers are curated manually. In *R. sativus*, 5260 markers [CAPS (7 markers), InDel (8 markers), SCAR (3 markers), SNP (921 markers), SSR (4032 markers), and Other (289 markers)] have been released. In the DNA marker DB, marker names, marker types, primers, population, chromosome numbers of linkage groups, and publication information (DOI or PMID) in each marker have been released. This database also presents the QTL information, QTL name, trait name, chromosome number and linkage group, significant markers, LOD, LOD peak, location, DOI or PMID. It integrates the DBs related to the metabolome, such as KNApSAck (http://kanaya.naist.jp/KNApSAcK/) and MassBase (http://webs2.kazusa.or.jp/massbase/). In KNApSAck Core, 74 metabolites including gibberellin, brassinolide, rphanusanin, have been registered for radish. In Skewed KNApSAcK, information about the geographic usage, biological activity, and the use of Kampo and Jamu formulas has been released. In MassBase, mass spectral tag archives for metabolomics in 12 entries have been released.

4.3.7 Kazusa Marker DataBase

The Kazusa Marker DataBase (http://marker.kazusa.or.jp) (Shirasawa et al. 2014) is a portal site for DNA markers developed at the Kazusa DNA Research Institute (KDRI). Currently, 14 plant species (*Solanum lycopersicum*, *Arachis hypogaea*, *R. sativus*, *Glycine max*, *Trifolium repens*, *Trifolium pratense*, *Lotus japonicus*, *Eucalyptus camaldulensis*, *Capsicum annuum*, *Fragaria* × *ananassa*, *Vicia faba*, *Vigna angularis*, *Hieracium* species, and *Trifolium subterraneum*) are registered. In *R. sativus*, the DNA markers (EST-SSR, isozyme marker, morphological marker, R-EST (disease resistance genes), RAPD, RFLP, SSR, STS) and linkage map of GHRI (RIL: *R. sativus* 'GSK3-1' × *R. sativus* 'HA2') have been released. In this DB, the files related to gene, genome, linkage map, and markers are available in the tsv format.

Note Added in Proof

The genome sequence of Chinese radish cv. XYB36-2 (Radish_v2.20) has been released from Institute of Vegetables and Flowers, Chinese Academy of Agricultural Science under the NCBI accession number GCA_002197605.1 (Zhang et al. 2015, Horticultural Plant Journal, 1, 155-164). However the database of the cultivar has not been published, therefore the description of the cultivar was excluded in this review.

References

Altschul SF, Madden TL, Schäffer AA, Zhang J, Zhang Z, Miller W, Lipman DJ (1997) Gapped BLAST and PSI-BLAST: a new generation of protein database search programs. Nucleic Acids Res 25:3389–3402

Asamizu E, Ichihara H, Nakaya A, Nakamura Y, Hirakawa H, Ishii T, Tamura T, Fukami-Kobayashi K, Nakajima Y, Tabata S (2014) Plant Genome DataBase Japan (PGDBj): a portal website for the integration of plant genome-related databases. Plant Cell Physiol 55:e8

Conesa A, Götz S, García-Gómez JM, Terol J, Talón M, Robles M (2005) Blast2GO: a universal tool for annotation, visualization and analysis in functional genomics research. Bioinformatics 21:3674–3676

Eddy SR (2011) Accelerated profile HMM searches. PLoS Comput Biol 7:e1002195

Grabherr MG, Haas BJ, Yassour M, Levin JZ, Thompson DA, Amit I, Adiconis X, Fan L, Raychowdhury R, Zeng Q, Chen Z, Mauceli E, Hacohen N, Gnirke A, Rhind N, di Palma F, Birren BW, Nusbaum C, Lindblad-Toh K, Friedman N, Regev A (2011) Full-length transcriptome assembly from RNA-Seq data without a reference genome. Nat Biotechnol 29:644–652

Haas B, Salzberg S, Zhu W, Pertea M, Allen J, Orvis J, White O, Buell C, Wortman J (2008) Automated eukaryotic gene structure annotation using EVidenceModeler and the program to assemble spliced alignments. Genome Biol 9:R7

Jeong YM, Kim N, Ahn BO, Oh M, Chung WH, Chung H, Jeong S, Lim KB, Hwang YJ, Kim GB, Baek S, Choi SB, Hyung DJ, Lee SW, Sohn SH, Kwon SJ, Jin M, Seol YJ, Chae WB, Choi KJ, Park BS, Yu HJ, Mun JH (2016) Elucidating the triplicated ancestral genome structure of radish based on chromosome-level comparison with the Brassica genomes. Theor Appl Genet 129:1357–1372

Kitashiba H, Li F, Hirakawa H, Kawanabe T, Zou Z, Hasegawa Y, Tonosaki K, Shirasawa S, Fukushima A, Yokoi S, Takahata Y, Kakizaki T, Ishida M, Okamoto S, Sakamoto K, Shirasawa K, Tabata S, Nishio T (2014) Draft sequences of the radish (*Raphanus sativus* L.) genome. DNA Res 21:481–490

Korf I (2004) Gene finding in novel genomes. BMC Bioinformat 5:59

Li L, Stoeckert CJ Jr, Roos DS (2003) OrthoMCL: identification of ortholog groups for eukaryotic genomes. Genome Res 13:2178–2189

Li W, Godzik A (2006) Cd-hit: a fast program for clustering and comparing large sets of protein or nucleotide sequences. Bioinformatics 22:1658–1659

Llorens C, Futami R, Covelli L, Domínguez-Escribá L, Viu JM, Tamarit D, Aguilar-Rodríguez J, Vicente-Ripolles M, Fuster G, Bernet GP, Maumus F, Munoz-Pomer A, Sempere JM, Latorre A, Moya A (2011) The Gypsy Database (GyDB) of mobile genetic elements: release 2.0. Nucleic Acids Res 39 (Database Issue):D70–D74

Lowe TM, Eddy SR (1997) tRNAscan-SE: a program for improved detection of transfer RNA genes in genomic sequence. Nucleic Acids Res 25:955–964

Mitsui Y, Shimomura M, Komatsu K, Namiki N, Shibata-Hatta M, Imai M, Katayose Y, Mukai Y, Kanamori H, Kurita K, Kagami T, Wakatsuki A, Ohyanagi H, Ikawa H, Minaka N, Nakagawa K, Shiwa Y, Sasaki T (2015) The radish genome and comprehensive gene expression profile of tuberous root formation and development. Sci Rep 5:10835

Moghe GD, Hufnagel DE, Tang H, Xiao Y, Dworkin I, Town CD, Conner JK, Shiu SH (2014) Consequences of whole-genome triplication as revealed by comparative genomic analyses of the wild radish *Raphanus raphanistrum* and three other Brassicaceae species. Plant Cell 26:1925–1937

Mulder NJ, Apweiler R, Attwood TK, Bairoch A, Bateman A, Binns D, Bork P, Buillard V, Cerutti L, Copley R, Courcelle E, Das U, Daugherty L, Dibley M, Finn R, Fleischmann W, Gough J, Haft D, Hulo N, Hunter S, Kahn D, Kanapin A, Kejariwal A, Labarga A, Langendijk-Genevaux PS, Lonsdale D, Lopez R, Letunic I, Madera M, Maslen J, McAnulla C, McDowall J, Mistry J, Mitchell A, Nikolskaya AN, Orchard S, Orengo C, Petryszak R, Selengut JD, Sigrist CJ, Thomas PD, Valentin F, Wilson D, Wu CH, Yeats C (2007) New developments in the InterPro database. Nucleic Acids Res 35 (Database Issue):D224–D228

Ogata H, Goto S, Sato K, Fujibuchi W, Bono H, Kanehisa M (1999) KEGG: Kyoto encyclopedia of genes and genomes. Nucleic Acids Res 27:29–34

Quevillon E, Silventoinen V, Pillai S, Harte N, Mulder N, Apweiler R, Lopez R (2005) InterProScan: protein domains identifier. Nucleic Acids Res 33(Web Server Issue):W116–W120

Rice P, Longden I, Bleasby A (2000) EMBOSS: the European molecular biology open software suite. Trends Genet 16:276–277

Shen D, Sun H, Huang M, Zheng Y, Li X, Fei Z (2013) RadishBase: a database for genomics and genetics of radish. Plant Cell Physiol 54:e3

Shirasawa K, Isobe S, Tabata S, Hirakawa H (2014) Kazusa marker DataBase: a database for genomics, genetics, and molecular breeding in plants. Breed Sci 64:264–271

Stanke M, Waack S (2003) Gene prediction with a hidden Markov model and a new intron submodel. Bioinformatics 19:ii215–ii225

Tatusov RL, Fedorova ND, Jackson JD, Jacobs AR, Kiryutin B, Koonin EV, Krylov DM, Mazumder R, Mekhedov SL, Nikolskaya AN, Rao BS, Smirnov S, Sverdlov AV, Vasudevan S, Wolf YI, Yin JJ, Natale DA (2003) The COG database: an updated version includes eukaryotes. BMC Bioinformat 4:41

Wu T, Nacu N (2010) Fast and SNP-tolerant detection of complex variants and splicing in short reads. Bioinformatics 26:873–881

Zhang X, Yue Z, Mei S, Qiu Y, Yang X, Chen X, Cheng F, Wu Z, Sun Y, Jing Y, Liu B, Shen D, Wang H, Cui N, Duan Y, Wu J, Wang J, Gan C, Wang J, Wang X, Li X (2015) A de novo genome of a Chinese radish cultivar. Hortic Plant J 1:155–164

Comparative Analysis of the Radish Genome with *Brassica* Genomes

5

Hee-Ju Yu, Young-Min Jeong and Jeong-Hwan Mun

Abstract

Raphanus sativus L. includes an annual root vegetable crop, radish, and diverse wild species. *R. sativus* has a long history of domestication, but its phylogenetic position in the tribe *Brassiceae* is controversial. A comprehensive analysis of the *R. sativus* genome will provide fundamental information about the structure of its genome, evolutionary features of polyploidy, and significant insight for phylogenetic delimitation of this species. Diverse genomic resources, including a high-density genetic map, clone libraries, cytogenetic data, and transcriptome data, have been developed to sequence the genome. Recently, the *R. sativus* cv. 'WK10039' ($2n = 18$, 510.8 Mb) genome was sequenced and assembled into nine chromosome pseudomolecules spanning >98% of the gene space. Comparative mapping of the tPCK-like ancestral genome based on conserved ortholog set markers and proteome comparison revealed that the *R. sativus* genome has intermediate characteristics between the *Brassica* A/C and B genomes with triplicated segments, suggesting an internal origin from the genus *Brassica*. The evolutionary characteristics shared between *R. sativus* and diploid *Brassica* species provide genomic evidence for species delimitation of *R. sativus* and reconstruction of the mesohexaploid ancestral genome.

5.1 Introduction

Raphanus sativus L. is a diploid ($2n = 18$) species in the tribe *Brassiceae* of the Brassicaceae family and includes a major commercial root vegetable crop, radish, important for agriculture and human nutrition. It has been defined as a distinct species of the genus *Raphanus* but is closely related to species in the genus *Brassica*.

H.-J. Yu · Y.-M. Jeong
Department of Life Science, The Catholic University of Korea, Bucheon 14662, Korea
e-mail: yuheeju@catholic.ac.kr

Y.-M. Jeong
e-mail: ymjeongbio@catholic.ac.kr

J.-H. Mun (✉)
Department of Bioscience and Bioinformatics, Myongji University, Yongin 17058, Korea
e-mail: munjh@mju.ac.kr

T. Nishio and H. Kitashiba (eds.), *The Radish Genome*,
Compendium of Plant Genomes, DOI 10.1007/978-3-319-59253-4_5

Genetics and breeding studies aiming to develop intergeneric hybrids between two genera established that *R. sativus* and three diploid *Brassica* species, including *B. rapa* (A genome), *B. nigra* (B genome), and *B. oleracea* (C genome), are close relatives sharing common ancestry in the tribe *Brassiceae*. Similar to the allotetraploid hybrids of *Brassica* such as *B. juncea* (AB genome), *B. napus* (AC genome), and *B. carinata* (BC genome) in U's triangle (U 1935), artificial hybridization between *R. sativus* and diploid *Brassica* species successfully produced allopolyploid hybrids (Arumugam et al. 2002; Lee et al. 2011; Matsuzawa and Sarashima 1986; Prakash et al. 2009). Meanwhile, phylogeny of the genus *Raphanus* in the tribe *Brassiceae* is still contentious. The tribe *Brassiceae* comprises approximately 240 species in 6–7 sub-tribes (Warwick and Sauder 2005), and the genus *Raphanus* is classified into sub-tribe Raphaninae, and the genus *Brassica* belongs to the sub-tribe *Brassicinae* based on morphological characteristics such as fruit structure. However, fruit structure has proven labile during evolution of Brassicaceae (Beilstein et al. 2006). Moreover, recent molecular phylogenetic studies have failed to provide obvious evidences for the division of *Raphanus* and *Brassica* into distinct sub-tribes suggesting that these two genera are paraphyletic. For example, molecular phylogenetic analysis of chloroplast DNA has placed *R. sativus* in the *rapa/oleracea* lineage of sub-tribe Brassicinae (Lü et al. 2008; Warwick and Black 1991, 1997; Yang et al. 2002), whereas independent studies using nuclear DNA markers have shown that *R. sativus* has a closer relationship with *B. nigra* than with *B. rapa/B. oleraceae* (Song et al. 1990, 1988; Thormann et al. 1994; Yang et al. 1998, 1999). These findings collectively suggest that *R. sativus* might have arisen by hybridization of a paternal *nigra/sinapis* lineage and a maternal *rapa/oleracea* lineage (Song et al. 1990; Yang et al. 2002).

Investigation into the speciation and evolution of *R. sativus* can be explored in detail by comparative analysis with the sequenced genomes of Brassicaceae. There are more than 10 sequenced genomes in Brassicaceae, including *Arabidopsis thaliana*, *A. lyrata*, *B. rapa*, *B. oleracea*, *Capsella rubella*, and *Thellungiella parvula*. In previous studies, genomic comparisons successfully elucidated dynamic patterns of genome evolution caused by duplications and hybridization in Brassicaceae. For example, *A. thaliana* has a small genome of five chromosomes; however, a remnant of a hexaploid genome has been found (The *Arabidopsis* Genome Initiative 2000). Comparative mapping and cytogenic analysis also suggest that the *A. thaliana* genome might have derived from an ancestral genome of eight chromosomes, similar to those of *A. lyrata* and *C. rubella* (Koch and Kiefer 2005; Lysak et al. 2006; Yogeeswaran et al. 2005). Based on that result, Schranz et al. (2006) proposed an Ancestral Crucifer Karyotype (ACK) with eight chromosomes composed of 24 ancestral karyotype (AK) genomic blocks (Fig. 5.1a). The ACK system has played a crucial role in understanding genome evolution in Brassicaceae species (Lysak et al. 2016; Murat et al. 2015). On the other hand, other species of the Brassicaceae family are believed to be descendants of an ancestor with seven chromosomes called the Proto-Calepineae Karyotype (PCK) (Mandakova and Lysak 2008). After translocation of whole chromosome arms, the PCK might have evolved into the translocation PCK (tPCK, $n = 7$; Fig. 5.1b) that can be seen in several species, such as *T. salsuginea* (Lysak and Koch 2011; Wu et al. 2012; Yang et al. 2013). The tPCK is considered to be an ancestral genome of the tribe *Brassiceae* that includes *Raphanus* and *Brassica* species (Cheng et al. 2013). In this regard, a detailed comparison of the *R. sativus* genome with other Brassicaceae genomes should provide significant insights into the origin, evolution, and even phylogenetic status of this species.

5.2 Characteristics of the *Brassica* Genomes

Construction of chromosome-scale draft genome assemblies for *B. rapa* (The *Brassica rapa* Genome Sequencing Project Consortium 2011), *B. oleracea* (Liu et al. 2014; Parkin et al. 2014),

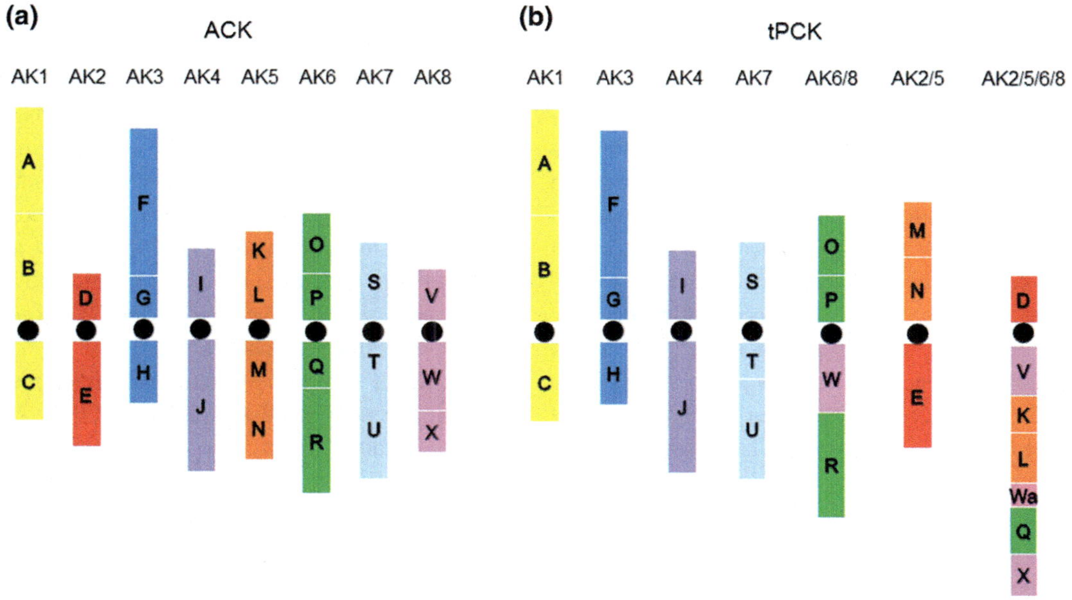

Fig. 5.1 Genome blocks of Ancestral Crucifer Karyotype (ACK) and translocation-Proto-Calepineae Karyotype (tPCK). **a** ACK with $n = 8$. **b** tPCK with $n = 7$. The *colors* of the 24 ancestral karyotype genomic blocks (*A–X*) indicate each ACK chromosome. *Black circles* represent centromeres

and *B. napus* (Chalhoub et al. 2014) has established detailed polyploid histories of *Brassica* species. *B. rapa* was the first sequenced genome in the genus *Brassica* and served as the fundamental basis for genome assembly and annotation and for studying the genomic structure and evolution of other *Brassica* species. *B. oleracea* was the second sequenced *Brassica* genome. *B. rapa* and *B. oleracea* are similar both in sequence and genomic organization (Parkin et al. 2014), as expected from their close phylogenetic relationship. In the allopolyploid *B. napus* genome, the overall genomic structures of the A and C sub-genomes have been retained, although several homologous exchanges between the sub-genomes were found (Chalhoub et al. 2014). These genomes showed dynamic evolutionary natures of the *Brassica* genomes; however, the *B. rapa* and *B. oleracea* genomes represent only the *rapa*/*oleracea* lineage of the tribe *Brassiceae*. To overview general patterns of genome evolution in the tribe, *Brassiceae* requires an examination of genomes in the *nigra*/*sinapis* lineage. The *B. nigra* genome has been analyzed based on a microsynteny comparison using bacterial artificial chromosome (BAC) sequences (Navabi et al. 2013) and comparative mapping (Panjabi et al. 2008; Paritosh et al. 2014). Unfortunately, whole-genome assembly of the B genomes is not yet available, and the genomic resources are also limited. Therefore, characterization of the B genome is a significant challenge in the genomic study of *Brassica*.

The most interesting characteristics of the diploid *Brassica* genomes are whole-genome triplication (WGT). Comparative chromosome painting using fluorescence *in situ* hybridization (FISH) suggests that WGT is shared in species of the tribe *Brassiceae* (Lysak et al. 2005). Moreover, the first released *B. rapa* genome enabled researchers to decipher the detailed nature of the triplicated genomic blocks (The *Brassica rapa* Genome Sequencing Project Consortium 2011). Interestingly, the triplicated *B. rapa* has only 1.4 times more genes than *A. thaliana*. This finding suggests that genome-wide diploidization is also involved in formation of the current *B. rapa* genome. A comparison of the microsynteny patterns of the triplicated genome blocks with the *A. thaliana* genome established detailed evolutionary changes of the *Brassica* genomes. In general, the duplicated genes in a duplicated genome can be freely mutated as long as one of the genes maintains its original function. Therefore, duplicated genes can be functionally diverse; otherwise, many of them are lost through the accumulation of null mutations (Lynch and Conery 2000). A tetraploid becomes a diploid as the copy number of genes decreases; however, gene loss is likely to occur non-randomly (Schnable et al. 2011; Cheng et al. 2012). Genes from one specific lineage can be lost more often than those from another lineage, which is called genome dominance. The differential gene loss of triplicated genome blocks is evident in both the *B. rapa* and *B. oleracea* genomes (Cheng et al. 2012; Parkin et al. 2014). The triplicated genome blocks of *B. rapa* and *B. oleracea* can be classified into least fractionated (LF), moderately fractionated (MF1), and most fractionated (MF2) sub-genome blocks according to the degree of gene retention. The overall gene retention rate of the LF, MF1, and MF2 sub-genome blocks in *B. rapa*, compared to those in *A. thaliana*, is 70, 46, and 36%, respectively (The *Brassica rapa* Genome Sequencing Project Consortium 2011).

Identification of three differentially fractioned sub-genomes suggests that WGT in the *Brassica* genome resulted from two rounds of polyploidization (Fig. 5.2; Cheng et al. 2012; Cheng et al. 2014). The first tetraploid was likely

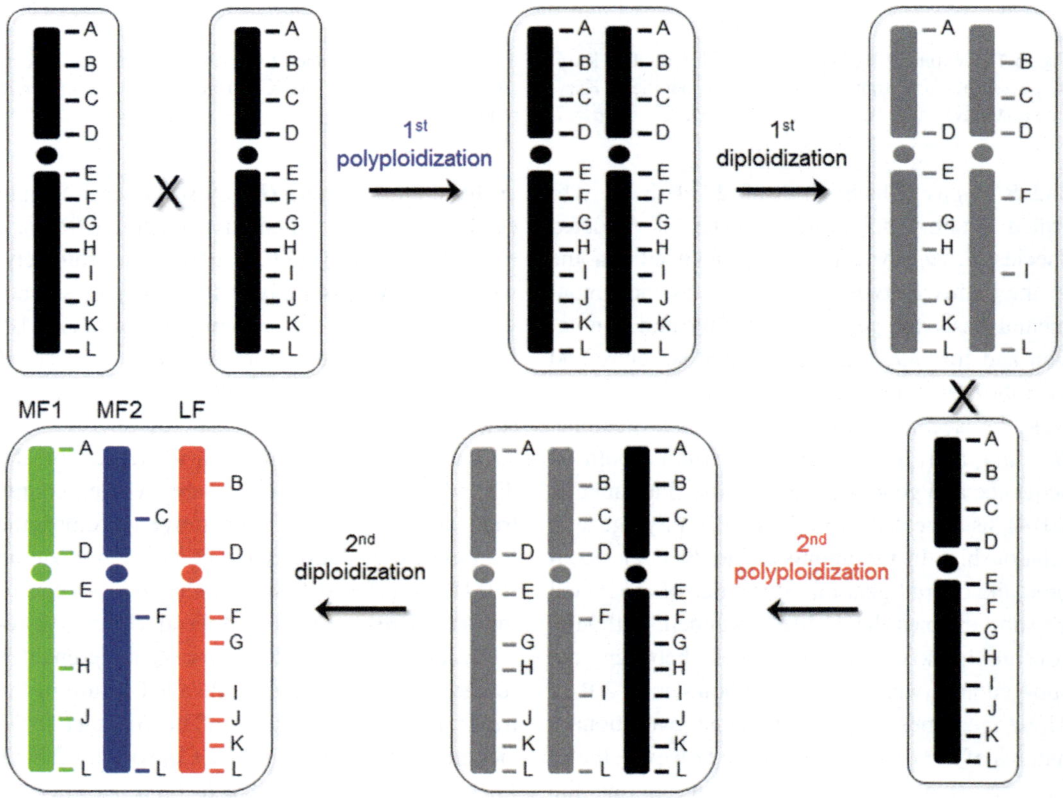

Fig. 5.2 A model of whole-genome triplication by two rounds of polyploidization. The current triplicated genomes in the tribe *Brassiceae* were likely formed through two rounds of polyploidization: auto- or allopolyploidization of tPCK species followed by differential gene loss resulted in tetraploid sub-genomes with unequal gene density. A second hybridization of a tetraploid species and a tPCK species formed a hexaploid genome harboring three sub-genomes. Diploidization including gene loss formed the triplicated genome of the progenitor of tribe Brassiceae. Because the tPCK genome (LF) introduced in the second tetraploidization experienced only one diploidization event, its gene density is higher than that of the other sub-genomes (MF1 and MF2)

formed either by autopolyploidization or allopolyploidization through hybridization of ancestral plants with tPCK genomes. Differential gene loss due to genome dominance might have followed during the first diploidization. The second hybridization of the tetraploid genome with another tPCK genome and diploidization and rearrangement after that might have resulted in the production of the ancestral hexaploid genome with the triplicated sub-genome blocks of *Brassica* species. This two-round polyploidization hypothesis explains differential gene loss in each sub-genome. The tPCK genome recently introduced into the hexaploid ancestral *Brassica* genome by the second polyploidization corresponds to the LF sub-genome in *B. rapa* and *B. oleracea,* because it has experienced a relatively short period of diploidization compared with the other sub-genomes. Likewise, the MF1 and MF2 sub-genomes originated from the first polyploidization. However, the chromosome number of the hexaploid common ancestor of the tribe *Brassiceae* after the second polyploidization remains uncertain.

5.3 Comparative Mapping of the Radish Genome Based on a *Brassica* Conserved Orthologous Set

The extensive genetic and sequence information of a sequenced genome can be easily transferred to closely related crop species using comparative genome mapping. In the case of a genome comparison of diploid species of *Brassica* (such as *B. rapa* and *B. oleracea*) with *A. thaliana*, the triplicated sub-genome blocks of each genome can be analyzed based on syntenic counterparts in the *A. thaliana* genome. On the other hand, a comparative analysis between triplicated genomes of the tribe *Brassiceae* that have not been sequenced is difficult due to multiplied genes. Theoretically, one *R. sativus* gene shows sequence similarity to three *B. rapa* genes and *vice versa*; therefore, additional examination is required to identify orthologous gene pairs between the genomes. Furthermore, gene loss at each lineage after speciation hinders determination of orthologous relationships because paralogous genes inherited from different ancestors are sufficiently similar at the sequence level (Fig. 5.3a). Therefore, sequence similarity is not enough to distinguish orthologous (allelic) gene pairs from paralogous (non-allelic) variants between species in the tribe *Brassiceae.* One possible approach to overcoming the paralogy issue is the development of conserved ortholog set (COS) markers. A COS is defined as a collection of genes that are conserved in sequence and copy number among multiple species, but practically it is used to find orthologous genes that exist in a single or low copy between related species. Since the first COS was developed from the Solanaceae family and COS markers were successfully applied to comparative mapping of *Solanum* species (Fulton et al. 2002), COSs have been identified from diverse plant species, including cereals (Quraishi et al. 2009), Rosaceae (Cabrera et al. 2009; Bushakra et al. 2011), Asteraceae (Timms et al. 2006; Chapman et al. 2007), and Pinaceae (Krutovsky et al. 2004; Liewlaksaneeyanawin et al. 2008).

In Brassicaceae, a *Brassica* COS was developed based on whole proteome comparison of *A. thaliana*, *B. rapa*, and *B. oleracea* (Jeong et al. 2014). A total of 1194 *Brassica* COS genes were mapped to the genetic map of *R. sativus* constructed using an F_2 population from a maternal 'WK10039' and paternal 'WK10024' cross (Fig. 5.3b). The *Brassica* COS markers provided an overview of the *R. sativus* genome before whole-genome sequencing. For example, the R1 chromosome of *R. sativus* has a structure similar to the AK1 of ACK. An AK1 chromosome is also in *Sinapis alba* in the tribe *Brassiceae* (Nelson et al. 2011), suggesting that AK1 of ACK is retained in R1 with little fractionation. In the *B. rapa* genome, most COS genes mapped to R1 were located in LF sub-genomic regions, indicating that the sub-genomic identity of COS genes is similar between *R. sativus* and *B. rapa.* Overall, the COS markers in the radish genetic map and the corresponding COS genes in *B. rapa* showed a one-to-one correspondence (Fig. 5.4a). Therefore, COS markers correctly interconnect

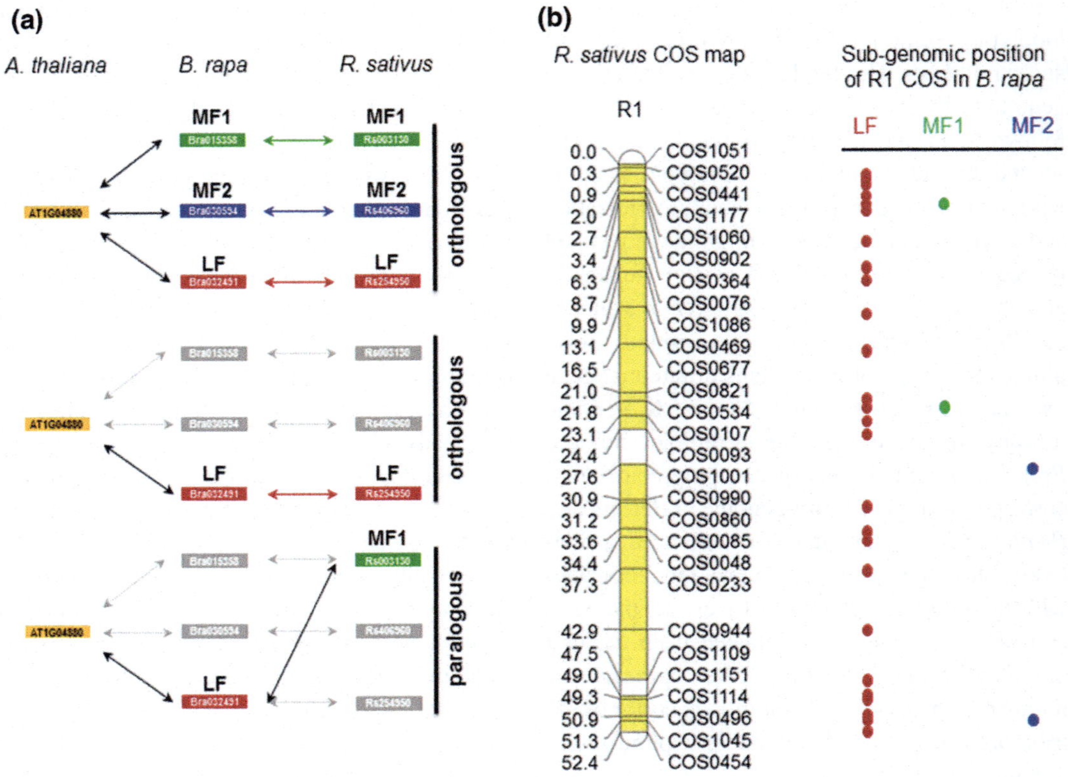

Fig. 5.3 Orthologous and paralogous relationships between *R. sativus* and *B. rapa.* **a** Orthologous relationship among triplicated genes. If gene loss occurred in the same sub-genome blocks, the remaining genes can be orthologous. If gene loss occurred in different sub-genome blocks, the remaining genes can be paralogous. **b** Genetic position and sub-genomic identity of COS genes in the radish R1 linkage group. Sub-genomic positions of radish COS genes were determined by those of counterpart COS genes in the *B. rapa* genome. *Red*, *green*, and *blue dots* indicate LF, MF1, and MF2 sub-genomic identities, respectively

the *R. sativus* genome blocks with the corresponding *B. rapa* genome. Orthologous correspondence of genome blocks based on COS genes between the genomes can be evaluated by whole proteome comparison using the chromosome pseudomolecule sequences of *R. sativus* (Jeong et al. 2016). Proteome comparison of *R. sativus* and *B. rapa* enables researchers to identify homologs of protein-coding genes in each genome. Figure 5.4 shows an example of orthologous counterparts of the *B. rapa* genome for genes in the R1 chromosome of *R. sativus.* The best hits from the BLASTP search using *R. sativus* genes in R1 as a query against the *B. rapa* genome are positioned throughout the *B. rapa* genome (Fig. 5.4b). To establish confident orthologous relationships, paralogous pairs should be removed. Considering that R1 consists of the A, B, and C blocks of tPCK AK1, noisy BLAST hits can be removed using AK block information from *B. rapa* (Fig. 5.4c). Additional paralogous relationships are filtered out using the sub-genomic characteristics of *B. rapa* genes. AK1 blocks in A6, A9, and A10 are orthologous to those in R1 because R1 has an LF sub-genome (Fig. 5.4d). The resulting relationship is fundamentally similar to that of COS genes between the genomes (Fig. 5.4e). Therefore, comparative mapping using *Brassica* COS can be an efficient way to elucidate the genome structure of any species in the tribe *Brassiceae* for which researchers lack detailed genomic information.

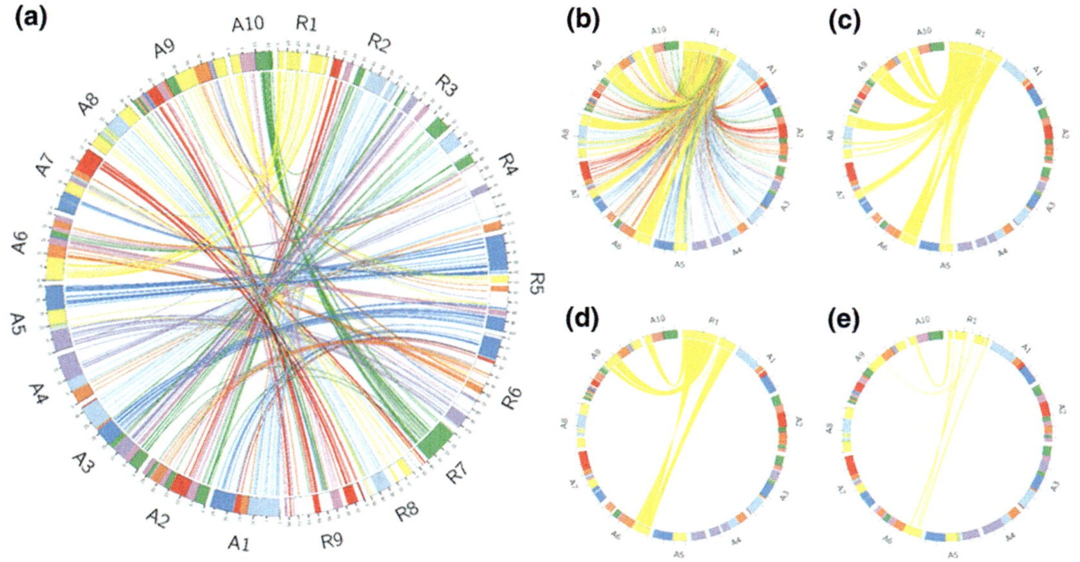

Fig. 5.4 Correspondence of COS genes between the *R. sativus* genetic map and the *B. rapa* genome. **a** COS genes in the *R. sativus* genetic map and the *B. rapa* genome are linked by *lines*. *Colors* indicate 24 AK genome blocks. *B. rapa* chromosomes (*A1–A10*) and *R. sativus* linkage groups (*R1–R9*) are indicated *outside of the circles*. **b** Genomic positions of best-hit *B. rapa* homologs searched using the genes in *R1* of *R. sativus* as queries. **c**, **d** Selection of genes in the AK1 (**c**) and LF sub-genomes (**d**). **e** Genomic positions of COS genes that belong to the R1 and $AK1_{LF}$ block of *B. rapa*. *Colors* indicate AK genome blocks. Chromosome and linkage groups are indicated outside the *circles*

5.4 Triplicated Genome Structure of Radish

A chromosome-scale genome assembly that covers the entire euchromatic gene space is a crucial resource to elucidate the genomic structure of *R. sativus*. Several independent genome sequencing efforts have been made for cultivars (Kitashiba et al. 2014; Mitsui et al. 2015) and wild radish (Moghe et al. 2014) that resulted in draft scaffold or contig sequences. Recently, nine chromosome pseudomolecule sequences of an Asian big radish cultivar 'WK10039' ($2n = 18$, 510 Mb) have been published (Jeong et al. 2016). Briefly, whole-genome shotgun sequencing of the 'WK10039' genome was performed using a combination of the 454, Illumina, and PacBio platforms for hybrid assembly. The *R. sativus* genome was assembled into 426.2 Mb spanning >98% of the gene space, of which 344.0 Mb were organized into nine chromosome pseudomolecules by mapping to the reference genetic map (Mun et al. 2015). The structural characteristics of the *R. sativus* genome were investigated in detail based on genome-wide proteome comparison (Jeong et al. 2016). Self-comparison of the *R. sativus* genome identified 128 syntenic blocks, many of which were members of triplicated fragments. In addition, genome-wide comparative analysis of *R. sativus* using BLASTP identified 326 genome blocks syntenic to their counterparts in the *A. thaliana*, *B. rapa*, and *B. oleracea* genomes. These blocks show a three-to-one or three-to-three syntenic match of long-range syntenic blocks between chromosomes of *R. sativus* and *A. thaliana* or *Brassica*s, respectively, demonstrating a whole-genome triplication of the *R. sativus* genome. Comparative mapping of 24 AK genome building blocks on the *R. sativus* genome found that three copies of the AK genome blocks were organized in nine chromosomes of *R. sativus* except for the D, G, and S blocks (Fig. 5.5). Differential gene loss among triplicated sub-genome blocks was evident as presented

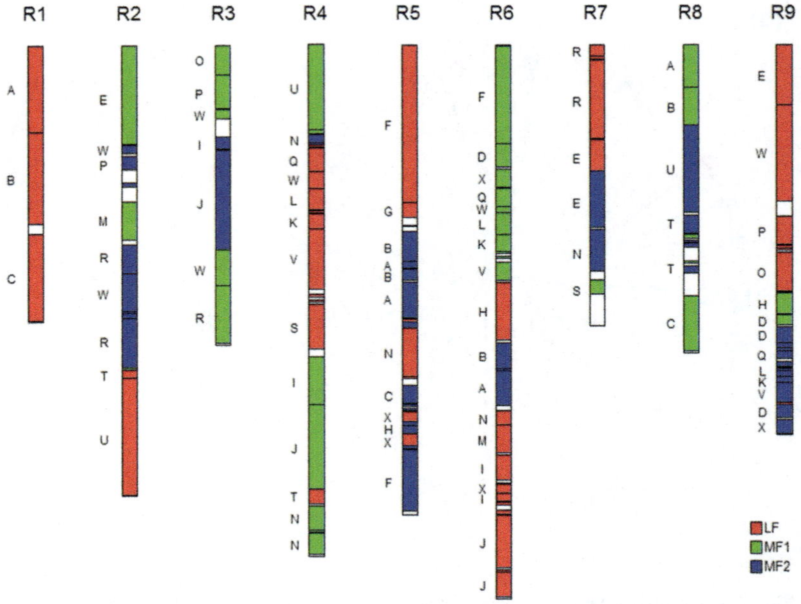

Fig. 5.5 Organization of sub-genome blocks in the *R. sativus* genome. AK block name is presented at the *left* side of each block. The three sub-genomes are *red* (LF), *green* (MF1), and *blue* (MF2)

from the *B. rapa* and *B. oleracea* genomes (Liu et al. 2014; Parkin et al. 2014; The *Brassica rapa* Genome Sequencing Project Consortium 2011). On average, the retention rate of non-redundant orthologous genes in sub-genome blocks compared to *A. thaliana* was 41% for LF blocks, 25% for MF1 blocks, and 22% for MF2 blocks, similar to the case of *B. rapa*. Considering the close phylogenetic relationship between the genera *Raphanus* and *Brassica* in the tribe *Brassiceae*, the three-to-three chromosome-level long-range syntenic matches for both the *R. sativus-B. rapa* and *R. sativus-B. oleracea* comparisons and the differentially fractionated sub-genomes (LF vs. MF1 and MF2) suggest that WGT of *R. sativus* might have occurred via two rounds of whole-genome duplication followed by fractionation, as suggested in the *B. rapa* genome (Cheng et al. 2012; The *Brassica rapa* Genome Sequencing Project Consortium 2011).

Characterization of the triplicated sub-genome blocks provides information on the ancestral genomic structure of *R. sativus*. As shown in Figs. 5.1 and 5.5, AK2/5 and AK2/5/6/8 are tPCK-specific chromosomes. In AK2/5, the M_{MF1}-E_{MF1} (R2) and N_{MF2}-E_{MF2} (R7) blocks are found in a single chromosome, whereas the N_{LF}-E_{LF} block is recombined in three chromosomes (R5, R7, and R9). Likewise, $AK2/5/6/8_{MF1}$ and $AK2/5/6/8_{MF2}$ are in R6 and R9, respectively, and a fragmented $AK2/5/6/8_{LF}$ is found in R4 and parts of R5 and R6. Therefore, the existence of D-V and M-E block associations with triplicated patterns in the *R. sativus* chromosomes is strong evidence that *R. sativus* is a descendent of an ancestor with a triplicated tPCK genome. Other AK sub-genome blocks are also reconstructed in the triplicated *R. sativus* genome, as shown in Fig. 5.6. Moreover, three copies of each AK block were not always identified in the *R. sativus* genome. Similar to the case of *B. rapa* (Cheng et al. 2013), the D_{LF} and G_{MF1} blocks were mostly fragmented in *R. sativus*, suggesting that the fragmentation occurred in their common ancestor. In addition, tPCK chromosomes showed diverse rearrangements and translocation patterns, and large parts of the G_{LF}, G_{MF2}, I_{MF2}, S_{MF1}, and T_{MF1} blocks were lost in the *R. sativus* genome. At the chromosomal level, one tPCK chromosome is separated and dispersed over 1–3 radish chromosomes. The most conserved tPCK chromosome is $AK1_{LF}$, which is retained in R1. In $AK4_{LF}$, $AK1_{MF1}$, all AK4, $AK6/8_{MF1}$, $AK6/8_{MF2}$, and $AK2/5/6/8_{MF2}$, all AK blocks exist in single chromosome of *R. sativus*, with or without block rearrangements. In contrast, $AK2/5_{LF}$ shows

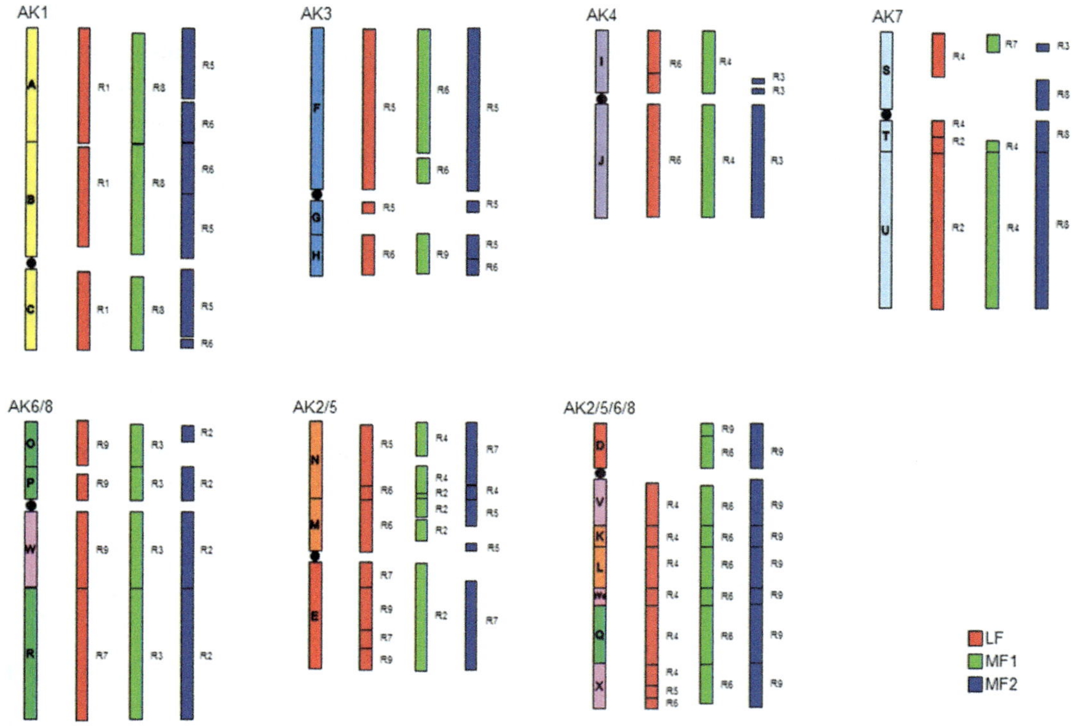

Fig. 5.6 Reconstruction of triplicated sub-genome blocks of *R. sativus*. The tPCK chromosomes are indicated at the *left* side of the three sub-genomes of *R. sativus*. The chromosomal positions of the sub-genome blocks are indicated at the *right* side. The three sub-genomes are colored *red* (LF), *green* (MF1), and *blue* (MF2)

several internal block breakages and dispersion in 4 chromosomes. Overall, the translocation patterns of the sub-genome blocks are similar, but the number of translocations is smaller than those of *B. rapa*.

5.5 Conserved Genome Blocks Between Radish and *Brassica* Genomes

Comparison of the *R. sativus* and *B. rapa* genomes showed chromosome range synteny between the two species, as well as species-specific differences. Therefore, genome-to-genome comparison of *R. sativus* with other genomes in the tribe *Brassiceae* will enable reconstruction of the three ancestral sub-genomes of *R. sativus*. However, only three chromosome-level genome assemblies of diploid species in the tribe *Brassiceae*, *B. rapa*, *B. oleracea*, and *R. sativus*, have been reported. Genomic information about distantly related species, such as *B. nigra* and *Sinapis alba* in the *nigra/sinapis* lineage, is also important to identify the conserved genome structure of the tribe *Brassiceae*. Currently, genetic maps available for *B. juncea* (Panjabi et al. 2008; Paritosh et al. 2014) and *S. alba* (Nelson et al. 2011) can provide genomic information about the *nigra/sinapis* lineage through genetic markers. In addition, expressed sequence tags (ESTs) from *B. juncea* originating from the B genome and several BAC sequences from *B. nigra* are also available (Navabi et al. 2013; Paritosh et al. 2014). Using those genomic resources, comparative genomic analyses were performed using two independent methods. To compare the *R. sativus* and *Brassica* A/C genomes, orthologous gene pairs were used to define chromosome-scale syntenic regions and infer triplicated sub-genome blocks based on gene retention. To compare *R. sativus* and *B. nigra*, marker sequences of the B genome in *B. juncea* genetic linkage groups were mapped onto the *R. sativus* genome assembly, and syntenic regions were defined. In *S. alba*, overall AK block association patterns were compared with those of *R. sativus* due to a lack of sub-genomic position information. In the results, a total of 12 chromosomal regions of the *R. sativus* genome exhibited chromosome-level synteny with the *Brassica* and *S. alba* genomes (Fig. 5.7).

Chromosomes R1, R2, R3, R8, and R9 had conserved AK block structures with single counterpart chromosomes in *B. rapa*, *B. nigra*, *B. oleracea*, or *S. alba*, suggesting a recently shared common ancestry. Although most LF sub-genomes were rearranged in the genomes investigated, R1 and S5 retained the original $AK1_{LF}$ structure of tPCK, indicating that these chromosomes are an ancient form. Interestingly, an R8-like chromosome structure was found in all species examined, and it existed as a single chromosome in *B. rapa* and *S. alba*. Given that R8 consists of $AK1_{MF1}$ and $AK7_{MF2}$, an R8-like chromosome might have existed before the introduction of the LF sub-genome at the second polyploidization event. A similar situation was found in R3, which is composed of MF1 and MF2 sub-genome blocks. Homologous regions of R3

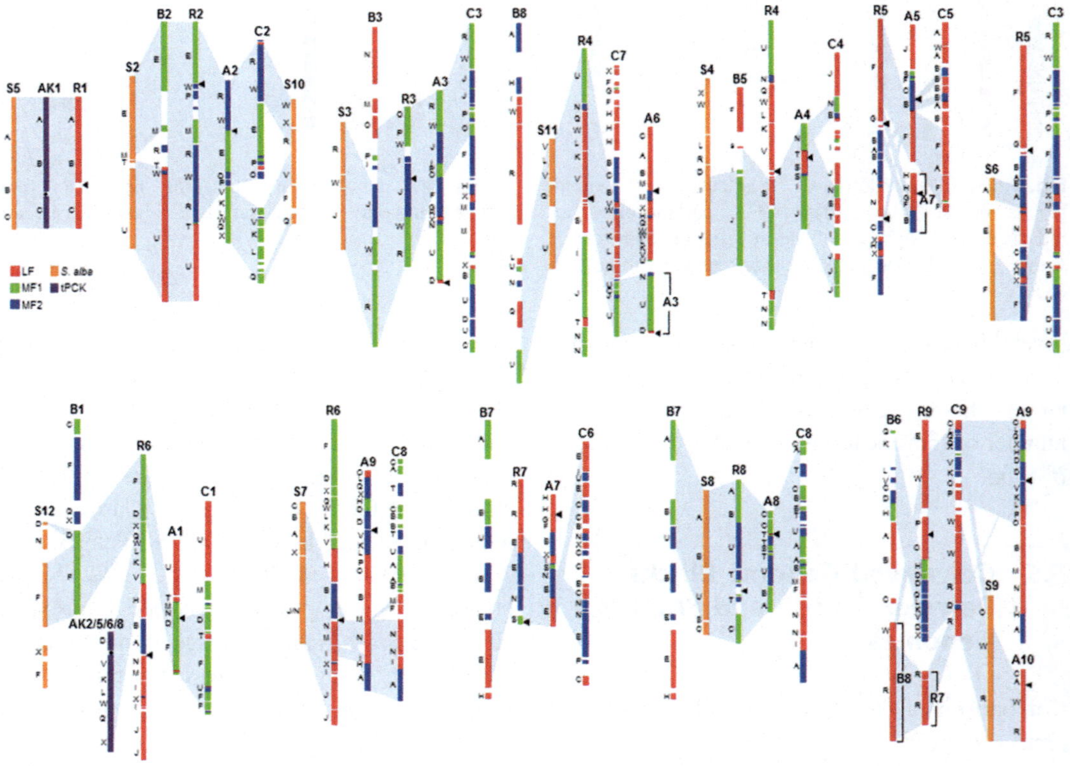

Fig. 5.7 The 12 chromosomal regions conserved in *R. sativus*, diploid *Brassica* species, and *S. alba* (Jeong et al. 2016). Conserved chromosomal regions in *R. sativus*, *B. rapa*, *B. oleracea*, *B. nigra*, and *S. alba* were identified based on the sub-genomic identity or similarity of AK block associations. The three sub-genomes are *red* (LF), *green* (MF1), and *blue* (MF2). *Orange* and *purple* indicate genome blocks of *S. alba* and tPCK, respectively. The names of the AK blocks are shown on the *left* side of each block. Regions conserved between species are connected by *gray shading*

are parts of single chromosomes in *Brassica* species. Because the conserved block associations found in A3, B3, C3, and S3 have an R3-like structure, it is anticipated that R3 is also an ancient chromosome retained in the genomes of the tribe *Brassiceae* before the second polyploidization event. In contrast, R4, R5, R6, and R7 were syntenic to more than two independent counterpart chromosomes of *Brassica* and *S. alba*, suggesting that these chromosomes were formed by unique recombination of the ancestral tPCK chromosomes. In detail, the sub-genome block association of R4 is largely divided into two parts. The upper part of R4 shows an association with U_{MF1}-N_{MF2} and part of the $[Q\text{-}W\text{-}L\text{-}K\text{-}V]_{LF}$ of AK2/5/6/8 that is also found in B8 and C7. In *B. rapa*, a similar block association is found in A6 and A3. The entire S11 region also shows an association with the U block and part of the AK2/5/6/8 tPCK chromosome. Meanwhile, the bottom part of R4 is almost homologous to the whole A4 chromosome. Moreover, C4 and B5 have A4-like block associations with rearrangements. Thus, R4 was likely formed by fusion of two tPCK chromosomes from a *Brassica–Raphanus* common ancestor. The upper part of the R6 chromosome has an association of F_{MF1} and the $[D\text{-}V\text{-}K\text{-}L\text{-}W\text{-}Q\text{-}X]_{MF1}$ of AK2/5/6/8 in an inverted arrangement. A retained MF1 sub-genome of AK2/5/6/8 has not been observed in other *Brassica* species, but F_{MF1} and D_{MF1} are commonly associated in the *R. sativus* and *Brassica* genomes. This finding suggests a possible association of AK3 and AK2/5/6/8 in the common ancestral lineage of *Brassica–Raphanus*. The conserved region in R7 consists of AK2/5 (E-N-M), and an association of E_{MF2}-E_{LF} blocks is an interesting structural characteristic of this chromosome that is also found in A7 and B7. However, the upper R_{LF} blocks have no synteny with *Brassica* and *S. alba* chromosomes, suggesting the occurrence of lineage- or species-specific rearrangement. The most divergent block association pattern is in R5, where $AK3_{LF}$, $AK3_{MF2}$, and $AK1_{MF2}$ are associated, indicating recombination of AK1 and AK3. Meanwhile, the C9 chromosome showed a syntenic relationship with two independent chromosomes from *B. rapa*, *B. nigra*, and *R. sativus*. The C9 chromosome was likely formed by the fusion of two tPCK chromosomes; therefore, it retains the ancient form of the ancestral chromosome.

Considering all the evidence together, *R. sativus* has rather simple and conserved AK block structures compared to the *Brassica* species. Overall, *R. sativus* shows AK block association patterns similar to those found in the *rapa/oleracea* lineage. However, specific chromosomes, such as R1 and R2, are highly similar to chromosomes in *B. nigra* and *S. alba* that belong to the *nigra/sinapis* lineage. Phylogenetic analysis based on sequence divergence of orthologs between *R. sativus* and *Brassica* species showed that *R. sativus* is positioned between the *rapa/oleracea* and *nigra/sinapis* genomes (Fig. 5.8a, b). Therefore, these results collectively suggest that the *R. sativus* genome has intermediate characteristics between the *Brassica* A/C and B genomes, rather than an ancient form, with a partly retained structure of the tPCK genome. Presumably, the ancestral triplicated *R. sativus* genome had 12 chromosomes and culminated in the current form of nine chromosomes through rearrangement and recombination of chromosomes. The divergence time of the *R. sativus* genome from the *Brassica* genomes can be deduced based on the synonymous substitution rate (Ks) of homologous genes (Fig. 5.8c). The age distribution patterns of paralogs in *R. sativus*, *B. rapa*, and *B. oleracea* are essentially identical, demonstrating at least three rounds of polyploidy events. The peak at $Ks = 0.3$ for each genome indicates a WGT event, estimated as 10–21 million years ago (Mya). The split of the *R. sativus* genome from the *Brassica* genome is estimated at $Ks = 0.2$ because the ortholog comparison between *B. rapa* and *B. oleracea* had a major peak at $Ks = 0.0$ to 0.1. This finding indicates that the *R. sativus* genome splits from the *Brassica* genome 7–14 Mya, soon after the shared WGT in their common ancestor.

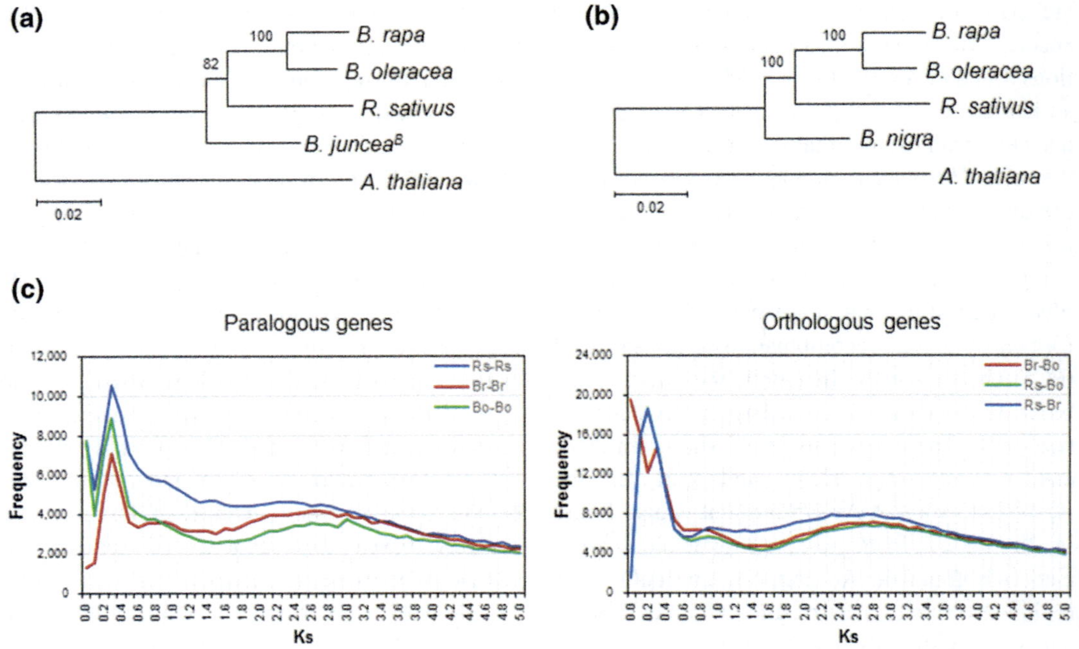

Fig. 5.8 Divergence of the *R. sativus* genome from *Brassica* genomes. **a** Maximum likelihood tree based on 229 *B. juncea* EST sequences (*B. juncea*^{B}) originating from the B genome (Paritosh et al. 2014) and their orthologs in the *B. rapa*, *B. oleracea*, *R. sativus*, and *A. thaliana* genomes. **b** Maximum likelihood tree based on 65 orthologous gene pairs among *B. rapa*, *B. nigra*, *B. oleracea*, *R. sativus*, and *A. thaliana*. Trees were generated using MEGA6, and the stability of tree nodes was tested by bootstrap analysis with 1000 replicates. Bootstrap values are indicated on the branches, and branch length represents the estimated number of substitutions per 100 sites. **c** The distribution of Ks values obtained from comparisons of sets of paralogous genes in *R. sativus*, *B. rapa*, and *B. oleracea* or orthologous genes between *R. sativus*, *B. rapa*, and *B. oleracea*. The *vertical axes* denote the frequency of paired sequences, and the *horizontal axes* indicate Ks values with an interval of 0.1

5.6 Reconstruction of the Ancestral Radish Genome

The genome sequences of *R. sativus* and diploid *Brassica* species provide a substantial amount of information for deducing the organization and reconstructing of the ancestral genome. Phylogenetic and cytological studies of the Brassicaceae family suggest that tPCK with 7 chromosomes might have been triplicated to form the ancestor of the tribe *Brassiceae* (Cheng et al. 2013, 2014; Lysak and Koch 2011; Mandakova and Lysak 2008). However, the chromosome number of the hexaploid ancestor after WGT is uncertain because the basic chromosome number of the tribe *Brassiceae* varies between species. The chromosome number of species in the tribe *Brassiceae* can be divided into two groups. Diploid species have 7–13 chromosomes, and polyploid species contain more than 13 chromosomes (Warwick and Al-Shehbaz 2006). A comparative analysis between the *R. sativus* genome and three *Brassica* species and *S. alba* identified 12 chromosomal regions. In most cases, those regions contain at least one centromeric sequence of *R. sativus* or *B. rapa*. This result suggests that a common ancestor of *R. sativus* and *Brassica* species might have triplicated genome consisting of 12 chromosomes. This finding shows good agreement with a previous cytogenetic study suggesting that the genomes of *Brassica* and *Raphanus* species might have arisen through a duplication of an ancestral genome with six chromosomes (Kaneko et al. 2007).

Based on a hypothesis that the 12 chromosomal regions are derived from a triplicated ancestral genome, one scenario for reconstruction of the *R. sativus* chromosomes is presented in Fig. 5.9. The *R. sativus* genome appears to have been derived from two rounds of polyploidization of tPCK chromosomes. Considering the AK block configuration and sub-genome composition with MF1 and MF2 blocks, the R3 and R8 chromosomes were likely formed at the first polyploidization event by the recombination of AK4-AK6/8 and AK1-AK7, respectively. Likewise, the recombination of AK2/5, AK3, AK7, and AK2/5/6/8 might have resulted in recombinant chromosomes T1 to T4 of a tetraploid progenitor. $AK1_{MF2}$, $AK3_{MF2}$, and $AK4_{MF1}$ seem to have maintained their tPCK structure without significant chromosome rearrangement or recombination. At the second polyploidization event, $AK1_{LF}$ and $AK3_{LF}$, R3, AK3, R8, and T3 with the MF1 or MF2 sub-genomes of the ancestral genome were likely to be retained. Meanwhile, the H1 to H6 chromosomes of a hexaploid progenitor might have

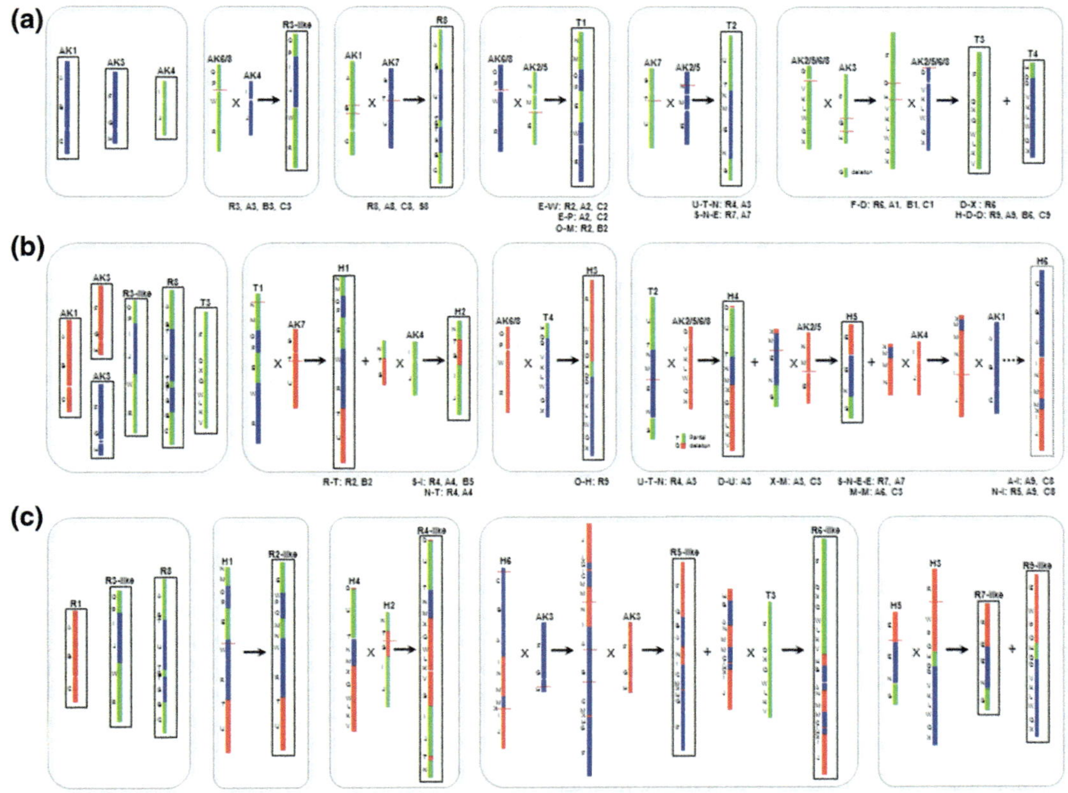

Fig. 5.9 Hypothetical model for evolution of the *R. sativus* chromosomes (Jeong et al. 2016). *R. sativus* chromosomes are thought to be formed by two rounds of polyploidization of tPCK chromosomes. Fusion and recombination after polyploidization might have formed 12 chromosomes of a hexaploid progenitor. **a** Hypothetical tetraploid ancestral genome of radish after first polyploidization. **b** Hypothetical hexaploid ancestral genome of radish after second polyploidization. Genome block associations in chromosomes of *Brassicas* (*A*, *B*, and *C*), *R. sativus* (R), and *S. alba* (S) that support the hypothetical ancestral chromosomes are indicated under the boxes. T and H chromosomes indicate hypothetical chromosomes of the tetraploid and hexaploid progenitors, respectively. **c** Reduction of 12 ancestral chromosomes into nine chromosomes of *R. sativus*. Sub-genome blocks are indicated by *red* (LF), *green* (MF1), and *blue* (MF2). The putative positions of recombination break points are indicated by *red dotted lines*. *Solid circles* are centromeres of the tPCK chromosomes. Putative ancestral chromosomes are indicated by *black rectangles*

been formed by recombination of the LF sub-genomes of the tPCK chromosomes (AK4, AK7, AK2/5, and AK6/8) with the T1, T2, and T4 chromosomes. Additional recombination and rearrangement of the hexaploid ancestral chromosomes seem to have resulted in reduction of the basic chromosome number of *R. sativus* to nine.

5.7 Conclusion

The chromosome-scale genome assembly of radish provides excellent information for understanding the genetic system of *R. sativus* and evolutionary changes in its genome. Comparative mapping of the radish genome using *Brassica* COS markers enabled the identification of a direct link between the *Raphanus* and *Brassica* genomes and the elucidation of the AK block structure in each triplicated genome, in which three sub-genomes with different fractionation were determined. A chromosome-level synteny comparison with sequenced genomes from the Brassicaceae family identified a tPCK-like ancestral genome of *Raphanus* that has intermediate structural characteristics between the *B. rapa*/*B. oleracea* and *B. nigra* genomes. A sequence-level comparison also showed that radish genes have an intermediate range of sequence divergence between *B. rapa*/*B. oleracea* and *B. nigra*. Evolutionary characteristics shared between the *R. sativus* and *Brassica* genomes provide evidence that the current form of nine chromosomes in *R. sativus* was rearranged from 12 chromosomes of a paleohexaploid progenitor through post-polyploid changes. The most significant beneficiaries of radish genome sequences will be *Raphanus* crop or weed researchers, from plant taxonomists and biologists to breeders. Intermediate characteristics of the *R. sativus* genome compared to diploid *Brassica* genomes will contribute to delimitation of the genera *Raphanus* and *Brassica*, allowing for taxonomic treatment of *R. sativus* and *Brassica* species. Moreover, the mesohexaploidization and subsequent rearrangement of the radish genome might have contributed to genomic selection of agronomic traits, including swollen flesh taproot, or weed traits, such as frequent outcrossing and interspecific hybridization. Therefore, with an increased number of fully sequenced genomes from the Brassicaceae family and re-sequenced data from diverse radish accessions, comparative genomic analysis will more fully reveal the complex and dynamic history of the genome evolution of *R. sativus*. A comprehensive analysis of evolutionarily selected genes will eventually serve as a fundamental resource for molecular-genetic and breeding studies of radish.

References

Arumugam N, Mukhopadhyay A, Gupta V, Sodhi Y, Verma J, Pental D, Pradhan A (2002) Synthesis of somatic hybrids (RCBB) by fusing heat-tolerant *Raphanus sativus* (RR) and *Brassica oleracea* (CC) with *Brassica nigra* (BB). Plant Breed 121:168–170

Beilstein M, Al-Shehbaz I, Kellogg E (2006) Brassicaceae phylogeny and trichome evolution. Am J Bot 93:607–619

Bushakra JM, Sargent DJ, Cabrera A, Crowhurst R, Girona EL, Velasco R, Symonds VV, Knaap E, Troggio M, Gardiner SE, Chagné D (2011) Rosaceae conserved orthologous set (RosCOS) markers as a tool to assess genome synteny between *Malus* and *Fragaria*. Tree Genet Genomes 8:643–658

Cabrera A, Kozik A, Howad W, Arus P, Iezzoni AF, van der Knaap E (2009) Development and bin mapping of a Rosaceae Conserved Ortholog Set (COS) of markers. BMC Genom 10:562

Chalhoub B, Denoeud F, Liu SY, Parkin IAP, Tang HB, Wang XY, Chiquet J, Belcram H, Tong CB, Samans B, Correa M, Da Silva C, Just J, Falentin C, Koh CS, Le Clainche I, Bernard M, Bento P, Noel B, Labadie K, Alberti A, Charles M, Arnaud D, Guo H, Daviaud C, Alamery S, Jabbari K, Zhao MX, Edger PP, Chelaifa H, Tack D, Lassalle G, Mestiri I, Schnel N, Le Paslier MC, Fan GY, Renault V, Bayer PE, Golicz AA, Manoli S, Lee TH, Thi VHD, Chalabi S, Hu Q, Fan CC, Tollenaere R, Lu YH, Battail C, Shen JX, Sidebottom CHD, Wang XF, Canaguier A, Chauveau A, Berard A, Deniot G, Guan M, Liu ZS, Sun FM, Lim YP, Lyons E, Town CD, Bancroft I, Wang XW, Meng JL, Ma JX, Pires JC, King GJ, Brunel D, Delourme R, Renard M, Aury JM, Adams KL, Batley J, Snowdon RJ, Tost J, Edwards D, Zhou YM, Hua W, Sharpe AG, Paterson AH, Guan CY, Wincker P (2014) Early allopolyploid evolution in the post-Neolithic *Brassica napus* oilseed genome. Science 345:950–953

Chapman MA, Chang J, Weisman D, Kesseli RV, Burke JM (2007) Universal markers for comparative mapping and phylogenetic analysis in the Asteraceae (Compositae). Theor Appl Genet 115:747–755

Cheng F, Mandáková T, Wu J, Xie Q, Lysak M, Wang X (2013) Deciphering the diploid ancestral genome of the mesohexaploid *Brassica rapa*. Plant Cell 25:1541–1554

Cheng F, Wu J, Fang L, Sun S, Liu B, Lin K, Bonnema G, Wang X (2012) Biased gene fractionation and dominant gene expression among the subgenomes of *Brassica rapa*. PLoS ONE 7:e36442

Cheng F, Wu J, Wang X (2014) Genome triplication drove the diversification of *Brassica* plants. Hortic Res 1:14024

Fulton TM, Van der Hoeven R, Eannetta NT, Tanksley SD (2002) Identification, analysis, and utilization of conserved ortholog set markers for comparative genomics in higher plants. Plant Cell 14:1457–1467

Jeong Y-M, Chung W-H, Chung H, Kim N, Park B-S, Lim K-B, Yu H-J, Mun J-H (2014) Comparative analysis of the radish genome based on a Conserved Ortholog Set (COS) of *Brassica*. Theor Appl Genet 127:1975–1989

Jeong Y-M, Kim N, Ahn B, Oh M, Chung W-H, Chung H, Jeong S, Lim K-B, Hwang Y-J, Kim G-B, Baek S, Choi S-B, Hyung D-J, Lee S-W, Sohn S-H, Kwon S-J, Jin M, Seol Y-J, Chae W, Choi K, Park B-S, Yu H-J, Mun J-H (2016) Elucidating the triplicated ancestral genome structure of radish based on chromosome-level comparison with the *Brassica* genomes. Theor Appl Genet 129:1357–1372

Kaneko Y, Kimizuka-Takagi C, Bang S, Matsuzawa Y (2007) Radish. In: Kole C (ed) Vegetables. Springer, Berlin, pp 141–160

Kitashiba H, Li F, Hirakawa H, Kawanabe T, Zou Z, Hasegawa Y, Tonosaki K, Shirasawa S, Fukushima A, Yokoi S, Takahata Y, Kakizaki T, Ishida M, Okamoto S, Sakamoto K, Shirasawa K, Tabata S, Nishio T (2014) Draft sequences of the radish (*Raphanus sativus* L.) genome. DNA Res 21:481–490

Koch MA, Kiefer M (2005) Genome evolution among cruciferous plants: a lecture from the comparison of the genetic maps of three diploid species—*Capsella rubella*, *Arabidopsis lyrata* subsp. *petraea*, and *A. thaliana*. Am J Bot 92:761–767

Krutovsky KV, Troggio M, Brown GR, Jermstad KD, Neale DB (2004) Comparative mapping in the Pinaceae. Genetics 168:447–461

Lee S-S, Lee S-A, Yang J, Kim J (2011) Developing stable progenies of × *Brassicoraphanus*, an intergeneric allopolyploid between *Brassica rapa* and *Raphanus sativus*, through induced mutation using microspore culture. Theor Appl Genet 122:885–891

Liewlaksaneeyanawin C, Zhuang J, Tang M, Farzaneh N, Lueng G, Cullis C, Findlay S, Ritland CE, Bohlmann J, Ritland K (2008) Identification of COS markers in the Pinaceae. Tree Genet Genomes 5:247–255

Liu S, Liu Y, Yang X, Tong C, Edwards D, Parkin I, Zhao M, Ma J, Yu J, Huang S, Wang X, Wang J, Lu K, Fang Z, Bancroft I, Yang T, Hu Q, Wang X, Yue Z, Li H, Yang L, Wu J, Zhou Q, Wang W, King G, Pires J, Lu C, Wu Z, Sampath P, Wang Z, Guo H, Pan S, Yang L, Min J, Zhang D, Jin D, Li W, Belcram H, Tu J, Guan M, Qi C, Du D, Li J, Jiang L, Batley J, Sharpe A, Park B, Ruperao P, Cheng F, Waminal N, Huang Y, Dong C, Wang L, Li J, Hu Z, Zhuang M, Huang Y, Huang J, Shi J, Mei D, Liu J, Lee T, Wang J, Jin H, Li Z, Li X, Zhang J, Xiao L, Zhou Y, Liu Z, Liu X, Qin R, Tang X, Liu W, Wang Y, Zhang Y, Lee J, Kim H, Denoeud F, Xu X, Liang X, Hua W, Wang X, Wang J, Chalhoub B, Paterson A (2014) The *Brassica oleracea* genome reveals the asymmetrical evolution of polyploid genomes. Nat Commun 5:3930

Lü N, Yamane K, Ohnishi O (2008) Genetic diversity of cultivated and wild radish and phylogenetic relationships among *Raphanus* and *Brassica* species revealed by the analysis of *trnK/matK* sequence. Breed Sci 58:15–22

Lynch M, Conery JS (2000) The evolutionary fate and consequences of duplicate genes. Science 290:1151–1155

Lysak MA, Berr A, Pecinka A, Schmidt R, McBreen K, Schubert I (2006) Mechanisms of chromosome number reduction in *Arabidopsis thaliana* and related Brassicaceae species. Proc Natl Acad Sci USA 103:5224–5229

Lysak MA, Koch MA (2011) Phylogeny, genome, and karyotype evolution of crucifers (Brassicaceae). In: Schmidt R, Bancroft I (eds) Genetics and genomics of the Brassicaceae. Springer, New York, pp 1–31

Lysak MA, Koch MA, Pecinka A, Schubert I (2005) Chromosome triplication found across the tribe *Brassiceae*. Genome Res 15:516–525

Lysak MA, Mandakova T, Schranz ME (2016) Comparative paleogenomics of crucifers: ancestral genomic blocks revisited. Curr Opin Plant Biol 30:108–115

Mandakova T, Lysak MA (2008) Chromosomal phylogeny and karyotype evolution in x = 7 crucifer species (Brassicaceae). Plant Cell 20:2559–2570

Matsuzawa Y, Sarashima M (1986) Intergeneric hybridization between *Raphanus sativus* L. and *Brassica nigra* Koch. and alloplasmic radish derivative. Breed Sci 36:122–130

Mitsui Y, Shimomura M, Komatsu K, Namiki N, Shibata-Hatta M, Imai M, Katayose Y, Mukai Y, Kanamori H, Kurita K, Kagami T, Wakatsuki A, Ohyanagi H, Ikawa H, Minaka N, Nakagawa K, Shiwa Y, Sasaki T (2015) The radish genome and comprehensive gene expression profile of tuberous root formation and development. Sci Rep 5:10835

Moghe G, Hufnagel D, Tang H, Xiao Y, Dworkin I, Town C, Conner J, Shiu S (2014) Consequences of whole-genome triplication as revealed by comparative genomic analyses of the wild radish *Raphanus raphanistrum* and three other Brassicaceae species. Plant Cell 26:1925–1937

Mun J-H, Chung H, Chung W-H, Oh M, Jeong Y-M, Kim N, Ahn B, Park B-S, Park S, Lim K-B, Hwang Y-J, Yu H-J (2015) Construction of a reference genetic map of *Raphanus sativus* based on genotyping by whole-genome resequencing. Theor Appl Genet 128:259–272

Murat F, Louis A, Maumus F, Armero A, Cooke R, Quesneville H, Crollius HR, Salse J (2015) Understanding Brassicaceae evolution through ancestral genome reconstruction. Genome Biol 16:262

Nagaharu U (1935) Genome analysis in *Brassica* with special reference to the experimental formation of *B. napus* and peculiar mode of fertilization. Jpn J Bot 7:389–452

Navabi Z-K, Huebert T, Sharpe A, O'Neill C, Bancroft I, Parkin IA (2013) Conserved microstructure of the *Brassica* B genome of *Brassica nigra* in relation to homologous regions of *Arabidopsis thaliana*, *B. rapa* and *B. oleracea*. BMC Genom 14:250

Nelson M, Parkin I, Lydiate D (2011) The mosaic of ancestral karyotype blocks in the *Sinapis alba* L. genome. Genome 54:33–41

Panjabi P, Jagannath A, Bisht NC, Padmaja KL, Sharma S, Gupta V, Pradhan AK, Pental D (2008) Comparative mapping of *Brassica juncea* and *Arabidopsis thaliana* using Intron Polymorphism (IP) markers: homoeologous relationships, diversification and evolution of the A, B and C *Brassica* genomes. BMC Genom 9:113

Paritosh K, Gupta V, Yadava S, Singh P, Pradhan A, Pental D (2014) RNA-seq based SNPs for mapping in *Brassica juncea* (AABB): synteny analysis between the two constituent genomes A (from *B. rapa*) and B (from *B. nigra*) shows highly divergent gene block arrangement and unique block fragmentation patterns. BMC Genom 15:396

Parkin IAP, Koh C, Tang HB, Robinson SJ, Kagale S, Clarke WE, Town CD, Nixon J, Krishnakumar V, Bidwell SL, Denoeud F, Belcram H, Links MG, Just J, Clarke C, Bender T, Huebert T, Mason AS, Pires JC, Barker G, Moore J, Walley PG, Manoli S, Batley J, Edwards D, Nelson MN, Wang XY, Paterson AH, King G, Bancroft I, Chalhoub B, Sharpe AG (2014) Transcriptome and methylome profiling reveals relics of genome dominance in the mesopolyploid *Brassica oleracea*. Genome Biol 15:R77

Prakash S, Bhat S, Quiros C, Kirti P, Chopra V (2009) *Brassica* and its close allies: cytogenetics and evolution. In: Jules J (ed) Plant breed reviews, vol 31. Wiley, London, pp 21–187

Quraishi UM, Abrouk M, Bolot S, Pont C, Throude M, Guilhot N, Confolent C, Bortolini F, Praud S, Murigneux A, Charmet G, Salse J (2009) Genomics in cereals: from genome-wide conserved orthologous set (COS) sequences to candidate genes for trait dissection. Funct Integr Genomics 9:473–484

Schnable JC, Springer NM, Freeling M (2011) Differentiation of the maize subgenomes by genome dominance and both ancient and ongoing gene loss. Proc Natl Acad Sci USA 108:4069–4074

Schranz ME, Lysak MA, Mitchell-Olds T (2006) The ABC's of comparative genomics in the Brassicaceae: building blocks of crucifer genomes. Trends Plant Sci 11:535–542

Song KM, Osborn TC, Williams PH (1988) Brassica taxonomy based on nuclear restriction fragment length polymorphisms (RFLPs): 2. Preliminary analysis of subspecies within *B. rapa* (syn. campestris) and *B. oleracea*. Theor Appl Genet 76:593–600

Song K, Osborn TC, Williams PH (1990) *Brassica* taxonomy based on nuclear restriction fragment length polymorphisms (RFLPs): 3. Genomic relationships in *Brassica* and related genera and the origin of *B. oleracea* and *B. rapa* (syn. *campestris*). Theor Appl Genet 79:497–506

The Arabidopsis Genome Initiative (2000) Analysis of the genome sequence of the flowering plant *Arabidopsis thaliana*. Nature 408:796–815

The Brassica rapa Genome Sequencing Project Consortium (2011) The genome of the mesopolyploid crop species *Brassica rapa*. Nat Genet 43:1035–1040

Thormann CE, Ferreira ME, Camargo LE, Tivang JG, Osborn TC (1994) Comparison of RFLP and RAPD markers to estimating genetic relationships within and among cruciferous species. Theor Appl Genet 88:973–980

Timms L, Jimenez R, Chase M, Lavelle D, McHale L, Kozik A, Lai Z, Heesacker A, Knapp S, Rieseberg L, Michelmore R, Kesseli R (2006) Analyses of synteny between *Arabidopsis thaliana* and species in the Asteraceae reveal a complex network of small syntenic segments and major chromosomal rearrangements. Genetics 173:2227–2235

Warwick S, Al-Shehbaz I (2006) Brassicaceae: chromosome number index and database on CD-Rom. Plant Syst Evol 259:237–248

U N (1935) Genome analysis in Brassica with special reference to the experimental formation of B. napus and peculiar mode of fertilization. Jpn J Bot 7:389-452

Warwick S, Black L (1991) Molecular systematics of *Brassica* and allied genera (subtribe Brassicinae, Brassiceae)-chloroplast genome and cytodeme congruence. Theor Appl Genet 82:81–92

Warwick S, Black L (1997) Phylogenetic implications of chloroplast DNA restriction site variation in subtribes Raphaninae and Cakilinae (Brassicaceae, tribe Brassiceae). Can J Bot 75:960–973

Warwick S, Sauder C (2005) Phylogeny of tribe Brassiceae (Brassicaceae) based on chloroplast restriction site polymorphisms and nuclear ribosomal internal transcribed spacer and chloroplast *trnL* intron sequences. Can J Bot 83:467–483

Wu HJ, Zhang Z, Wang JY, Oh DH, Dassanayake M, Liu B, Huang Q, Sun HX, Xia R, Wu Y, Wang YN, Yang Z, Liu Y, Zhang W, Zhang H, Chu J, Yan C, Fang S, Zhang J, Wang Y, Zhang F, Wang G, Lee SY, Cheeseman JM, Yang B, Li B, Min J, Yang L, Wang J, Chu C, Chen SY, Bohnert HJ, Zhu JK, Wang XJ, Xie Q (2012) Insights into salt tolerance from the genome of *Thellungiella salsuginea*. Proc Natl Acad Sci USA 109:12219–12224

Yang R, Jarvis DE, Chen H, Beilstein MA, Grimwood J, Jenkins J, Shu S, Prochnik S, Xin M, Ma C, Schmutz J, Wing RA, Mitchell-Olds T, Schumaker KS, Wang X (2013) The reference genome of the halophytic plant *Eutrema salsugineum*. Front Plant Sci 4:46

Yang Y, Tai P, Chen Y, Li W (2002) A study of the phylogeny of *Brassica rapa*, *B. nigra*, *Raphanus sativus*, and their related genera using noncoding regions of chloroplast DNA. Mol Phylogenet Evol 23:268–275

Yang YW, Lai KN, Tai PY, Ma DP, Li WH (1999) Molecular phylogenetic studies of *Brassica*, *Rorippa*, *Arabidopsis* and allied genera based on the internal transcribed spacer region of 18S-25S rDNA. Mol Phylogenet Evol 13:455–462

Yang YW, Tseng PF, Tai PY, Chang CJ (1998) Phylogenetic position of *Raphanus* in relation to *Brassica* species based on 5S rRNA spacer sequence data. Bot Bull Acad Sin 39:153–160

Yogeeswaran K, Frary A, York TL, Amenta A, Lesser AH, Nasrallah JB, Tanksley SD, Nasrallah ME (2005) Comparative genome analyses of *Arabidopsis* spp.: inferring chromosomal rearrangement events in the evolutionary history of *A. thaliana*. Genome Res 15:505–515

DNA Polymorphism and Genetic Diversity in *Raphanus* Accessions

6

Li Zhang and Qingbiao Wang

Abstract

Radish is an important vegetable crop that is widely distributed throughout the world. Because of the long cultivation history, radish is morphologically diverse and a large number of cultivars with widely varying morphology, especially in root shapes, sizes, skin, and flesh colors, are produced. According to the morphology of its edible root and its intended uses, cultivated radishes are classified into five main varieties. Radish has a wide range of genetic diversity which provides a tremendous scope for genetic improvement of economic traits. Studies on DNA polymorphisms and genetic diversity of radish help us to determine the genetic relationship between radish cultivars and to understand the origin and evolution of this species. In the past two decades, various molecular tools provide easy, less laborious means for identifying the DNA polymorphisms and genetic diversity of radish. Here, we review recent research on the DNA polymorphisms, including cytoplasmic and nuclear genomes, summarize the genetic relationship between wild and cultivated radishes, and discuss the origin and evolution of *Raphanus* according to their genetic diversity.

6.1 Introduction

Radish (*Raphanus sativus* L., $2n = 18$) is a member of the Brassicaceae family and includes both cultivated and wild species. The cultivation history of this crop dates back to 4000 years ago in Egypt, where it was grown as an oilseed crop (Becker 1962). Currently, radish is grown worldwide, especially in China, Japan, and Korea. In China, this crop was cultivated in arable lands of 1.2 million ha, covering 6% of the

L. Zhang (✉) · Q. Wang
National Engineering Research Center for Vegetables, 50 Zhanghua Street, Haidian, Beijing 100097, People's Republic of China
e-mail: zhangli@nercv.org

L. Zhang · Q. Wang
Beijing Vegetable Research Center, Beijing Academy of Agriculture and Forestry Sciences, Beijing 100097, People's Republic of China

T. Nishio and H. Kitashiba (eds.), *The Radish Genome*,
Compendium of Plant Genomes, DOI 10.1007/978-3-319-59253-4_6

whole cultivated acreages for vegetables, and producing 38.8 million tons in 2014.

Radish is used for many purposes, for example, as food, forage, cover crop, and for processing and medical purposes. The root is the most commonly eaten part, although the leaf and the fresh pod are also used for cooking or eaten raw. The roots are processed into various pickles in China, Japan, and Korea. Some special varieties have been developed for this purpose. In the mid-Atlantic region, radish is used as a cover crop that helps inhibit undesirable winter annual weed growth and improve the physical structure of the soil (Lawley et al. 2011). Radish seeds, which are called Laifuzi in traditional Chinese medicine, are used to treat digestive and respiratory diseases in China (Tang et al. 2003). Wild radish, distributed worldwide, is one of the most troublesome weeds for crop production, colonizing not only new regions, but also causing yield losses (Sahli et al. 2008a, b). For example, in Australia, the weedy *R. raphanistrum* is estimated to cause an average yield loss of 2 ton ha^{-1} (Streibig et al. 1989). Some useful agricultural traits have also been found in wild radish, such as cytoplasmic male sterility (CMS), tolerance to saline soil, and resistance to blackleg (Kaneko et al. 2007; Ogura 1968; Salisbury 1987; Warwick 1993). According to the morphology of its edible root and its intended uses, cultivated radishes are classified into five main varieties, that is, *R. sativus* var. *hortensis* Becker (Asian big radish), var. *sativus* (European small radish), var. *niger* Kerner (black radish), var. *oleiferus* Metzg (oil radish), and var. *caudatus* Hooker and Anderson (rat's tail radish; Kitamura 1958; Lü et al. 2008; Wang et al. 2015). Wild radishes, including *Raphanus raphanistrum*, *R. landra*, and *R. maritimus*, are distributed throughout the coastal areas of the Mediterranean and Black Seas (Kaneko et al. 2007; Lewis-Jones et al. 1982). Another wild *Raphanus* species, *R. sativus* var. *raphanistroides* Makino (East Asian wild radish), is distributed in sandy coastal areas, and, occasionally inland, in Japan, Korea, and China (Huh and Ohnishi 2002; Makino 1961); however, most radish specialists are not convinced that it is a true wild species, but might be an escaped type of cultivated radish (Kitamura 1958).

The abundant radish germplasms, including cultivated and wild radishes, provide good materials for breeding new varieties. The variation within or among radish accessions is the basis for its selection and improvement. Studies on DNA polymorphisms and genetic diversity of radish help us to determine the genetic relationship among *Raphanus* and to understand the origin and evolution of this species. These results could also provide useful information for efficient use. In this chapter, we focus on DNA polymorphisms and the genetic diversity of radish, including the latest information in this field. First, we introduce the different radish varieties around the world and their various morphological traits. We then review the DNA polymorphisms, including cytoplasmic and nuclear genomes. Third, we summarize the genetic relationship between wild and cultivated radishes. Finally, the origin and evolution of *Raphanus* is discussed, based on their genetic diversity. In addition, the influence of breeding practice on the diversity of radish is also discussed.

6.2 Diversity Based on Morphological Traits, Geographical Distribution, and Intended Use

6.2.1 Cultivated Radish

The extensive crossing and selection of radish, especially in East Asian countries, have produced a large number of cultivars with widely varying morphology, especially in root shapes, sizes, and skin and flesh colors. The abundant morphological traits of radish provided genetic resources for breeding and will be very useful for the use and improvement of radish germplasms. Among the five cultivated varieties, accessions of the var. *hortensis* (Asian big radish) are further split into several subgroups, indicating a greater genetic diversity compared with other radish types. This type is mainly distributed in China, Japan, and Korea.

Upon the introduction of radish to China about 2700 years ago, a huge diversity of roots was created through traditional breeding, exhibiting differences in skin and flesh colors and in terms of their sizes and shapes. Radish genetic resources are very abundant in China. To date, about 2100 radish accessions have been conserved in the National Mid-term Gene-bank for Vegetables, located in the Institute of Vegetables and Flowers, Chinese Academy of Agricultural Sciences. The useful agricultural traits in Chinese radish germplasms are summarized in Table 6.1. These include late bolting, tolerance to high temperature, resistance to disease, high fresh-edible quality, processing use, and male sterility. These accessions are distributed in seven cultivation districts, according to geographical distribution and climatic conditions (Table 6.2). Generally speaking, red- and green-skinned types are grown in the North of China, and white long types are grown in the South. About 45% of radish accessions are distributed in District 3, suggesting that this district is the original center of Chinese radish (Wang and He 2005). The cropping types of Chinese radish are classified into spring, summer–autumn, autumn–winter, and winter. Some varieties can be planted all year-round. The large ones with red or green skin usually are planted in

Table 6.1 Radish genetic resources in China

Types of radish genetic resources	Characteristics	Representative varieties
Late bolting	Not bolt under 10 °C for 15 days	Shuiluobo, Hongdanzi, Yanghua
Tolerance to high temperature	Grow normally under 30 °C	Duanye13, Rebai, Zhongqiuhong
Resistance to diseases	Resistance to *TuMV*, black rot, and downy mildew	Rebai, Xiangyanghong, Qiukuiqing
Fruit used	Flesh is crisp, sweet, and juicy	Xinlimei, Weixianqing, Xiguahong
Processing	Thin skin, compact texture, low water content	Yadantou, Yanzhihong, Weiqing
Male sterility	No functional pollen, but retain normal female fertility	Jinhuatai, 48A, 262A

Table 6.2 Distribution of Chinese radish accessions in seven cultivation districts

No. of Districts	Locations	Represent cultivars
1	Liaoning, Jilin, Heilongjiang	Daliangxiaowuying, Hongfeng, Qiaotouqing
2	Beijing, Tianjin, Hebei, Inner-Mogolia	Mantanghong, Weiqing, Xiangyabai, Jinghong No. 4
3	Shandong, Jiangsu, Anhui, Henan	Weixianqing, 791, Luluobo No. 2, Yanghualuobo, Zhongqiuhong
4	Gansu, Qinghai, Shanxi, Ningxia, Tibet	Qingtouluobo, Guoguang, Yinchuanqingpi, Qinghailvtoudong, Fengguan, Fengqiao
5	Shanghai, Zhejiang, Jiangxi, Fujian, Hunan, Hubei	Panxiluobo, Zhedachang, Yidianhong, Chunhong, Qiyehong, Xiakang 40, Sanbai
6	Guangdong, Guangxi, Taiwan	Duanye13, Nanpanzhou, Baishamaer, Ronganwanluobo,
7	Sichuan, Yunnan, Guizhou	Yuanbai, Chunbulao, Yanzhihong, Quanshenhong

autumn, and the small ones in spring or in all four seasons. Chinese radish can be used for cooking, processing, and in salads. Some radish cultivars with green skin and green or scarlet flesh can be consumed as fruit, whose flavors are sweet and crisp. The root shape of Chinese radish includes round, flat round, conical, spindle, and cylindrical, and the size varies from 10 g to 2 kg per root. The skin can be white, scarlet, pink, green, purple, and bicolor, such as green to white and red to white. The flesh is commonly white, green, scarlet, red, or purple. Most accessions have white flesh, and some accessions with green skin have green, scarlet, red, or purple flesh (Fig. 6.1). The Chinese small radish has the same shape and color as European radish, but it is not clear whether the two types have a close phylogenetic relationship.

The Korean and Japanese varieties are believed to have originated from South China (Aoba 1979), and most of these varieties have large long roots. Japanese cultivated radishes are divided into 16 groups according to their morphological traits and cropping systems (Kumazawa 1961). 'Moriguchi-daikon' and 'Sakurajima-daikon', the longest and the largest radishes in the world, typify the morphological variation in Japanese radish. European small radish (*R. sativus* var. *sativus*), which is commonly cultivated in the temperate regions of

Fig. 6.1 Various Chinese cultivated radish accessions **a** Gangshui-luobo, **b** Weiqing, **c** Fengguang, **d** Jingyanzidan, **e** Xinlimei, **f** Chuanxinghong, **g** Duanye13, **h** Manshenhong, **i** Jinghong No.4, **j** Nanxiang-luobo, **k** Chunbulao, **l** Yanzhihong, **m** Xiaowuying, **n** Jingcui No.1, **o** Fengrun, **p** Jinghong No.5

Europe, USA, and as far as the Arctic under protection, has small roots with red, white, purple, and yellow skins and globular or cylindrical shapes (George and Evans 1981). Black radish (*R. sativus* var. *niger* Kerner) has black skin and a round or cylindrical shape. Oil radish (*R. sativus* var. *oleiferus* Metzg), which does not form an expanded fleshy root, can be used to extract oil and for fodder. Rat's tail radish (*R. sativus* var. *caudatus* Hooker and Anderson), which has early-bolting traits, is not an edible root variety but is used for its slender and edible seedpods.

6.2.2 Wild Radish

Wild radish, growing as the dominant weed around the coastal areas, exhibits polymorphisms with respect to leaf shape and flower color. Yamaguchi (1987) investigated the spatial pattern of variation in the leaf shape and size of Japanese wild radish. A high level of diversity of leaf shape was recognized in populations from different locations. Three leaf forms (lyrate, pinnatifid, and pinnatisect) were observed, and the polymorphism level increased with rising elevation. Wild radish exhibits polymorphism with respect to flower color, and four petal color morphs (yellow, pink, white, and bronze) have been found. Pollinator-mediated natural selection has been shown to act on phenotypic variation in floral morphology.

6.3 DNA Polymorphisms and Genetic Diversity

The traditional methods to assess the diversity of radish were based on morphological observation (particularly agro-morphological), geographical distribution, and intended use. In the past two decades, DNA polymorphisms and genetic diversity have been extensively studied with the use of various molecular markers such as enzyme electrophoresis, randomamplified polymorphic DNA (RAPD), amplified fragment length polymorphism (AFLP), sequence-related amplified polymorphism (SRAP), simple sequence repeat (SSR), and single nucleotide polymorphism (SNP) (Curtis 2011). These molecular markers avoid the influence of the environment, and provide better sampling of the genome, thus increasing the resolution of measurements of genetic variation. These assessments not only help our efforts to conserve the germplasm, but also provide guidance for better germplasm use for genetic improvement.

6.3.1 Polymorphisms and Genetic Diversity of Mitochondrial and Chloroplast DNA

Mitochondrial and chloroplast genomes, which are characterized by maternal inheritance; a slow frequency of gene silencing and replacement; and the evolutionary rate of base substitutions, have been used to analyze many plant species. Analysis of polymorphisms in plants organelle genomes is useful for defining cytoplasmic DNA variation and phylogenetic relationships (Arrieta-Montiel and Mackenzie 2011; Tanaka et al. 2012; Wolfe et al. 1987).

6.3.1.1 Polymorphisms of Mitochondrial DNA

Compared with animal mitochondrial genomes, plant mitochondrial genomes are extremely complex and have characteristic features, including large genome size, frequent reorganization, and incorporation of foreign DNA. Until now, four mitochondrial genomes (Ogura CMS, normal, DCGMS, and 'Chuanxinhong') have been sequenced and assembled to approximately 240,000-bp circular sequences (Chang et al. 2013; Park et al. 2013; Tanaka et al. 2012). Analyses of these genomes showed that repeat sequences account for 10.2% of the radish mitochondrial genome (Chang et al. 2013), which cause rearrangements and are responsible for extensive structural changes. About 7% of Ogura and 3% of normal sequences are unique to each mitochondrial genome, and the Ogura-type mitochondrial genome contains four unique regions (Tanaka et al. 2012). Moreover, comparative analysis of three mitotypes of DCGMS, normal, and Ogura CMS showed that short-repeat sequences could be identified in almost all junctions between the 17 syntenic sequence blocks and the position and orientation of the sequence blocks were highly rearranged (Park et al. 2013).

In addition to the mitochondrial genome rearrangements, different regions of mitochondrial DNA (mtDNA) were studied and polymorphisms were found in wild and cultivated radishes. Yamagishi and Terach (2003) analyzed the structural variations in the mitochondrial *cox I* regions by PCR, using three wild and one cultivated species. Five (I–V) and four (I–IV) types of mtDNA were identified for *R. raphanistrum* and *R. maritimus*, respectively. Five types were also found in the cultivated radish, which showed that mtDNA variations were common to wild and cultivated radishes. The polymorphisms of mitochondrial *orfB* were assessed by restriction fragment length polymorphism (RFLP) analysis in wild and cultivated radishes (Yamagishi 2004). All three types of *orfB* were found in *R. raphanistrum* and Japanese wild radish, while the cultivated radish contained only one type, indicating that the wild radish had more polymorphic mtDNA than the cultivated ones. Lee et al. (2009) divided 120 radish germplasms from diverse countries into four mitotypes (Ogura, DCGMS, DBRMF1, and DBRMF2) using molecular markers in the *atp6-nad3-rps12* region (Kim et al. 2007; Lee et al. 2009).

6.3.1.2 Polymorphisms of Chloroplast DNA

Different regions of the chloroplast DNA (cpDNA) and chloroplast simple sequence repeats (cpSSRs) were studied, and polymorphisms were found in wild and cultivated radishes. Three haplotypes (designated A–C) were recognized in 17 accessions, based on sequence variation of the 5′-*matK*, and two haplotypes (designated D, E) were identified in 60 accessions based on PCR-RFLP in the *rpoC3-trnC* region (Yamane et al. 2005). In another region of the cpDNA, *trnK/matK*, nucleotide diversity was calculated in cultivated and wild radish species (Lü et al. 2008). The result showed two insertions/deletions and nine substitutions among 18 *Raphanus* accessions. The value of the diversity was found to be higher in cultivated radish than in the wild species, which indicated that cultivated radish had larger variations than wild one. This conclusion was inconsistent with the result obtained from mtDNA, which suggested that the wild radish had larger variations than cultivated one (Yamagishi 2004).

However, only three sites were involved in the above study, which did not provide sufficient information. CpSSRs show a higher level of intraspecific variation and are readily analyzed by PCR. In *Raphanus*, 20 haplotypes were detected in 82 accessions, including four cultivated radish species (var. *hortensis*, var. *niger*, var. *caudatus*, and var. *sativus*) and three wild ones (var. *raphanistroides*, var. *raphanistrum*, and var. *landra*), using 25 cpSSR loci. As shown in Table 6.3, 20 haplotypes were detected, 13 haplotypes in cultivated radish and 13 in wild radish. Only one haplotype was found in accessions of cultivated radish from the Mediterranean regions and two in European small radish. By contrast, 11 haplotypes were identified in Asian big radish. The Asian cultivated radish, which has the largest morphological variation among all the radish accessions, had higher cpSSR diversities than the Mediterranean cultivated or wild radishes. These findings are consistent with the diversity of agronomic traits in cultivated radish, i.e., cultivars from East Asia show greater diversity than cultivars from the Mediterranean region (Crisp 1995).

In summary, polymorphisms in organelle genomes are common in wild and cultivated radishes, which provide molecular proof of the diversity of radish accessions. Four mitotypes Ogura CMS, normal, DCGMS, and ‘Chuanxinhong’ were sequenced and revealed many large and multiple short repeats causing genome rearrangement and polymorphisms of mitochondrial genomes. Many mitotypes and cpDNA haplotypes were identified using sequence analysis of specific genes or intergenic spacers from the

Table 6.3 Number of accessions and haplotypes of cultivated and wild *Raphanus* species and varieties (Yamane et al 2009)

Species and varieties	Number of accessions	Number of haplotypes
cultivated *Raphanus*		
R. sativus var. *hortensis* (Asian big radish)	48	11
R. sativus var. *niger* (black radish)	2	1
R. sativus var. *caudatus* (rat's tail radish)	1	1
R. sativus var. *sativus* (European small radish)	8	2
total number of haplotypes in cultivated *Raphanus*	59	13
wild *Raphanus*		
R. sativus var. *raphanistroides* (=East Asian wild radish)	10	6
R. raphanustrum	4	4
R. landra	9	7
total number of haplotypes in wild *Raphanus*	23	13

mitochondrial and chloroplast genomes. These studies provide available information for the phylogenetic assignment of radish cultivars.

6.3.2 Nuclear DNA Polymorphisms and Genetic Diversity

Recently, the draft genome of *R. sativus* var. *hortensis*, a Japanese radish with long and thick roots, was published. Two radish genomes were sequenced and assembled to approximately 402 Mb and 383.3 Mb by the two groups independently (Kitashiba et al. 2014; Mitsui et al. 2015). Mitsui et al. (2015) estimated the genome size as 574 Mb using flow cytometry, which was similar to previous results by Johnston et al. (2005). With the same Brassicaceae crop (Wang et al. 2011), partial whole genome triplication (WGT) triggered diversification may also have occurred in *R. sativus*. After WGT, chromosome rearrangements might have occurred many times by the time *R. sativus* was established. Subsequently, a number of repetitive sequences were detected, and the frequency and classes of repeat sequences in *R. sativus* (36.61%) were similar to those in *B. rapa* (39.51%) (Kitashiba et al. 2014; Mitsui et al. 2015). These will result in more abundant DNA polymorphisms in the radish genome.

Before the genomic era, nuclear DNA polymorphisms and genetic diversity among wild and cultivated radishes were studied using various molecular markers and *Raphanus* accessions from different geographical locations (Table 6.4). Among these markers, RAPD, AFLP, and SRAP markers have been used for cultivar fingerprinting, genetic diversity analysis, and genetic map construction (Huh and Ohnishi 2002; Madhou et al. 2005; Muminović et al. 2005). Recently, both expressed sequence tag (EST) and genomic sequences have been accumulated rapidly with the development of sequencing technology, and have led to the discovery of genome-wide SSRs and SNP variations in several vegetable crop species, including radish (Bae et al. 2015; Jiang et al. 2012; Li et al. 2011; Shen et al. 2013a; Shirasawa et al. 2011; Zhai et al. 2013). Using different markers and various materials, nuclear DNA polymorphisms were found in both wild and cultivated radishes. The differing levels of polymorphisms observed most likely result from differing marker selection and sampling strategies between the studies (Table 6.4).

6.3.2.1 Random Amplified Polymorphic DNA and Amplified Fragment Length Polymorphisms

RAPD and AFLP markers were first applied to DNA polymorphisms and genetic diversity analysis in cultivated and wild radishes over 10 years ago. Yamagishi et al. (1998) compared

Table 6.4 Studies on genetic diversity in radish using molecular markers

Molecular marker used	Reference	Radish accessions studied	Total no. of bands/primers	No of polymorphic bands/primers (rate of polymorphism)
RAPD	Pradhan et al. (2004)	Australian radish cultivars	52	46 (88.5%)
	Madhou et al. (2005)	European wild radish	118	97 (82%)
	Liu et al. (2008)	Asian big radish	412	352 (85.4%)
	Kong et al. (2011)	Asian big radish; European small radish; black radish	109	72 (66.1%)
AFLP	Huh and Huh (2001)	East Asian wild radish	308	180 (58.4%)
	Huh and Ohnishi (2002)	East Asian wild radish; European wild radish; Asian big radish	315	230 (72.8%)
	Muminović et al. (2005)	European small radish; black radish; Chinese wild radish; European wild radish	349	267 (76.5%)
	Wang et al. (2008)	Asian big radish; European small radish; South Asian radish;	478	221 (46.2%)
	Kong et al. (2011)	Asian big radish; European small radish; black radish	327	128 (39.1%)
SRAP	Liu et al. (2008)	Asian big radish	233	199 (85.41%)
ISSR	Muminović et al. (2005)	European small radish; black radish; Chinese wild radish; European wild radish	116	45 (38.8%)
	Liu et al. (2008)	Asian big radish	250	213 (85.2%)
SSR	Ohsako et al. (2010)	East Asian wild radish	78 primers	8 (10.3%)
	Wang et al. (2007)	Asian big radish	14 primers	8 (57.14)
	Nakatsuji et al (2011)	Rat's tail radish and Japanese cultivar 'Harufuku'	256 primers	130 (50.8%)
	Shirasawa et al. (2011)	'GSK3-1' and 'HA2'	3800 primers	642 (16.9%)
	Jiang et al. (2012)	Asian big radish	58 primers	28 (48.3%)
	Shen et al. (2013a)	Cultivated and wild radish	13,570	–
	Zhai et al. (2013)	Asian big radish	67 primers	39 (58.2%)
	Wang et al. (2015)	Asian big radish; European wild radish; Asian big radish; European small radish; black radish; oilseed radish; rat's tail radish	626 primers	559(89.3%)
	Bae et al. (2015)	Cultivated radish	184 primers	27 (14.7%)

(continued)

Table 6.4 (continued)

Molecular marker used	Reference	Radish accessions studied	Total no. of bands/primers	No of polymorphic bands/primers (rate of polymorphism)
SNP	Li et al. (2011)	'Sayatori 26704' and 'Aokubi S-h'	2290 SNPs	1465 (64%)
	Shen et al. (2013a, b)	Cultivated and wild radish	RadishBase	28,758

the similarity ratio of RAPD within and between radish varieties. They found that the commercial varieties showed high similarity ratio values (above 80) within a variety. The native varieties and the two wild radishes had low similarity ratio values, suggesting that larger genetic variations exist within the wild populations than within the cultivated ones: Artificial selection improved the uniformity of the commercial varieties. Using RAPD markers, the genetic diversity among cultivated and wild Korean radishes was studied (Huh and Ohnishi 2003). The index of diversity was 1.20 and 1.15, on average, for wild and cultivated radishes, respectively. The wild populations had higher diversity than the cultivated ones, suggesting that the domestication process has reduced the levels of genetic diversity via artificial selection. Results using AFLP markers further confirmed the high diversity in Koran wild radish, with an average of 58.4% polymorphic, and the populations at different areas had various polymorphisms within the population. The Seongsanpo population at Cheju-do showed the highest within population variability, whereas the population at Gyeongsangbuk-do exhibited the lowest genetic diversity of all the Korean wild radish populations (Huh and Ohnishi 2001).

Using AFLP and inter-simple sequence repeat (ISSR) analyses, the genetic diversity of cultivars grown in Europe countries, including var. *sativus* and var. *niger*, was investigated (Muminović et al. 2005). These accessions consisted of open-pollinated varieties, hybrids, and inbred lines. Of 349 bands tested with AFLP primers, 267 (76.5%) bands were polymorphic, and the average polymorphic information content (PIC) values ranged from 0.18 to 0.29. Applying ISSR primers, 45 of 116 bands (38.8%) were polymorphic, and the PIC values were between 0.14 and 0.43. The open-pollinated varieties showed larger diversity than the F_1 hybrids and inbred lines. Wang et al. (2008) analyzed AFLP and genetic diversity of 65 cultivated accessions from 21 countries from Europe, the Middle East, South Asia, and East Asia. Among 478 AFLP bands, 221 (46.2%) presented distinct polymorphisms and the largest rate of polymorphisms was 80.8%. The total number of polymorphic AFLPs for the cultivars from the same area (Europe, the Middle East, South Asia, and East Asia) ranged from 174 (36.4%) to 190 (39.7%). No significant difference was found in the average genetic diversity among East Asian radishes, although these accessions showed great diversity of morphological types. In that study, only three Chinese radish cultivars were included, which may have led to the low genetic diversity among East Asian radishes.

RAPD and AFLP marker systems were integrated to examine the genetic diversity among 56 cultivated radish accessions representing three groups, var. *hortensis*, var. *sativus*, and var. *niger* (Kong et al. 2011). Among 109 bands generated by RAPD primers, 72 were polymorphic, and the proportions ranged from 40.0 to 88.9%, with the mean polymorphic proportion of 66.1%. Meanwhile, 39.1% of bands generated by AFLP assay were polymorphic. The East Asian radishes were divided into six subgroups, and each subgroup had distinct agro-morphological traits. Of the three accessions, var. *hortensis* Becker germplasms had the largest genetic diversity. This conclusion is inconsistent with a previous report (Wang et al. 2008). Perhaps the different materials used led to the inconsistence. The genetic diversity of RAPD and AFLP markers was also found in European, Australian, and East Asian wild radish accessions (Huh and Ohnishi 2002; Madhou et al. 2005; Pradhan et al. 2004).

6.3.2.2 Development of SSR Markers and DNA Polymorphism Analysis

SSRs, EST-SSR, or microsatellites comprise 2–6 nucleotide repeat units and are abundant throughout the genome. These markers have been used to study genetic diversity, genetic map construction, and the identification of genes associated with agriculturally important traits in radish because of the advantages of multi-allelic detection, high variability, ease of detection, codominant inheritance, high reproducibility, and locus specificity (Bae et al. 2015; Hashida et al. 2013; Shirasawa et al. 2011; Wang et al. 2015).

For SSR marker development in radish, Wang et al. (2007) firstly identified a set of polymorphic SSRs using an EST database that contained only 409 radish EST sequences from GenBank (2006). With these ESTs, 14 SSR primer pairs were designed, eight primer pairs were able to amplify a clear band, and polymorphisms were estimated among 43 inbred lines of Japanese radish cultivars. However, many ESTs and genomic sequences have accumulated rapidly following the advent of improved sequencing technology. Up to 2013, 314,799 *Raphanus* ESTs had been collected from the National Center of Biotechnology Information (NCBI) dbEST database. These ESTs were generated from 22 different *Raphanus* cDNA libraries derived from 18 different accessions (Shen et al. 2013a).

During this period, a number of radish SSR markers have been developed using EST databases and next-generation sequencing (NGS). Nakatsuji et al. (2011) developed 256 SSR markers (156 genomic and 100 EST-SSRs) using three SSR-enriched libraries (a rat's tail radish, a Japanese cultivar 'Harufuku,' and their F_1 plant) and cDNA contig data in the RadishDB, respectively, of which 130 markers were polymorphic between the two parental lines. A total of 26,606 ESTs have been collected from cDNA libraries of 'GSK3-1' (Shirasawa et al. 2011). These EST sequences were used to design 3800 SSR markers and then 630 polymorphic EST-SSRs were mapped into nine linkage groups. Cui et al. (2012) downloaded 287,349 radish ESTs from the NCBI database and observed 3718 SSRs. Out of 183 primers designed, 64 pairs were polymorphic among the 12 typical radish germplasms. In the study of Jiang et al. (2012), a taproot cDNA library was constructed and a large number of high-quality ESTs were generated. Out of 125 primer pairs designed, 110 (88%) could generate unambiguous amplification products. Twenty-eight EST-SSRs were used for genetic diversity analysis in 32 radish genotypes. Wang et al. (2012) identified a large number of assembled unigenes (61,554) and detected SSR markers (14,641) for gene discovery and molecular marker development using high-throughput transcriptomic sequencing in the radish cultivar 'Weixianqing.. Using a similar method, Zhai et al. (2013) produced 73,084 unigene sequences using NGS of the transcriptome of the radish cultivar 'NAU-RG,' with a white global taproot. Furthermore, 5503 genic-SSR primers were developed, among which 1052 SSR primers were synthesized, and a subset of 823 (78.23%) primers could generate stable bands. Bae et al. (2015) developed genomic SSR markers using a SSR-enriched library and investigated genetic diversity in a collection of 144 radish cultivars. In that study, 237 primer pairs for SSRs were designed and 184 (77.6%) primer pairs produced PCR amplicons.

A number of SSR markers have been developed in radish; however, only a few SSR polymorphism information (e.g., multi-allele, PIC) was investigated in some radish genetic resources. Wang et al. (2007) developed eight EST-SSR markers using an EST database from the NCBI. The number of alleles per locus ranged from two to five among 43 cultivars of Japanese radish, and the PIC value ranged from 0.171 to 0.575. Ohsako et al. (2010) studied the genetic variability and its spatial structure within and between East Asian wild radishes from seven regions using microsatellite markers. Ten subpopulations within a local population were also investigated. They drew the conclusion that significant genetic differentiation existed within and among the populations of East Asian wild radishes. A positive correlation between spatial distance and genetic differentiation was found at regional and local levels, which suggested that East Asian wild radish is genetically structured and that each population retains a unique genetic composition. Polymorphisms between the rat's tail radish and Japanese cultivars were studied using EST-SSR markers (Nakatsuji et al. 2011). Of 81 SSR primer pairs, 48 (59.3%) showed polymorphisms in the rat's tail radish, and 65.4% of primer pairs had polymorphisms in the Japanese cultivated radish. Of the markers tested, 16 primer pairs were amplified using 16 radish cultivars and four other Brassicaceae species. These markers detected 3–15 alleles (average = 9.6), and the PIC values ranged from 0.54 to 0.92. Moreover, 28 and 67 selected informative SSR markers were used to determine the genetic diversity of 32 radish genotypes by two groups, respectively (Jiang et al. 2012; Zhai et al. 2013), in which the PIC values ranged from 0.49 to 0.89 and ranged from 0.00 to 0.825. The 32 accessions were grouped into three main clusters at a similarity index value of 0.60, which was mainly in accordance with the different biological characteristics of the accessions. In the study of Bae et al. (2015), 27 SSR markers (14.7%) were selected and used to assess the genetic relationships between 144 radish varieties. The number of alleles per marker ranged from two to 18, with an average of 7.77 alleles, and the PIC values ranged from 0.491 to 0.906. Genetic relationship analysis showed significant genetic differentiation between the five market classes of 135 radish cultivars (74 big, 15 small, 29 young, 12 Altari, and five processing).

Although a number of SSR markers have been developed and used to evaluate genetic diversity in radish, the above studies have a number of deficiencies: (1) the morphological traits of the materials used are not clear; (2) some cultivars with different names from different countries are the same types, and these cultivars are introduced to the other areas and are called new names according to the local habits; and (3) some studies did not include sufficient types of cultivated and wild radishes.

Using EST-SSR markers, Wang et al. (2015) analyzed the genetic diversity within and among *Raphanus* species, including all the five cultivated types (var. *sativus*, var. *hortensis Becker*, var. *niger Kerner*, var. *oleiferus* Metzg, and var. *caudatus* Hooker and Anderson) and four wild types (var. *raphanistroides* Makino, *R. raphanistrum*, *R. landra*, and *R. maritimus*) . The cultivated accessions included 83 varieties from 16 countries, representing nearly all typical radish germplasms. Different from the previous reports, the morphological information such as root shape, root color, leaf shape, and flower color was investigated in this study. Of 626 EST-SSR markers, 559 (accounting for 89.3%) were polymorphic, and produced amplified fragments with sizes corresponding to those of Shirasawa et al. (2011). The number of alleles for each primer varied from one (dominant marker) to seven, with an average of 3.03 polymorphic bands. The average PIC value per primer was 0.39, ranging from 0.07 to 0.73. Furthermore, the genetic diversity among *Raphanus* species and subspecies was assessed with 50 selected core markers. A total of 254 alleles at 52 loci were detected, with an average of 4.88 alleles per locus. The mean PIC value was 0.57. The wild radish has the largest range of genetic variation (Table 6.5). Among the cultivated radish subspecies, the Asian big radish and the European small radish possess the highest morphological diversity, which is reflected in a higher number of alleles (3.42, 3.35), and higher PIC value (0.46, 0.39) compared with other subspecies (Table 6.5).

6.3.2.3 SNPs and Genetic Diversity Analysis

SNPs are the most abundant type of DNA polymorphism in genomes. However, only a few studies on genetic mapping, genetic diversity, and mapping of genes associated with

Table 6.5 Information of amplification in radish subspecies/varieties by 50 core markers (Wang et al. 2015)

Parameters	Total	Cultivated radish	Asian big radish	European small radish	Black radish	Oil radish	Rat's tail radish	Wild radish
No. of samples	139	98	52	24	5	13	4	41
No. of makers	50	50	50	50	50	50	50	50
No. of locus	52	52	52	52	52	52	52	52
No. of variation (allele)	254	218	178	174	135	142	107	235
Mean number of alleles per marker	4.88	4.19	3.42	3.35	2.60	2.73	2.06	4.52
No. of frequent alleles[a]	1	1	1	4	8	4	18	1
No. of middle frequency alleles[a]	194	181	157	137	127	132	89	199
No. of rare alleles[a]	59	36	20	33	0	6	0	35
No. of absent alleles	–	36	76	80	119	112	147	19
No. of unique alleles	–	19	3	1	2	0	0	36
Mean polymorphic information content (PIC)	0.57	0.53	0.46	0.39	0.37	0.31	0.28	0.56

[a]Signal presence with the frequency f: $0 < f \leq 0.05$—rate alleles, $0.05 < f < 0.95$—middle frequency alleles, $0.95 \leq f < 1$—frequent alleles

agriculturally important traits using SNP markers have been published for radish (Li et al. 2011; Shen et al. 2013a; Zou et al. 2013).

The research groups of Nishio T. (Graduate School of Agricultural Science, Tohoku University) developed large-scale SNP markers, constructed a genetic map with high-density EST-based SNP markers, and identified candidate genes controlling 4-methylthio-3-butenyl glucosinolate contents in radish roots (Kitashiba et al. 2014; Li et al. 2011; Zou et al. 2013). In these studies, researchers used EST sequences to design primer pairs to amplify specific genes, and identified SNPs to produce EST-based SNP markers by determining the nucleotide sequences of two parental lines. In the study of Li et al. (2011), 2290 sequences, covering 791.9 kb, were obtained. Among them, 1465 (64.0%) DNA fragments showed SNPs or indel polymorphisms between the parental lines. In the study of Zou et al. (2013), the comparison of the sequence data of 1777 fragments between 'TBS' and 'AZ26H' revealed 2655 possible SNPs. By assembling the sequence data, 1953 SNPs in 750 DNA fragments were identified, 437 of which had been mapped previously on a linkage map. Recently, draft sequences of the radish genome were published (Kitashiba et al. 2014). The 'Aokubi' reference genome, a Japanese radish with a long and thick root, was carried out using de novo assembly. Then, SNPs were surveyed by mapping reads of 'Sayatori' to the 'Aokubi' reference sequences. Multiplex PCR fragments from four inbred lines of 'Yumehomare,' 'Sakurajima,' 'N1-3,' and 'Nishimachi-Risou' were sequenced. Combined with their previous studies (Zou et al. 2013), SNPs were surveyed between all eight inbred lines and the results for the number of SNPs and the number of common amplicons containing SNPs between different lines are available at http://radish.kazusa.or.jp/.

Another strategy for SNP development is the use of databases. During the past several years, large-scale EST and genomic sequence data have accumulated rapidly. To date, >300,000 radish ESTs have been deposited in the GenBank dbEST database (Shen et al. 2013a, b). In the study of Shen et al. (2013b), 314,799 ESTs derived from 22 *Raphanus* cDNA libraries from 18 different genotypes were analyzed, and then clustered and assembled into 85,083 unigenes: 28,758 high-quality SNPs were detected in 4764 unigenes. Among them, 15,029 were transitions, 10,051 were transversions, and 3678 were single-base indels (Table 6.6). In addition, RadishBase (http://bioinfo.bti.cornell.edu/radish) has been developed. The identified EST-derived SNPs and their associated information are available for radish breeding and research (Shen et al. 2013b).

In summary, DNA polymorphisms were surveyed in wild and cultivated radishes using different molecular markers. These results provide molecular evidence of the genetic diversity of radish accessions. As shown in Table 6.7, the index of DNA polymorphism information showed a large variation among different studies. For example, in the study of Wang et al. (2015), DNA fragments of all 93 cultivated and wild radish accessions were amplified by PCR using 50 core EST-SSR markers. An average of 4.88 (ranging from two to seven) alleles per locus was detected, and the mean PIC value was 0.57 (ranging from 0.07 to 0.73). The number of alleles and the PIC value were lower than those obtained using SSR analysis (Bae et al. 2015; Nakatsuji et al. 2011), and higher than those generated by AFLP analysis (Muminović et al. 2005) and other EST-SSR analyses (Wang et al. 2007), but were similar with the result of Jiang et al. (2012) and Zhai et al. (2013).

The differing PIC values and polymorphism levels most likely result from differences of marker selection, allele detection methods, and sampling strategies between the studies. Generally, SNPs are the most abundant type of DNA polymorphism in genomes (Shen et al. 2013a, b); SSR markers have the advantage of being multi-allelic and show the highest variability (Wang et al. 2015); however, RAPD, AFLP, RFLP, and SRAP markers have some defects, such as poor stability and reproducibility, and anonymity (Table 6.7). In any allele detection method, the most sensitive is sequencing of the PCR products, second is the Genetic Analyzer platform with fluorescent labels (Bae et al. 2015; Nakatsuji et al. 2011), and the least sensitive is gel electrophoresis (Wang et al. 2015). The genetic variation of cultivated species is generally considered to be less than that of their wild ancestors. Thus, the use of only cultivated inbred lines in a study might result in these relatively low levels of genetic parameters. In theory, the comparison of complete genome sequences would properly reflect each species' exact specificity although that is very difficult. So in the near future, a series of cultivated and wild *Raphanus* species will be genome-resequenced. These data will undoubtedly lead to increased understanding of the DNA polymorphisms and genetic diversity of radish.

Table 6.6 Statistics of radish single nucleotide polymorphisms (SNPs, Shen et al. 2013a)

SNP	No. SNPs	Type	Total
A ↔ G or C ↔ T	15,029	Transition	15,029
A ↔ C	2433	Transversion	10,051
A ↔ T	2828		
C ↔ G	2276		
G ↔ T	2514		
T ↔ -	1288	Indel	3678
A ↔ -	1172		
C ↔ -	611		
G ↔ -	607		

Table 6.7 Differences of DNA polymorphism between the various molecular markers in radish

Marker types	Materials	No. of primers	No. of alleles (Avg.)	Information of DNA polymorphism*	References
RAPD	Three wild radish populations	9	6–17 (11)	–	Madhou et al. (2005)
	35 cultivars	35	4–16 (10)	*Rp* (0.857–7.143)	Liu et al. (2008)
	56 cultivars	12	2–13 (6)	–	Kong et al. (2011)
AFLP	13 East Asian wild radish, one wild radish and two cultivars populations	8	33–48(39.4)	P_p (65.9–80%)	Huh and Ohnishi (2002)
	Seven East Asian wild radish populations	8	30–48(38.5)	P_p (40–67.5%)	Huh and Huh (2001)
	68 cultivars, one Chinese wild radish and one wild radish	12	11–42 (22.3)	*PIC* (0.18–0.29)	Muminović et al. (2005)
	65 cultivars from around the world	14	5–30 (15.8)	*Nei's* (0.36)	Wang et al. (2008)
	56 cultivars	8	10–32 (16)	–	Kong et al. (2011)
SRAP	35 cultivars	17	6–17 (11.76)	*Rp* (2.229–8.457)	Liu et al. (2008)
ISSR	68 cultivars, one Chinese wild radish and one wild radish		2–7(4.5)	*PIC* (0.14–0.43)	Muminović et al. (2005)
	35 cultivars	22	6–14 (7.61)	*Rp* (2.057–7.6)	Liu et al. (2008)
SSR	43 cultivars	8	2–5(3.13)	*PIC* (0.171–0.575)	Wang et al. (2007)
	Seven East Asian wild radish populations	8	4–9(6.13)	H_T (0.418–0.731)	Ohsako et al. (2010)
	16 cultivars and four other Brassicaceae species	16	3–15(9.6)	*PIC* (0.54–0.92)	Nakatsuji et al. (2011)
	32 cultivars	28	1–7(3.5)	*PIC* (0.00–0.825)	Jiang et al. (2012)
	32 cultivars	39	1–10(3.49)	*PIC* (0.49–0.89)	Zhai et al. (2013)
	37 cultivars	559	1–7(3.03)	*PIC* (0.07–0.73)	Wang et al. (2015)
	83 cultivars, three East Asian wild radish, and 10 wild radish	50	2–7 (4.88)	*PIC* (0.32–0.77)	Wang et al. (2015)
	144 cultivars from nine companies	27	2–17(7.7)	*PIC* (0.5–0.914)	Bae et al. (2015)
SNP	18 cultivars and four wild radish	28,758	28,758	–	Shen et al. (2013a, b)

Note: H_T, overall gene diversity; *PIC*, polymorphism information content; *Nei's*, Nei's similar index; P_p, percentage of polymorphism; R_p, resolving power

6.4 Genetic Relationships Based on the Nucleotide Diversity

Long cultivation history and wide distribution have led to cultivated radish being classified into many varieties according to the morphology of its edible root (i.e., shape, size, and color) and its intended uses. Abundant morphological diversity also makes it difficult to study the genetic structure and relationships within and among cultivated and wild radishes. In the past two decades, the relationships between different genera or species of *Raphanus* have been studied using various molecular markers.

The relationship between radish cultivars from five regions (Europe, North Africa, Middle East, South Asia, and East Asia) was studied using AFLP markers, (Wang et al. 2008). The cultivars were classified into four groups. The European accessions and East Asian accessions had a long genetic distance, occupying the two sides of the neighbor-joining tree. The South Asian accessions were closely related to the East Asian ones, and the accessions from the Middle East had a close relationship with the European ones. Concerning the genetic relationship between *R. raphanistrum* (European wild radish), East Asian wild radish, and East Asian cultivated radish, Yamagishi et al. (1998) suggested that the Japanese wild radish is closely related to Japanese and Chinese cultivated radishes, and *R. raphanistrum* was clearly separated from East Asian wild radish. Huh and Ohnishi (2002) showed that the East Asian wild radish populations could be further subdivided into Japanese and Korean subgroups, and that *R. raphanistrum* was found unexpectedly to be closely related to cultivated and wild East Asian populations in Kazakhstan. Among cultivated radish varieties, Kong et al. (2011) suggested that the black radish had a closer relationship to European radish than to Asian big radish. Among Asian big radish varieties, six subgroups were divided and those accessions showing heat tolerance and late-bolting accessions could be distinguished from the others.

In the study of Wang et al. (2015), genetic relationships of nearly all typical *Raphanus* germplasms, including five cultivated types and four wild radishes, were assessed with 50 core EST-SSR markers. Unweighted pair group method with arithmetic mean cluster analysis classified the 93 germplasms into four groups (I, II, III, and IV; Fig. 6.2). Group I included 86 cultivated radish accessions (*R. sativus*) and the East Asian wild radish (*R. sativus* var. *raphanistroides*) , while seven wild *Raphanus* accessions were separated into the remaining groups (*R. maritimus*, *R. raphanistrum* and *R. landra*). Group I could be further subdivided into two subgroups (A and B), mainly cultivated in Asia and Europe, respectively. Subgroup A included *R. sativus* var. *hortensis* (A1), var. *raphanistroides* Makino (A2), and var. *caudatus* (A3). *R. sativus* var. *hortensis* A1 has the most cultivar types mainly from China, Japan, and Korea, including long white radish, red skin radish, green skin radish, and 'Xinlimei' radish accessions. Subgroup B was divided into four types, as follows: *R. sativus* var. *sativus* (B1 and B2), var. *niger* Kerner (B3), and var. *oleiferus* Metzg (B4) (Fig. 6.2).

This UPGMA cluster analysis of the radish varieties was almost in accord with the phylogenetic results using a subset of 1800 SNPs (Shen et al. 2013a). The eight accessions were clearly separated into two groups: The first group included four accessions belonging to cultivated radish (*R. sativus* var. *caudatus*, var. *hortensis*, var. *radiculus*, and var. *oleiformis*,) and the second group included four wild radish accessions (*R. raphanistrum*, subsp. *Raphanistrum*, subsp. *maritimus*, and subsp. *landra*) (Fig. 6.2).

6.5 The Origin and Evolution Revealed by Genetic Diversity

The molecular biology studies on the genetic diversity also revealed the origin and evolution of radish. However, the origin of cultivated radish and the details of wild radish domestication are unclear. Presently, there are three different views: 1) most reports suggest that cultivated radish originated from *R. raphanistrum* L. or *R. maritimus* Smith (Henslow 1898; Kobabe 1959);

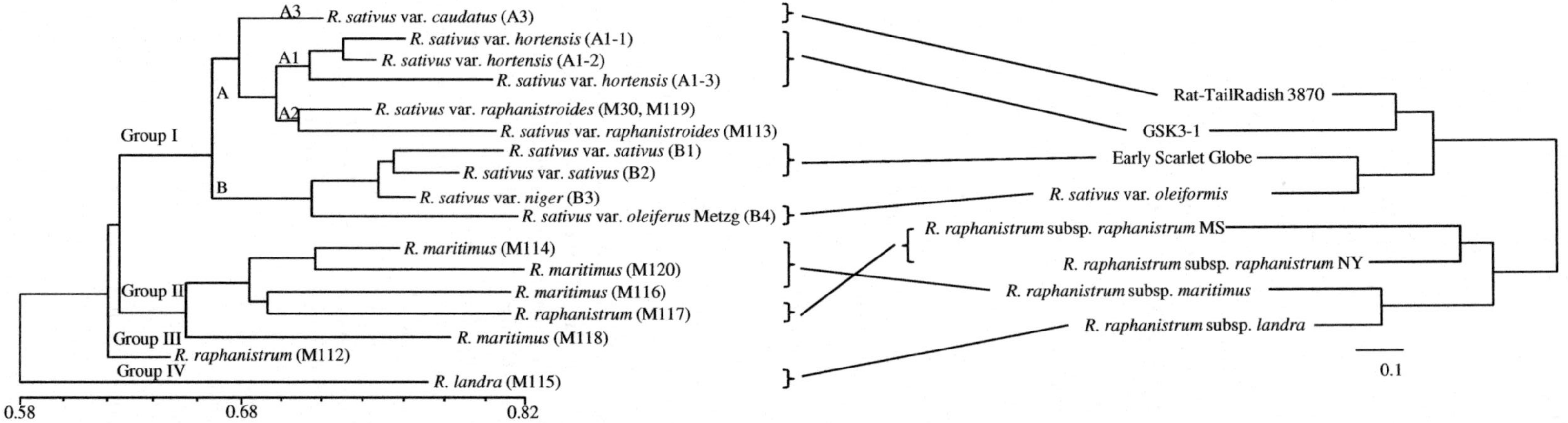

Fig. 6.2 Compare UPGMA cluster dendrogram with neighbor-joining tree of radish. *Left* UPGMA dendrogram of 137 radish accessions based on cluster analysis with 50 core EST-SSR markers (Wang et al. 2015). *Right* Neighbor-joining tree of radish by 1800 SNPs (Shen et al, 2013a). Rat-Tail 3870: *var. caudatus*, GSK 3-1: *var. hortensis*, Early Scarlet Globe: *var. radiculus*

2) *R. sativus* L. resulted from interspecific hybridization between *R. maritimus* and *R. landra* or a variety of parental species (Lewis-Jones et al. 1982); and 3) cultivated radishes had multiple ancestors (Huh and Ohnishi 2002; Kumazawa 1961).

6.5.1 Cultivated Radish Originated from Multiple Wild Radish Ancestors

To date, many studies related to genetic diversity and the relationship between cultivated and wild radishes have provided indirect proof of the origin and evolution of this genus. Chloroplast and mitochondrial genome sequence variation in cultivated and wild radishes reflect the fact that cultivated radish originated from multiple wild radish ancestors. Yamagishi and Terach (2003) analyzed the mitochondrial *coxI* and *orfB* regions of three European wild radishes (*R. raphanistrum*, *R. landra*, and *R. maritimus*) , and cultivars from Europe and Asia. All of five types of mtDNA were found in the wild species *R. raphanistrum*, while four were found in *R. maritimus* and one in *R. landra*. The same mtDNA variations were present in the cultivated radish, indicating that the diversity of the mtDNA was maintained during the domestication of wild radish. This conclusion supported indirectly the view that the three wild species would be the potential candidates for the ancestor of the cultivated radish and that modern cultivated radishes had multiple origins.

The sequences of the 5′-*matK* region in the cpDNA were investigated to elucidate the evolution of cultivated radish; however, this analysis produced a different conclusion on the origin of radish (Yamane et al. 2005). Using a combination of chloroplast *matK* sequences and PCR-RFLP, four haplotypes were found in wild and cultivated radishes. *R. raphanistrum* had a unique chloroplast haplotype that was not found in cultivated radish, implying that *R. raphanistrum* was not the maternal ancestor of cultivated radish. The East Asian wild radish had three haplotypes and one of them was predominant in East Asian cultivated radishes. The close relationship between East Asian wild radish and East Asian cultivated radish indicated that cultivated East Asian radishes might have originated from the East Asian wild radish.

Yamane et al. (2009) showed that wild radish in Japan had higher levels of chloroplast SSR variation, and was a valuable resource for the extension of the genetic diversity of cultivars by comparing chloroplast DNA variability. They suggested that the East Asian wild radish contributed to the establishment of the Asian big radish. Lü et al. (2008) investigated nucleotide sequence variation in the chloroplast *trnK/matK* region of cultivated and wild radishes; the results suggested the existence of multiple independent origins for the varieties studied.

The East Asian wild radish is found in sandy coastal areas, and, occasionally inland, in Japan, Korea, and China. It does not develop an edible succulent root, and bears white to purple flowers (Makino 1961; Ohsako et al. 2010). However, the origin of the East Asian wild radish, and its role and position in the evolution of cultivated radish, is still debated. Kitamura (1958) believed that the East Asian wild species was not a true wild species, rather it was an escaped form of a cultivated radish. However, many reports have shown that Japanese wild radish is a descendant from *R. raphanistrum* (Terachi et al. 2001; Yasumoto et al. 2008). The *orf138* and nuclear *Rf* genes are distributed widely in Japanese wild radish; however, most of the Japanese cultivated radishes do not have these genes, which cast doubt on the escaped cultivated radish theory (Yasumoto et al. 2008).

6.5.2 Cultivated Radish Originated from a Single Wild Raphanus Species or from a Hybridization of Wild Species

Many studies showed that wild radish had larger variation than cultivated species. In the study of Wang et al. (2015), wild *Raphanus* species contain 36 unique alleles which are not present in any species of cultivated radish, and the wild radish has a higher PIC values and more alleles per locus than cultivated ones. In the study of Huh and Ohnishi (2002), the values of diversity parameters H_o, H_{EP} of *R. raphanistrum* were higher than those of East Asian wild radish, although only one population of *R. raphanistrum* was investigated. This result was in accord with Lewis-Jones et al. (1982) based on allozyme analysis. On the other hand, in the UPGMA cluster analysis (Fig. 6.2), the position of wild *Raphanus* species strongly supports the idea that all six radish cultivars share the same ancestor, and might have originated from one wild *Raphanus* species or from an interspecific hybridization (Shen et al. 2013a; Wang et al. 2015). However, further studies are required to definitively establish the phylogenetic relationship between cultivated and wild radishes.

European radish showed a closer genetic relationship to wild *Raphanus* than to radish varieties mainly planted in Asia (Huh and Ohnishi 2002; Wang et al. 2008, 2015) and that wild radish is widely distributed in Europe. We assumed three wild *Raphanus* species were involved in the evolution of *R. sativus*, and that this probably took place somewhere in the Eastern Mediterranean region toward the Caspian Sea, or even further east. Eventually, cultivated radish was introduced into Asia along the Silk Road (Banga 1976; Kitamura 1958; Wang et al. 2008, 2015).

6.6 The Impact of Modern Breeding Programs on Radish Genetic Diversity

Genetic variation within a population is the basis for selection and plant improvement. Thus, conservation of plant genetic diversity is essential for present and future human well-being (Rao and Hodgkin 2002). Modern breeding has had a profound impact on crop production, such as improvements in yield, uniformity, resistance to biotic and abiotic stresses, and edible quality. However, it is generally considered that continuous selection among crosses of genetically related cultivars has led to a narrowing of the genetic base of the crops, contributing to the genetic erosion of the crop gene pools on which breeding is based (Plucknett et al. 1987).

To date, about 200 radish hybrid cultivars have been released in China, and more than 90% of cultivars used in production are F_1 hybrids. The same situation occurred in Japan and Korea. These F_1 hybrids have been developed relying on limited genetic resources, and local cultivars with abundant useful genes are ignored, resulting in an increasingly narrow genetic base for radish germplasms. Meanwhile, the extensive use of F_1 hybrids has led to losses of radish genetic resources. For example, domestic varieties with red skin named 'Wuyingluobo' were cultivated widely in spring in Northern China in the past; however, late-bolting F_1 hybrids with white skin imported from Korea are being planted in those regions, replacing the indigenous 'Wuyingluobo,' leading to the gradual disappearance of this domestic variety. Following the disappearance of domestic varieties, some good properties such as disease resistance, adaptation, and edible quality are at risk of disappearing.

The wild radish accessions, which have good agro-morphological traits, such as male sterility and stress tolerance, are useful breeding materials to introduce these good traits into cultivars. However, they have been poorly incorporated into the breeding of cultivated radish. Thus, wild radish should be further studied and used. The rapid development of molecular biological technologies provides new methods to study the diversity of radish. Through improved characterization and development of core collections, based on genetic diversity information, it will be possible to better exploit the available resources.

References

Aoba T (1979) Varieties and breeding in Japanese radish. The Iden 33:55–60 (in Japanese)

Arrieta-Montiel MP, Mackenzie SA (2011) Plant mitochondrial genomes and recombination. In: Kempken F (ed) Plant mitochondria. Springer, New York, pp 65–82

Bae KM, Sim SC, Hong JH, Choi KJ, Kim DH, Kwon YS (2015) Development of genomic SSR markers and genetic diversity analysis in cultivated radish (*Raphanus sativus* L.). Hortic Environ Biotechnol 56: 216–224

Banga O (1976) Radish, *Raphanus sativus* (Cruciferae). In: Simmonds NW (ed) Evolution of crop plants. Longman, London, pp 60–62

Becker C (1962) Rettish und radies (*Raphanus sativus*). Handbuch der Pflanzenzuchtung 6:23–78

Chang S, Chen J, Wang Y, Gu B, He J, Chu P, Guan R (2013) A mitochondrial genome of *Raphanus sativus* and gene evolution of Cruciferous mitochondrial types. J Genet Genomics 40:117–126

Crisp P (1995) Radish-*Raphanus sativus* (Cruciferae). In: Smartt J, Simmonds NW (eds) Evolution of crop plants. Longman Scientific & Technical, Longman House, Harlow, pp 86–89

Cui N, Qiu Y, Li X, Di Shen, Wang H, Song J (2012) Data mining for SSRs in EST rescources and EST-SSR markers development in radish. Acta Hortic Sinica 39:1303–1312 (in Chinese)

Curtis IS (2011) Genetic engineering of radish: current achievements and future goals. Plant Cell Rep 30:733–744

George RAT, Evans DR (1981) A classification of winter radish cultivars. Euphytica 30:483–492

Hashida T, Nakatsuji R, Budahn O, Schrader O, Peterka H, Fujimura T, Kubo N, Hirai M (2013) Construction of a chromosome-assigned, sequence-tagged linkage map for the radish, *Raphanus sativus* L. and QTL analysis of morphological traits. Breed Sci 63:218–226

Henslow G (1898) The history of the radish. Gard Chron 23:389

Huh MK, Huh HW (2001) Genetic diversity of *Raphanus sativus* var. *hortensis* f. *raphanistroides* in Korea using AFLP markers. Korean J Genet 23:45–53

Huh MK, Ohnishi O (2001) Allozyme diversity and population structure of Japanese and Korean populations of wild radish, *Raphanus sativus* var. *hortensis* f. *raphanistroides* (Brassicaceae). Genes Genet Syst 76:15–23

Huh MK, Ohnishi O (2002) Genetic diversity and genetic relationships of East Asian natural populations of wild radish revealed by AFLP. Breed Sci 52:79–88

Huh MK, Ohnishi O (2003) Genetic diversity and relationships among natural and cultivated populations of radish in Korea revealed by RAPD. Korean J Genet 25:119–125

Jiang L, Wang L, Liu L, Zhu X, Zhai L, Gong Y (2012) Development and characterization of cDNA library based novel EST-SSR marker in radish (*Raphanus sativus* L.). Sci Hortic 140:164–172

Johnston JS, Pepper AE, Hall AE, Chen ZJ, Hodnett G, Drabek J, Lopez R, Price HJ (2005) Evolution of genome size in Brassicaceae. Ann Bot 95:229–235

Kaneko Y, Kimizuka-Takagi C, Bang SW, Matsuzawa Y (2007) Radish. In: Kole C (ed) Genome mapping and molecular breeding in plants. Springer, Berlin Heidelberg, pp 141–160

Kim S, Lim H, Park S, Cho K, Sung S, Oh D, Kim K (2007) Identification of a novel mitochondrial genome type and development of molecular makers for cytoplasm classification in radish (*Raphanus sativus* L.). Theor Appl Genet 115:1137–1145

Kitamura S (1958) Varieties of radish and their transition. In: Nishiyama I (ed) Japanese radish. The Japan Science Society Press, Tokyo, pp 1–19

Kitashiba H, Li F, Hirakawa H, Kawanabe T, Zou Z, Hasegawa Y, Tonosaki K, Shirasawa S, Fukushima A, Yokoi S, Takahata Y, Kakizaki T, Ishida M, Okamoto S, Sakamoto K, Shirasawa K, Tabata S, Nishio T (2014) Draft sequences of the radish (*Raphanus sativus* L.) genome. DNA Res 21:481–490

Kobabe G (1959) Naturliche Einkreuzung von Hederich (*Raphanus raphanistrum* L.) in Radies (*Raphanus sativus* var. *radicula* DC.) und das Verhalten von Knollenform und Farbe in den nachfolgenden F-und R-Generationen. Zeitschrift fur Pflanzenzuchtung 42:1–10

Kong Q, Li X, Xiang C, Wang H, Song J, Zhi H (2011) Genetic diversity of radish (*Raphanus sativus* L.) germplasm resources revealed by AFLP and RAPD markers. Plant Mol Biol Rep 29:217–223

Kumazawa S (1961) Horticulture, detailed discussion of crops. Tokyo, Yokendo, p 637 (in Japanese)

Lawley YE, Weil RR, Teasdale JR (2011) Forage radish cover crop suppresses winter annual weeds in fall and before corn planting. Agron J 103(1):137–144

Lee Y, Kim S, Lim H, Ahn Y, Sung S (2009) Identification of mitochondrial genome rearrangements unique to novel cytoplasmic male sterility in radish (*Raphanus sativus* L.). Theor Appl Genet 118:719–728

Lewis-Jones LJ, Thorpe JP, Wallis GP (1982) Genetic divergence in four species of the genus *Raphanus*: Implications for the ancestry of the domestic radish *R. sativus*. Biol J Linn Soc 18:35–48

Li F, Hasegawa Y, Saito M, Shirasawa S, Fukushima A, Ito T, Fujii H, Kishitani S, Kitashiba H, Nishio T (2011) Extensive chromosome homoeology among Brassiceae species were revealed by comparative genetic mapping with high-density EST-based SNP markers in radish (*Raphanus sativus* L.). DNA Res 18:401–411

Liu LW, Zhao LP, Gong YQ, Wang MX, Chen LM, Yang JL, Wang Y, Yu FM, Wang LZ (2008) DNA fingerprinting and genetic diversity analysis of late-bolting radish cultivars with RAPD, ISSR and SRAP markers. Sci Hortic 116:240–247

Lü N, Yamane K, Ohnishi O (2008) Genetic diversity of cultivated and wild radish and phylogenetic relationships among *Raphanus* and *Brassica* species revealed by the analysis of *trnK/matK* sequence. Breed Sci 58:15–22

Madhou P, Wells A, Pang ECK, Stevenson TW (2005) Genetic variation in populations of Western Australian wild radish. Aust J Agric Res 56:1079–1087

Makino T (1961) Makino's new illustrated flora of Japan. Hokuryukan, Tokyo (in Japanese)

Mitsui Y, Shimomura M, Komatsu K, Namiki N, Shibata-Hatta M, Imai M, Katayose Y, Mukai Y, Kanamori H, Kurita K, Kagami T, Wakatsuki A, Ohyanagi H, Ikawa H, Minaka N, Nakagawa K, Shiwa Y, Sasaki T (2015) The radish genome and comprehensive gene expression profile of tuberous root formation and development. Sci Rep 5:10835

Muminović J, Merz A, Melchinger AE (2005) Genetic structure and diversity among radish varieties as inferred from AFLP and ISSR analyses. J Am Soc Hortic Sci 130(1):79–87

Nakatsuji R, Hashida T, Matsumoto N, Tsuro M, Kubo N, Hirai M (2011) Development of genomic and EST-SSR markers in radish (*Raphanus sativus* L.). Breed Sci 61:413–419

Ogura H (1968) Studies on the new male-sterility in Japanese radish, with special reference to the utilization of this sterility towards the practical raising of hybrid seeds. Mem Fac Agric Kagoshima Univ 6:39–78

Ohsako T, Hirai M, Yamabuki M (2010) Spatial structure of microsatellite variability within and among populations of wild radish *Raphanus sativus* L. var. *hortensis* Backer f. *raphanistroides* Makino (Brassicaceae) in Japan. Breed Sci 60:195–202

Park JY, Lee Y, Lee J, Choi B, Kim S, Yang T (2013) Complete mitochondrial genome sequence and identification of a candidate gene responsible for cytoplasmic male sterility in radish (*Raphanus sativus* L.) containing DCGMS cytoplasm. Theor Appl Genet 126:1763–1774

Plucknett DL, Smith NJH, Williams JT (1987) Gene bands and the world's food. Princeton University Press, Princeton

Pradhan A, Yan G, Plummer JA (2004) Development of DNA fingerprinting keys for the identification of radish cultivars. Aust J Exp Agric 44:95–102

Rao VR, Hodgkin T (2002) Genetic diversity and conservation and utilization of plant genetic resources. Plant Cell Tissue Cult 68:1–19

Sahli HF, Conner JK, Shaw FH, Howe S, Lale A (2008a) Adaptive differentiation of quantitative traits in the globally distributed weed, wild radish (*Raphanus raphanistrum*). Genetics 180:945–955

Salisbury PA (1987) Blacklet resistance in weedy crucifers. Crucif News 12:90

Sahli HF, Conner JK, Shaw FH, Howe S, Lale A (2008b) Adaptive differentiation of quantitative traits in the globally distributed weeds, wild radish (*Raphanus raphanistrum*). Genetics 180:945–955

Shen D, Sun H, Huang M, Zheng Y, Qiu Y, Li X, Fei Z (2013a) Comprehensive analysis of expressed sequence tags from cultivated and wild radish (*Raphanus spp.*). BMC Genom 14:721

Shen D, Sun H, Huang M, Zheng Y, Li X, Fei Z (2013b) RadishBase: a database for genomics and genetics of radish. Plant Cell Phys 54:e3

Shirasawa K, Oyama M, Hirakawa H, Sato S, Tabata S, Fujioka T, Kimizuka-Takagi C, Sasamoto S, Watanabe A, Kato M, Kishida Y, Kohara M, Takahashi C, Tsuruoka H, Wada T, Sakai T, Isobe S (2011) An EST-SSR linkage map of *Raphanus sativus* and comparative genomics of the Brassicaceae. DNA Res 18:221–232

Streibig JC, Combellack JH, Pritchard GH, Richardson RG (1989) Estimation of thresholds for weed control in Australian cereals. Weed Res 29:117–126

Tanaka Y, Tsuda M, Yasumoto K, Yamagishi H, Terachi T (2012) A complete mitochondrial genome sequence of Ogura-type male sterile cytoplasm and its comparative analysis with that of normal cytoplasm in radish (*Raphanus sativus* L.). BMC Genom 13:352

Tang J, Zhang L, Peng C (2003) Study on mechanism of semen *Raphanus* in moving qi and removing food retention. Chin J Integr Tradit West Med Dig 11 (5):287–289 (in Chinese)

Terachi T, Yamaguchi K, Yamaguchi H (2001) Sequence analysis on the mitochondrial *orfB* locus in normal and Ogura malesterile cytoplasms from wild and cultivated radishes. Curr Genet 40:276–281

Wang LZ, He QW (2005) Chinese radish. Scientific and Technical Document Publishing House, Beijing (in Chinese)

Wang N, Hu J, Ohsawa R, Ohta M, Fujimura T (2007) Identification and characterization of microsatellite markers derived from expressed sequence tags (ESTs) of radish (*Raphanus sativus* L.). Mol Ecol Notes 7:503–506

Wang N, Kitamoto N, Ohsawa R, Fujimure T (2008) Genetic diversity of radish (*Raphanus sativus*) germplasms and relationships among worldwide accessions analyzed with AFLP markers. Breed Sci 58:107–112

Wang QB, Zhang L, Zheng PJ (2015) Genetic diversity and evolutionary relationship analyses within and among *Raphanus* species using EST-SSR markers. Mol Breeding 35:62

Wang SF, Wang XF, He QW, Liu XX, Xu WL, Li LB, Gao JW, Wang FD (2012) Transcriptome analysis of the roots at early and late seedling stages using Illumina paired-end sequencing and development of EST-SSR markers in radish. Plant Cell Rep 31:1437–1447

Wang X, Wang H, Wang J, Sun R, Wu J, Liu S, Bai Y, Mun JH, Bancroft I, Cheng F et al (2011) The genome of the mesopolyploid crop species *Brassica rapa*. Nat Genet 43:1035–1139

Warwick SI (1993) Guide to the wild germplasm of *Brassica* and allied crops. Part IV. Wild species in the tribe Brassiceae (Cruciferae) as sources of agronomic traits. Tech Bull 17E:1–19

Wolfe KH, Li WH, Sharp PM (1987) Rates of nucleotide substitution vary greatly among plant mitochondrial, chloroplast, and nuclear DNAs. Proc Natl Acad Sci 84:9054–9058

Yamagishi H (2004) Assessment of cytoplasmic polymorphisms by PCR-RFLP of the mitochondrial *orfB* region in wild and cultivated radishes (*Raphanus*). Plant Breeding 123:141–144

Yamagishi H, Tateishi M, Terachi T, Murayama S (1998) Genetic relationships among Japanese wild radishes (*Raphanus sativus* f. *raphanistroides* Makino), cultivated radishes and *R. raphanistrum* revealed by RAPD analysis. J Jpn Soc Hortic Sci 67:526–531

Yamagishi H, Terach T (2003) Multiple origins of cultivated radishes as evidenced by a comparison of the structural variations in mitochondrial DNA of *Raphanus*. Genome 46:89–94

Yamaguchi H (1987) Latitudinal cline and intra populational differentiation in leaf shape of wild radish in Japan. Jpn J Breed 37:54–65 (in Japanese)

Yamane K, Lü N, Ohnishi O (2005) Chloroplast DNA variations of cultivated radish and its wild relatives. Plant Sci 168:627–634

Yamane K, Lü N, Ohnishi O (2009) Multiple origins and high genetic diversity of cultivated radish inferred from polymorphism in chloroplast simple sequence repeats. Breed Sci 59:55–65

Yasumoto K, Matsumoto Y, Terachi T, Yamagishi H (2008) Restricted distribution of *orf687* as the pollen fertility restorer gene for Ogura male sterility in Japanese wild radish. Breed Sci 58:177–182

Zhai L, Xu L, Wang Y, Cheng H, Chen Y, Gong Y, Liu L (2013) Novel and useful genic-SSR markers from de novo transcriptome sequencing of radish (*Raphanus sativus* L.). Mol Breeding 33:611–624

Zou Z, Ishida M, Li F, Kakizaki T, Suzuki S, Kitashiba H, Nishio T (2013) QTL analysis using SNP markers developed by next-generation sequencing for identification of candidate genes controlling 4-methylthio-3-butenyl glucosinolate contents in roots of radish, *Raphanus sativus* L. PLoS One 8(1):e53541

7 Cytoplasmic Male Sterility and Mitochondrial Genome Variations in Radish

Hiroshi Yamagishi and Toru Terachi

Abstract

The first CMS found in radish was Ogura CMS and it has been widely used in Brassicaceae crops. CMS named 'UK-1,' was reported, whereas more recently 'NWB' CMS was detected. A CMS radish that showed a similar phenotype to 'NWB' CMS was identified in an accession collected from Uzbekistan. This CMS was named as 'DCGMS.' 'DCGMS' differed from Ogura CMS in all the features investigated, whereas molecular studies revealed that 'DCGMS' was the same as 'NWB.' The gene *orf138* was responsible for Ogura CMS. Nine types of *orf138* sequences have been identified in wild and cultivated radishes. The *orf125* of 'Kosena' and H-type *orf138* of 'UK-1' are variants of *orf138*. Next-generation sequencing (NGS) has revolutionized studies on the plant mitochondrial genome, and the complete nucleotide sequences of the mitochondrial genomes of eight radish cultivars have been elucidated. The size of the radish mitochondrial genome ranged from 239,186 bp for 'DCGMS' to 258,426 bp for 'MS-gensuke', which contains the Ogura cytoplasm. Close examinations indicated that the eight radish mitochondrial genomes were classified into five types. Comparative genome sequencing unambiguously proved that the mitochondrial genomes of 'DCGMS' and 'MS-gensuke' differ from each other. Apart from 'Black radish,' no other radish contains a homologous sequence to *orf463*, which is responsible for CMS in 'DCGMS.' More NGS data will clarify evolutionary and phylogenetic relationships between the various mitochondrial genomes of radish.

H. Yamagishi · T. Terachi (✉)
Faculty of Life Sciences, Kyoto Sangyo University, Motoyama, Kamigamo, Kita-ku, Kyoto 603-8555, Japan
e-mail: terachi@cc.kyoto-su.ac.jp

T. Nishio and H. Kitashiba (eds.), *The Radish Genome*,
Compendium of Plant Genomes, DOI 10.1007/978-3-319-59253-4_7

7.1 Introduction

Radish is one of the most popular vegetables in Japan. It is regularly eaten throughout the year in many ways, i.e., raw, pickled, as sprouts, dried, and simmered. Many local varieties or cultivars have been developed in Japan since ancient times in order to fulfill this demand. The commercial production of radish now involves F_1 hybrids, and modern radish breeding depends on cytoplasmic male sterility (CMS). CMS is a well-known phenomenon in higher plants and is caused by the actions of specific genes (*orfs*) in the mitochondrial genome. In radish, Ogura CMS cytoplasm, found by a Japanese researcher (Ogura 1968), has been widely used in breeding programs; however, a high degree of cytoplasmic genetic uniformity may make the crop genetically vulnerable in terms of disease susceptibility. This has prompted us to discover and characterize new types of CMS. In this review, relationships between different CMS in radish are considered after looking back at the history of the discovery of CMS in radish. And based on the data generated by next-generation sequencing (NGS), sequence and structural variations among the mitochondrial genomes of eight different cultivars are discussed in detail.

7.2 Discovery of and Relationships Between CMS in Radish

7.2.1 History of Discoveries

Male sterility in radish was initially reported by Tokumasu (1951), who identified male sterile plants in the 'Tokinashi-Daikon' variety. Based on the findings of cross experiments, male sterility was suggested to be controlled by a single recessive gene. Paulmann and Röbbelen (1988) introduced the cytoplasm of the male sterile radish found by Tokumasu (1951) into *Brassica napus* using intergeneric hybridization and subsequent backcrossing. The desired male sterile alloplasmic *B. napus* that formed normal green leaves at low temperatures was obtained (Paulmann and Röbbelen 1988). Albaum et al. (1995) studied the mitochondrial *atp9* of the Tokumasu radish in detail. The Tokumasu radish had the *atp9* gene in two copies, whereas *B. napus* and *B. rapa* contained a single copy. However, the mitochondrial genomes of normal and Ogura CMS radishes also have two copies of *atp9*, and cytoplasmic differences between the Tokumasu radish and other CMS radishes currently remain unclear. Furthermore, it has not yet been established whether male sterility in the Tokumasu radish is controlled by a nuclear gene, as originally described by Tokumasu (1951), or if a mitochondrial gene is the causal factor.

The most well-known and widely used CMS in Brassicaceae crops was identified by Ogura (1968). In that study, a male sterile radish strain was obtained from an unidentified variety at Ohnejime, Kagoshima prefecture, Japan. Various crossing experiments were conducted in order to clarify the inheritance of the male sterile characteristic. Cross combinations included not only those among plants of the original male sterile strain composed of male sterile and normal fertile plants, but also reciprocal crossings with normal commercial varieties. Based on the segregation of male sterile plants, male sterility was concluded to result from the interaction between the cytoplasm and gene responsible for pollen degeneration (Ogura 1968). The genotypes of male sterile and fertile plants were designated in the original population, in which the male sterile mutants were detected as S msms (sterile plants) and S Msms (fertile plants) (Ogura 1968). This was the first study on CMS in radish.

Ikegaya (1986a, b) reported CMS in a Chinese radish and a Japanese local cultivar. Ikegaya (1986a) introduced an F_1 radish line from Taiwan and crossed it as a female parent with another Chinese radish cultivar. One plant in the progeny had rudimentary anthers that did not produce pollen. Test crosses were conducted using various radish lines, and the appearance of male sterile plants was noted. Based on the findings obtained, male sterility was attributed to the cytoplasm, and named 'UK-1' (Ikegaya 1986a). On the other hand, Ikegaya (1986b) observed segregation of male fertile and sterile plants in a local variety, 'Kosena,' introduced from Yamagata City, Japan. After open pollination in a population, in which fertile and sterile plants were present, male sterile plants produced male sterile progeny plants at very high frequencies (67–92%). Using test crosses, this male sterility was found to be CMS, and the maintainer lines for 'UK-1' also functioned as maintainers for CMS in 'Kosena.' The cytoplasm of 'Kosena' was named Kosena CMS cytoplasm (kos) (Sakai and Imamura 1992).

Nahm et al. (2005) more recently detected CMS that differs from Ogura in a Korea radish collection. This NWB CMS radish collected from South Korea had more yellowish anthers than Ogura CMS lines, and some pollen was visible. The amount of pollen per anther in the NWB CMS line was approximately half that of the male fertile lines; however, it was not viable, while Ogura CMS radish lines produced no pollen. Although Nahm et al. (2005) crossed the NWB CMS line with 58 breeding lines from various countries, they did not identify any restorer line. In NWB CMS, a DNA fragment was not amplified by PCR with a primer pair specific to the Ogura *orf138* gene. Therefore, they developed a specific PCR primer pair at the 3′ region of the *atp6* gene and 5′ region of the *nad3* gene to detect the NWB CMS line. They also found another type of male sterile line, 'Won-Baek,' in a China collection; however, a detailed study has not yet been conducted.

Lee et al. (2008) found the CMS radish that showed a similar phenotype to that of NWB CMS in an accession collected from Uzbekistan. They named this CMS Dongbu cytoplasmic and genic male sterility (DCGMS). DCGMS differs from Ogura CMS in all features of the morphological phenotype, fertility restoration spectrum, and molecular characteristics of the mitochondrial genome. After surveying diverse germplasms, they identified one accession collected from Russia that completely restored male fertility. When the molecular marker that exclusively detects NWB CMS (Nahm et al. 2005) was tested with DCGMS and diverse radish germplasms, positive bands appeared in almost all radish germplasms including DCGMS (Lee et al. 2008). These findings indicated that CMS is the same between NWB and DCGMS. Park et al. (2013) elucidated the complete mitochondrial genome sequence in DCGMS cytoplasm and identified a novel chimeric gene, *orf463*, as a causal factor for DCGMS. However, they did not investigate whether NWB contains this chimeric gene. Therefore, the relationship between NWB and DCGMS needs to be examined in more detail.

7.2.2 Relationships Between the Cytoplasms of Ogura, Kosena, and UK-1

We elucidated the sequences of the DNA fragments amplified by PCR with a primer pair specific to the Ogura CMS gene, *orf138*, for diverse wild and cultivated radishes (Yamagishi and Terachi 2001). Nine types of DNA sequences were found in *orf138* distributed in wild and cultivated radishes. The *orf125* gene named for the 'Kosena' CMS gene was found to have a deletion of 39 nucleotides from that of *orf138*. Thus, *orf125* is regarded as one of the nine variants of *orf138*. *orf125* was also detected in Japanese wild radish collected in several areas in the Tohoku region where 'Kosena' is cultivated. In addition, 'Kosena' has a unique characteristic as a cultivar in that the root is not swollen. Therefore, 'Kosena' is a local cultivar that appears to have been developed by the domestication of Japanese wild radish.

The cytoplasm of 'UK-1' was derived from 'Daibaika,' which is a Chinese radish grown in Taiwan (Ikegaya; personal communication). 'Daibaika' has the H type among the nine variants of *orf138* (Yamagishi and Terachi 2001). This finding demonstrates that 'UK-1' is also a variant of *orf138*. The CMS of H-type *orf138* is used in F_1 varieties bred in Taiwan, as reported by Yamagishi and Terachi (2001). We recently showed that this type is also used in Japanese F_1 varieties (Jikuya et al. 2015). In contrast to the *orf125* of 'Kosena,' H-type *orf138* was not found in Japanese wild radish (Yamagishi and Terachi 2001). A survey on the cytoplasm of East Asian wild radish growing in Taiwan and continental Asia is necessary for establishing the origin of this type.

7.2.3 Appearance of the Ogura Male Sterile Radish

Unlike the other CMS in radish, the variety name for radish in which Ogura (1968) found male sterility was not defined. He reported that male sterile plants were taken from an unknown variety of Japanese radish. Japanese cultivated and wild radishes have contrasting distributions of *orf138* and a fertility restorer gene. Most of the Japanese cultivars lack both *orf138* and the fertility restorer gene, while these two genes are widely distributed in Japanese wild radish. Based on these findings, we speculate that the process for the induction of Ogura male sterility was as follows: reproductive contacts occurred between a wild radish with the Ogura-type cytoplasm and the cultivar that lacks the restorer gene. In the progeny, male sterile segregants appeared and were found by Ogura (1968) 'under an escaped condition' (Yamagishi and Terachi 1994).

Including Ogura CMS, all male sterile radishes described above were identified as variants in radish species. CMS is induced under alloplasmic conditions in which the nuclear genome of a species is combined with the cytoplasm of other species. This typical CMS was obtained in an alloplasmic radish line carrying *Brassica maurorum* cytoplasm (Bang et al. 2011). However, radish contains large variations in the mitochondrial genome within a species, as discussed later in this chapter. Thus, the CMS lines obtained in radish hitherto are regarded as a type of alloplasmic lines even though they are within a species.

7.3 Structural and Sequence Variations in the Radish Mitochondrial Genome Revealed by NGS

7.3.1 Historical Studies on the Mitochondrial Genome of the Ogura Radish

After the pioneering work by Palmer and Herbon (1986), in which the tripartite structure of the radish mitochondrial genome was first reported, sequence and structural variations in radish mitochondrial genomes have since been studied. Previous studies mainly focused on differences between normal and Ogura male sterile radishes due to increasing interest in the molecular basis for male sterility caused by the Ogura cytoplasm (Makaroff and Palmer 1988). For example, Makaroff et al. (1989) investigated the *atp6* locus in the mitochondrial genomes of Ogura and normal radishes and found 5′ and 3′ rearrangements associated with short repeated sequences. Their group then studied and compared the structures of the *atpA* (Makaroff et al. 1990) and *coxI* (Makaroff et al. 1991) loci of Ogura and normal radishes and clarified the nature of rearrangements that had occurred nearby these genes. Although these structural variations or rearrangements appeared to be unrelated to Ogura male sterility, their efforts to reveal differences between the two mitochondrial genomes led to the finding that the gene *orf138* is responsible for male sterility in the Ogura cytoplasm (Krishnasamy and Makaroff 1993). Bonhome et al. (1991, 1992) also investigated a cytoplasmic male sterile *B. napus* cybrid containing the Ogura mitochondrial genome and identified the gene *orf138* as being responsible for its male sterility. A number of studies have since addressed questions on this Ogura (and its related Kosena) CMS system in radish and Brassicaceae crops. Some of the subjects extensively studied include the characterization of the ORF138 protein (Grelon et al. 1994; Krishnasamy and Makaroff 1994; Iwabuchi et al. 1999; Duroc et al. 2005), the origin and differentiation of *orf138* (Yamagishi and Terachi 1997, 2001; Giancola et al. 2007), and the fertility restoration of Ogura/Kosena male sterility (Brown et al. 2003; Desloire et al. 2003; Koizuka et al. 2003; Yasumoto et al. 2009).

7.3.2 Radish Mitochondrial Genomes Revealed by NGS

Similar to any fields in biological science, NGS has revolutionized studies on the mitochondrial genomes of higher plants. The complete nucleotide sequences of the mitochondrial genomes of eight radish cultivars have already been elucidated (Tanaka et al. 2012; Chang et al. 2013; Park et al. 2013; Jeong et al. 2016), and the data obtained has been deposited in the public database to facilitate detailed investigations on sequence and structural variations among radish mitochondrial genomes. These sequence data also may help to infer the evolutionary differentiation of radish mitochondrial genomes.

Table 7.1 lists radish cultivars for which a complete mitochondrial genome sequence was obtained. This table also lists the accession number, its genome size, and a reference for each cultivar for the convenience of readers. The sizes of radish mitochondrial genomes ranged from 239,186 bp for 'DCGMS' to 258,426 bp for 'MS-gensuke', falling within the range of the mitochondrial genome sizes of several Brassicaceae crops. We previously reported the first complete sequences of radish mitochondrial genomes using two isonuclear strains, 'MS-gensuke' and 'Uchiki-gensuke' (Tanaka et al. 2012). 'MS-gensuke' is an experimental strain containing the Ogura male sterile cytoplasm in the genetic background of 'Uchiki-gensuke.' 'Uchiki-gensuke' is a Japanese local cultivar belonging to var. *hortensis* (i.e., Asian big radish with a long and white root) and is used as a maintainer of Ogura male sterility in 'MS-gensuke.' Figure 7.1a, b shows the master circle molecule of mitochondrial genomes of both cultivars. The findings of the sequencing study clearly revealed the gene organization of the two radish mitochondrial genomes.

The 'Uchiki-gensuke' mitochondrial genome contained 33 protein-coding and three rRNA genes, while the 'MS-gensuke' genome contained the same genes and one additional *atp9* (*atp9*-2). Gene content was identically conserved between the two radish mitochondrial genomes and was also identical to that of rapeseed (Handa 2003), except for *atp9*-2 in 'MS-gensuke' and *orf222* in rapeseed genomes. The 'Uchiki-gensuke' genome contains a similar sequence to *atp9*-2; however, the 3′ part of this sequence lacks similarity to *atp9*-2. The *atp9*-2-like sequence in the 'Uchiki-gensuke' genome encodes a 70-amino acid protein that partially lacks the *atp9* domain. Eighteen out of the 33 conserved protein-coding genes produce components of the electron transport chain and ATP synthase: nine subunits of complex I (*nad1*, *nad2*, *nad3*, *nad4*, *nad4L*, *nad5*, *nad6*, *nad7*, and *nad9*), apocytochrome b (*cob*) of complex III, three subunits of complex IV (*cox1*, *cox2*, and *cox3*), and five subunits of complex V (*atp1*, *atp4*, *atp6*, *atp8*, and *atp9*). Five additional proteins are involved in the biogenesis of cytochrome c (*ccmB*,

Table 7.1 Radish cultivars for which the complete mitochondrial genome sequence was elucidated

Cultivar	Abbreviation	Accession no.	Genome size (in bp)	References
Uchiki-gensuke	UCG	AB694743.1	244,036	Tanaka et al. (2012)
MS-gensuke	MSG	AB694744.1	258,426	Tanaka et al. (2012)
Uchiki-gensuke OS40-type	OS40	AP012991.1	259,201	Tanaka et al. (unpubl.)
Black radish	BLR	AP012990.1	239,208	Tanaka et al. (unpubl.)
Aonaga	ANG	AP013077.1	239,717	Tanaka et al. (unpubl.)
WK10039	WK10039	KJ716484.1	244,054	Jeong et al. (2016)
Dongbu cytoplasmic and genic male sterility	DCGMS	KC193578.1	239,186	Park et al. (2013)
Chuanxinhong	CXH	JQ083668.1	239,723	Chang et al. (2013)

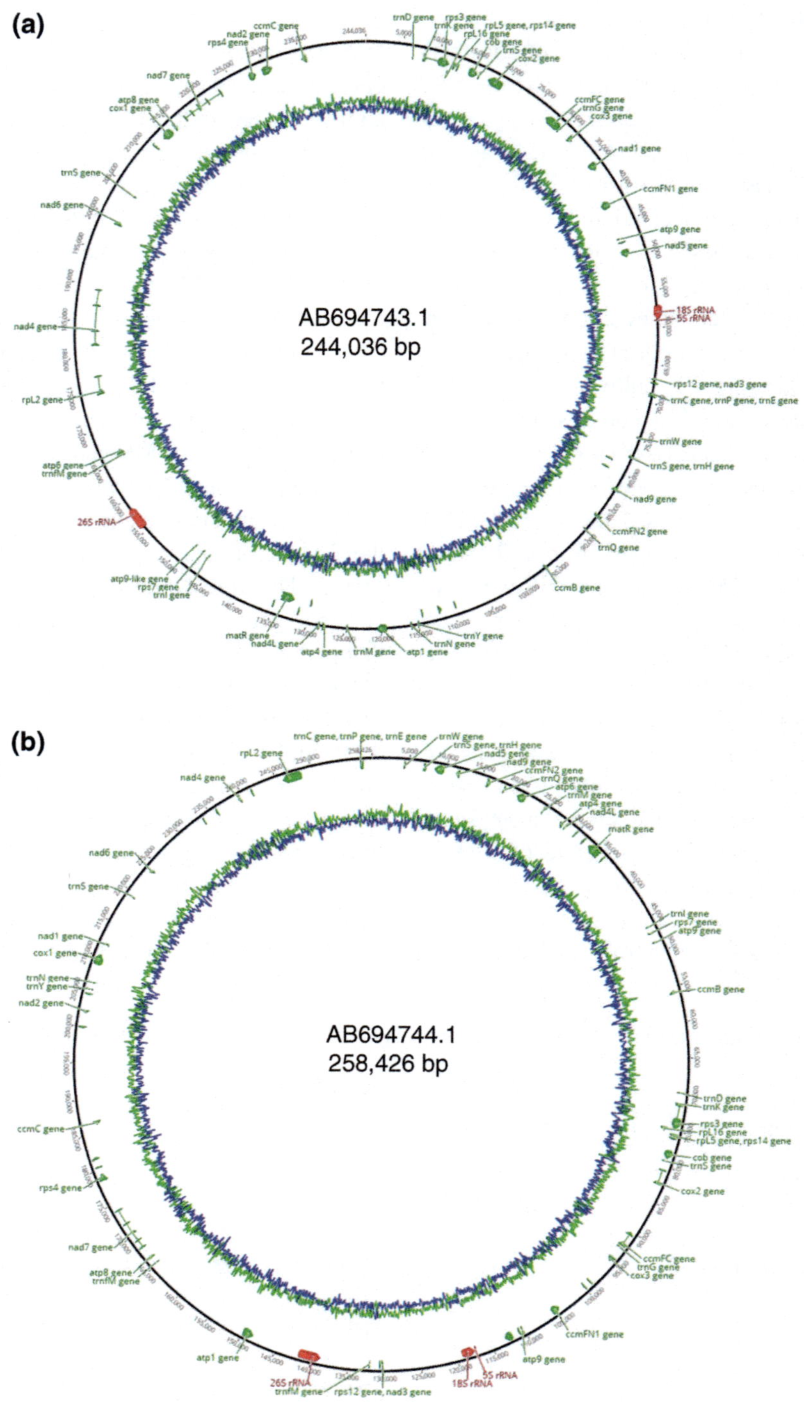

Fig. 7.1 Master circle molecules of mitochondrial genomes for 'Uchiki-gensuke' (**a**) and 'MS-gensuke' (**b**) predicted from complete nucleotide sequences of their mitochondrial DNAs. Protein-coding genes and tRNA genes are shown in *blue* and rRNA genes in *red*. An *inner circle* indicates GC (*blue*) and AT (*green*) contents

ccmC, *ccmFN1*/*ccmFN2*, and *ccmFC*). Another eight genes encode ribosomal proteins (*rpl2*, *rpl5*, and *rpl16*, and *rps3*, *rps4*, *rps7*, *rps12*, and *rps14*). The two remaining genes encode maturase (*matR*) and *orfX*, now named *mttB* (or *tatC*). The gene *ccmFN* is divided into two reading frames in the mitochondrial genomes of radish. Three highly conserved rRNA genes (*rrn5*, *rrn18*, and *rrn26*) are also present in the mitochondrial genomes of radish. The 5′ region of the gene *atp6* is highly divergent between these two radishes due to recombination in the locus, as demonstrated by Makaroff et al. (1989). The 3′ end of *mttB* of 'MS-gensuke' also lacks similarity to that of 'Uchiki-gensuke' due to a 48-bp repeated sequence that is regarded as a minisatellite. Eleven single nucleotide polymorphisms (SNPs) were identified in the nine genes: four were synonymous and seven were non-synonymous. As for tRNA, 'MS-' and 'Uchiki-gensuke' mitochondrial genomes both contain 17 tRNAs, whereas the 'MS-gensuke' genome has an additional *trnfM* (*trnfM*-2). *trnfM*-2, which is located close to the gene *orf138*, has a specific SNP from the other sequenced mitochondrial *trnfM*s in Brassicaceae.

7.3.3 Sequence Variations in Protein-Coding Genes

Table 7.2 summarizes nucleotide sequence variations found in the protein-coding genes of eight radish cultivars. Nineteen SNPs were identified in 13 different genes, i.e., *nad4*, *nad7*, *cox1*, *atp8*, *atp9*, *ccmC*, *ccmFC*, *rpl2*, *rpl16*, *rps3*, *rps4*, *rps12*, and *matR*. In addition to SNPs, a deletion and inversions were detected within the coding regions of the protein-coding genes: a 6-bp deletion in *rps3* of 'WK10039,' a 22-bp inversion in *nad5* of 'CHX,' and a 5-bp inversion in the *rpl2* of the two cultivars 'ANG' and 'CHX.' Two cultivars ('BLR' and 'DCGMS') contained *atp9*-2, which was found in 'MS-gensuke'; however, their *atp9*-2 had an SNP (A vs. G) at the 64th position similar to *atp9*-1. The two cultivars, 'OS40' and 'WK10039,' contained a truncated *atp9*-2 gene similar to 'Uchiki-gensuke,' and the remaining two cultivars, 'ANG' and 'CXH,' completely lacked *atp9*-2.

The rate of nucleotide substitutions markedly varies among mitochondrial, chloroplast, and nuclear DNAs (Wolfe et al. 1987). Except for a few cases, the plant mitochondrial genome evolved more slowly than chloroplast or nuclear genomes. Therefore, as expected, only a moderate number of SNPs (a total of 19) were observed in the coding regions of protein-coding genes. It is important to note that a small deletion and inversions have been identified in the coding regions of some protein-coding genes; however, the phylogenetic and functional significance of these types of intraspecific variations remains to be clarified.

7.3.4 Structures of Radish Mitochondrial Genomes Predicted by NGS

Although the in vivo molecular structure of plant mitochondrial DNA has been a long-standing source of controversy (Sloan 2013), the mitochondrial genomes of radish (and those of Brassicaceae crops) are traditionally depicted as a circular DNA molecule predicted by a master circle model (Palmer and Herbon 1986). In this model, the radish mitochondrial genome takes a tricircular organization, i.e., the genome consists of a 242-kb master circle and two subgenomic circles of 139 and 103 kb (Fig. 7.1b). These subgenomic circles are produced by intramolecular recombination through the 10-kb recombination repeats that are located in direct orientation relative to one another in the master circle (Palmer and Herbon 1986). Since the equimolar stoichiometry of four repeat-containing fragments, i.e., two in a master circle and one in each of the two subgenomic circles, has been observed, three circles are present at the same copy number.

The radish mitochondrial genome assembled by NGS does not appear to conflict with that predicted by the conventional master circle model. For example, a 258,426-bp master circle was assembled using NGS for 'MS-gensuke' with the Ogura-type mitochondrial genome, and

Table 7.2 Nucleotide sequence variations in mitochondrial protein-coding genes of eight radish cultivars

		Complex I			Complex IV		Complex V			
		nad4(1488)[a]	nad5(2010)	nad7(1185)	cox1(1584)		atp8(477)		atp9(225)	atp9-2
Abbreviation	Accession no.	1010	1447–1468	685	117	207	370	449	64	
UCG	AB694743.1	T	GATATGATGATTGGTTTAGGTA	G	G	C	A	T	A	G/truncated
MSG	AB694744.1	C	GATATGATGATTGGTTTAGGTA	G	T	C	C	C	G	G
OS40	AP012991.1	T	GATATGATGATTGGTTTAGGTA	G	G	C	A	T	A	G/truncated
BLR	AP012990.1	C	GATATGATGATTGGTTTAGGTA	A	G	C	A	T	A	G
ANG	AP013077.1	C	GATATGATGATTGGTTTAGGTA	G	T	T	A	T	A	–
WK10039	KJ716484.1	T	GATATGATGATTGGTTTAGGTA	G	G	C	A	T	A	G/truncated
DCGMS	KC193578.1	C	GATATGATGATTGGTTTAGGTA	A	G	C	A	T	A	G
CXH	JQ083668.1	C	TACCTAAACCAATCATCATATC	G	T	T	A	T	A	–

		Cytochrome C biogenesis			Ribosomal protein								
		ccmC(744)	ccmFC		rpl2(1050)			rpl16(540)	rps3(1665)		rps4(1089)		rps12(378)
Abbreviation	Accession no.	351	605(1329)	606	90–94	204	840	492	70–75	551	776	1052	9
UCG	AB694743.1	G	A	G	TTTTC	G	C	G	AGTGAG	G	C	C	G
MSG	AB694744.1	A	G	T	TTTTC	G	T	A	AGTGAG	G	T	C	G
OS40	AP012991.1	G	A	G	TTTTC	G	C	G	AGTGAG	G	C	C	G
BLR	AP012990.1	G	G	T	TTTTC	A	C	A	AGTGAG	A	C	C	G
ANG	AP013077.1	G	G	T	GAAAA	G	C	A	AGTGAG	G	C	T	A
WK10039	KJ716484.1	G	A	G	TTTTC	G	C	G	–	G	C	C	G

(continued)

Table 7.2 (continued)

		Cytochrome C biogenesis			Ribosomal protein								
		ccmC(744)	ccmFC		rpl2(1050)			rpl16(540)	rps3(1665)		rps4(1089)		rps12(378)
Abbreviation	Accession no.	351	605(1329)	606	90–94	204	840	492	70–75	551	776	1052	9
DCGMS	KC193578.1	G	G	T	TTTTC	A	C	A	AGTGAG	A	C	C	G
CXH	JQ083668.1	G	G	T	GAAAA	G	C	A	AGTGAG	G	C	T	A

Abbreviation	Accession no.	Others		No variation[b]	
		matR(2034)		cob(1182)	rps14(303)
		210	488	nad5(2010)	atp6(786)[c]
UCG	AB694743.1	C	C	nad4L(303)	
MSG	AB694744.1	C	C	cox3(798)	
OS40	AP012991.1	C	C	atp1(1524)	
BLR	AP012990.1	C	T	atp4(579)	
ANG	AP013077.1	T	C	ccmB(621)	
WK10039	KJ716484.1	C	C	ccmFN1(1146)	
DCGMS	KC193578.1	C	T	ccmFN2(636)	
CXH	JQ083668.1	T	C	rpL5(558)	

[a]Numerals in parentheses indicate the length of the coding region (in bp)
[b]nad1, cox2, and rps7 genes are excluded from this analysis because of mis-annotations of the genes in the deposited data for some cultivars
[c]A structural variation exists in the 5′ of atp6 in MSG

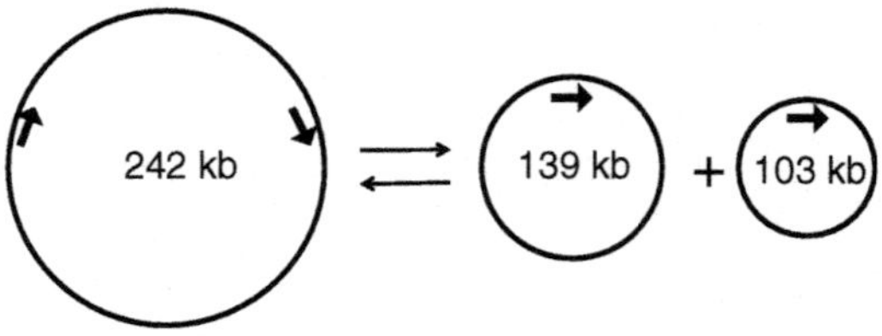

Fig. 7.2 Tripartite structure of the radish mitochondrial genome. *Arrows* indicate the position and relative orientation of recombination repeats. In this model, a 242-kb master circle interconverts with two subgenomic circles of 139 and 103 kb (after Palmer and Herbon 1986)

two subgenomic circles of 130,760 and 127,666 bp, respectively, were predicted from its nucleotide sequence. Similarly, a 244,036-bp master circle and two subgenomic circles of 139,398 and 104,638 bp, respectively, were expected for 'Uchiki-gensuke' (Tanaka et al. 2012). These findings coincide with the restriction map reported almost three decades ago (Makaroff and Palmer 1988). In that study, the 257-kb Ogura master circle was interconverted with subgenomic circles of 130 and 127 kb, and a 242-kb master circle coexisted in the normal radish with subgenomic circles of 139 and 103 kb (Makaroff and Palmer 1988). The radish is known to have one of the smallest mitochondrial genomes in higher plants, and only two direct repeated sequences in a master circle are involved in 'active' recombination that produces two subgenomic circles. These are the reasons why new NGS and old mapping data are consistent with each other. Although additional experiments, e.g., pulse field gel electrophoresis, are required in order to obtain conclusive evidence, the long-standing notion that radish mitochondrial genomes mainly have a tripartite structure currently appears to be accepted.

7.3.5 Relationships Between Eight Radish Mitochondrial Genomes

In order to clarify phylogenetic relationships between mitochondrial genomes of different radish cultivars, a multiple genome alignment analysis was conducted using the software Mauve 2.4.0 (Darling et al. 2010). Figure 7.2 shows sequence comparisons along the entire mitochondrial genomes of eight different cultivars. It is clear that large-scale evolutionary rearrangements such as inversions and translocations occurred among radish mitochondrial genomes. However, closer examinations indicate that the eight radish mitochondrial genomes may be classified into five types (Fig. 7.2). Type 1 consists of the mitochondrial genomes of 'Uchiki-gensuke,' 'WK10039,' and 'OS40'. Type 2 is composed of the mitochondrial genomes of 'Aonaga-daikon' and 'Chuanxinhong'. In this review, Types 3, 4, and 5 are assigned to each mitochondrial genome of 'BLR,' 'DCGMS', and 'MS-gensuke', respectively. Within Type 1, the genome sequence of 'Uchiki-gensuke' is virtually identical to that of 'WK10039.' Table 7.3 lists all sequence differences between the two mitochondrial genomes. Sequence differences were identified at 41 positions in the intergenic or noncoding regions of the genome. It is important to note that more than one half (26 positions) of these differences were detected in mononucleotide repeats (homopolymers). Since the majority of errors in Rosch/454 data are related to homopolymers, true biological variations must be carefully distinguished from simple sequencing errors. If the differences found in homopolymers are ignored, only ten base substitutions and five deletions of a single base are documented in the entire (>244 kb) mitochondrial genomes of the two cultivars. The other Type 1 genome of 'OS40' is 15,164 bp larger than that of 'Uchiki-gensuke.' The size difference between these two genomes is mainly caused by two pairs of direct repeated sequences of 9687 and 8367 bp found in the 'OS40' mitochondrial genome (Fig. 7.3). The 'OS40' mitochondrial genome was originally detected in the Japanese cultivar 'Oosaka-Shijunichi' when we performed a PCR analysis on the 5′ end of *coxI* (Yamagishi and Terachi 2003). The 'OS40' mitochondrial genome was subsequently identified in some individuals of 'Uchiki-gensuke,' which allowed us to consider the 'OS40' type as one of the common mitochondrial genome types found in Japanese radish cultivars.

The complete sequencing of the 'OS40' mitochondrial genome revealed that it contained two pairs of direct repeated sequences. This finding indicates a more complex configuration for the mitochondrial genome of 'OS40' than the tripartite structure; however, it currently remains unclear whether both direct repeated sequences are 'active' in recombination.

The coexistence of the 'Uchiki-gensuke' type and 'OS40' type mitochondrial genomes in the cultivar 'Uchiki-gensuke' suggests that unexpectedly large intra-cultivar or intra-varietal variations exist in Japanese radish cultivars, and a large-scale survey may lead to the discovery of a new mitochondrial genome type.

The mitochondrial genome of 'Aonaga-daikon' appears to be almost identical to that of 'Chuanxinhong'; only two base substitutions and two deletions of a single base are present if the eight differences in the homopolymers are excluded (Table 7.4). The material 'Chuanxinhong' used in the study by Chang et al. (2013) was described as 'the common radish that is the most commonly cultivated in Northern China.' Although we obtained 'Aonaga-daikon' in Japan, it is one of the Chinese varieties. Thus, the small number of differences detected between the two mitochondrial genomes is reasonable.

Regarding the types of mitochondrial genomes in radish, a large number of evolutionary rearrangements, e.g., inversions and translocations of syntenic regions, have been identified (Fig. 7.2). These rearrangements prevent the phylogenetic analysis between radish cultivars being deduced from the organization of the mitochondrial genome. As already described, the 'DCGMS' cytoplasm causes male sterility in radish, and its genetic properties such as the inheritance of fertility restoration and profiles of mitochondrial DNA-based molecular markers differ from those of Ogura CMS (Park et al. 2013). Comparative genome sequencing unambiguously proved that the mitochondrial genomes of 'DCGMS' and 'MS-gensuke' differ from each other, and sequencing data will provide the basis for designing new primer pairs that may be used to distinguish between male sterile cytoplasm by PCR. The gene *orf463*, which was previously reported to be responsible for CMS, consists of the 5′ part of *coxI* and a unique sequence of the 'DCGMS' mitochondrial genome (Park et al. 2013). No other radish mitochondrial genomes contain a homologous sequence to *orf463*, except for that of 'Black radish' (BLK). A previous study found 'DCGMS' in an accession collected from Uzbekistan (Lee et al. 2008). It will be of interest and importance to investigate whether our 'Black radish' ('Kuro-daikon' in Japanese) with the Type 3 mitochondrial genome induces male sterility in some radish cultivars.

7.4 Conclusions

Almost three decades ago, Palmer and Herbon (1986) published a study entitled 'Plant mitochondrial DNA evolves rapidly in structure but slowly in sequence' in which the tempo and mode of mitochondrial DNA evolution were examined in six crucifer species including radish. Even in the era of NGS, the meaning of this symbolic title holds true when considering the evolution of plant mitochondrial genomes. Many structural variations have been identified among the radish mitochondrial genomes sequenced to date. However, gene contents and nucleotide

Table 7.3 Nucleotide differences between 'Uchiki-gensuke' and 'WK10039' mitochondrial genomes

Abbreviation	Accession no.	6767[a]	6768	19617–19624	19938–19943	24409	32380	37611–37619	44348	44355	44459–44465
UCG	AB694743	G	A	C(7)[b]	T(5)	C	A	A(8)	C	A	A(6)
WK10039	KJ716484	A	G	C(8)	T(6)	G	–	A(9)	A	C	A(7)

Abbreviation	Accession no.	74844	74847	74852	74854	74862–74875	77031–77038	97262–97272	113215–113223	114887–114893	115110–115118
UCG	AB694743	T	A	G	T	A(13)	A(7)	T(10)	A(8)	A(6)	A(8)
WK10039	KJ716484	–	–	T	G	A(14)	A(8)	T(11)	A(9)	A(7)	A(9)

Abbreviation	Accession no.	116080	119054–119062	119199–119207	120937–120944	129146–129160	129692–129696	130229–130238	148635–148639	176557	178376–178383
UCG	AB694743	C	T(8)	T(8)	T(7)	A(14)	C(4)	A(9)	T(4)	T	C(7)
WK10039	KJ716484	A	T(9)	T(9)	T(8)	A(15)	C(5)	A(10)	T(5)	–	C(8)

Abbreviation	Accession no.	186505–186517	186522	187154–187159	204401–204408	205487–205496	210623–210629	211675	211676	220458–220464	220846–220854
UCG	AB694743	A(12)	A	G(5)	A(7)	T(9)	A(7)	T	A	T(6)	T(8)
WK10039	KJ716484	A(13)	–	G(6)	A(8)	T(10)	A(6)	A	T	T(7)	T(9)

Abbreviation	Accession no.	236123–236128
UCG	AB694743	A(5)
WK10039	KJ716484	A(6)

[a]The nucleotide positions in AP013077

[b]The numerals in parentheses indicate the number of mononucleotides

sequences are almost identical between genomes, with a few exceptions. One of the findings revealed by NGS is that the structural variations detected are not only as a consequence of simple genome shuffling such as inversions and translocations, but are often associated with the acquisition of a new sequence of unknown origin. This new sequence may be a good source for a new ORF related to male sterility; *orf138* in Ogura and *orf463* in 'DCGMS' are examples. Another interesting finding is that some radish cultivars may contain a mitochondrial genome with an exceptional structure. As described earlier, the master circle of the radish mitochondrial genome typically possesses a pair of direct repeated sequences and homologous recombination results in a tripartite structure. However, two pairs of direct repeated sequences are present in the master circle of 'OS40' (Fig. 7.3), and a pair of inverted repeated sequences in 'DCGMS'. In the former case, the presence of at least five circles (one master and four subgenomic circles) is expected, and, in the latter case, two master molecules differing in the relative orientation of their single copy sequences have been proposed; however, further studies are warranted to confirm this (Fig. 7.4).

Subliminary fragments or sublimons exist in plant mitochondria and play a significant role in the evolution of the plant mitochondrial genome (Small et al. 1987). Although previous findings demonstrated that sublimons exist in radish mitochondrial genomes (e.g., Lee et al. 2008), these minor molecules may not be reflected in the current master molecule. Since substoichiometric shifting in mitochondrial genomes is influenced by the nuclear genetic background (Abdelnoor et al. 2003), it will be interesting to clarify whether the configuration of mitochondrial genomes represented by the master circle is influenced by nuclear genotypes. As for the phylogenetic use of radish mitochondrial genomes, it will be important to distinguish 'true' evolutionary rearrangements from those caused by in vivo recombination. More NGS data will clarify the evolutionary and phylogenetic relationships between the various mitochondrial genomes of radish.

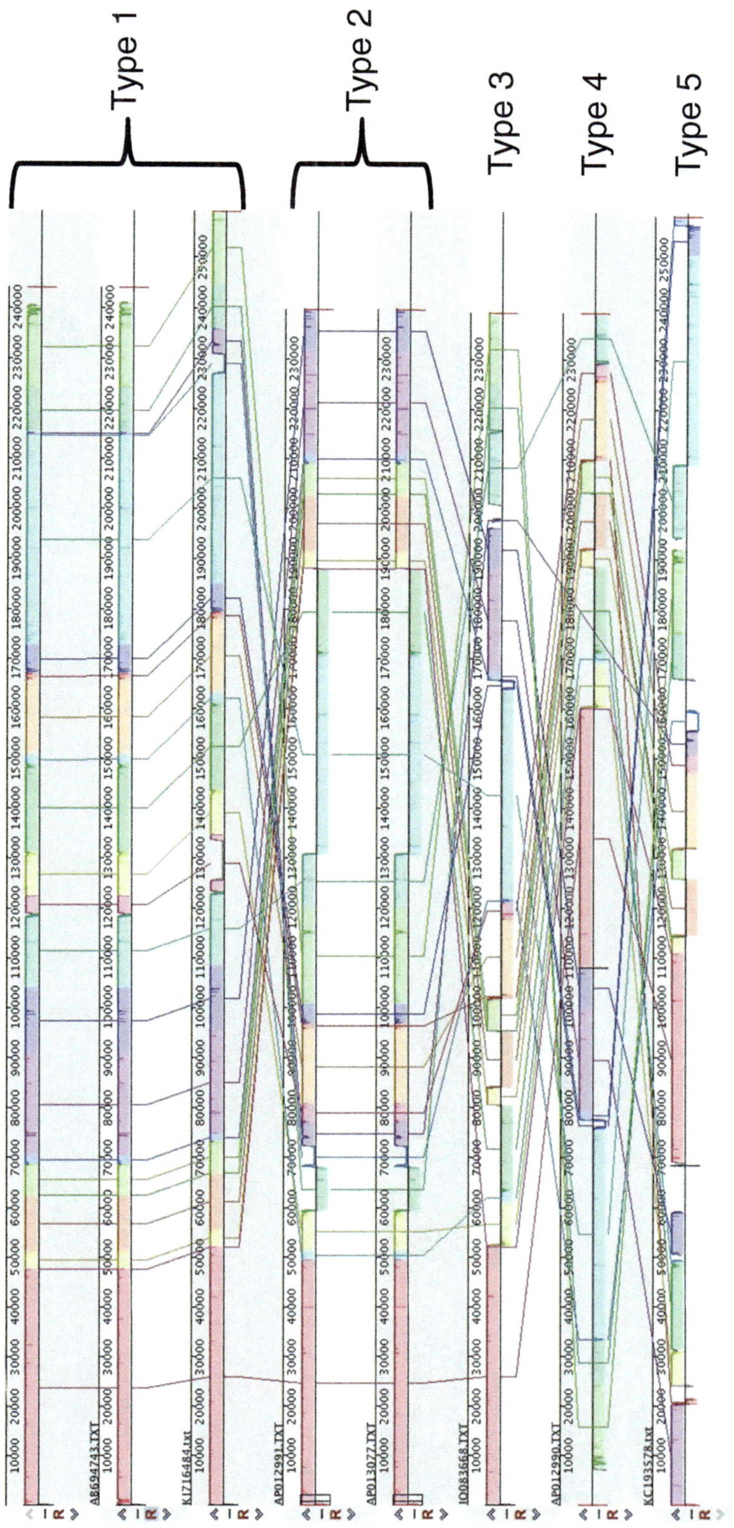

Fig. 7.3 Multiple alignment of radish mitochondrial genomes using the program Mauve. Eight radish mitochondrial genomes are aligned and grouped into five different types according to the degree of the colinearity of syntenic regions shown in the same color. The blocks located below the *middle lines* in type 2–5 genomes are in the reverse complement orientation relative to the reference genome

Table 7.4 Nucleotide differences between 'Aonaga' and 'Chuanxinhong' mitochondrial genomes

Abbreviation	Accession no.	8756–8766[a]	21929–21934	62129–62137	80771	130719	139797–139806	140885–140893	157451–157461	209845–209857	215466–215478	221885	237869
ANG	AP013077	A(10)[b]	T(5)	A(9)	T	–	A(9)	T(8)	A(10)	A(12)	A(12)	T	T
CXH	JQ083668	A(11)	T(6)	A(8)	C	A	A(10)	T(9)	A(11)	A(13)	A(13)	G	–

[a]The nucleotide positions in AP013077
[b]The numerals in parentheses indicate the number of mononucleotides

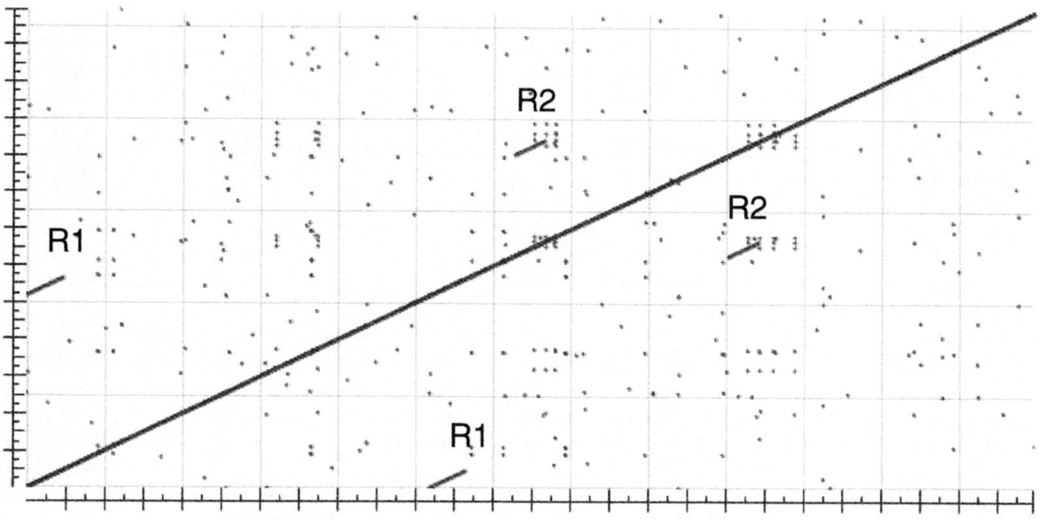

Fig. 7.4 Self-comparison of a mitochondrial genome sequence of 'OS40'. A dot matrix plot is made by nucleotide BLAST at https://blast.ncbi.nlm.nih.gov/Blast.cgi. Two direct repeated sequences are marked as *R1* and *R2*

References

Abdelnoor RV, Yule R, Elo A, Christensen AC, Meyer-Gauen G, Mackenzie SA (2003) Substoichiometric shifting in the plant mitochondrial genome is influenced by a gene homologous to *MutS*. Proc Natl Acad Sci USA 100:5968–5973

Albaum M, Lührs R, Trautner J, Abel WO (1995) The Tokumasu radish mitochondrial genome contains two complete *atp9* reading frames. Plant Mol Biol 29:179–185

Bang SW, Tsutsui K, Shim S, Kaneko Y (2011) Production and characterization of the novel CMS line of radish (*Raphanus sativus*) carrying *Brassica maurorum* cytoplasm. Plant Breed 130:410–412

Bonhomme S, Budar F, Férault M, Pelletier G (1991) A 2.5 kb *Nco*I fragment of Ogura radish mitochondrial DNA is correlated with cytoplasmic male-sterility in *Brassica* cybrids. Curr Genet 19:121–127

Bonhomme S, Budar F, Lancelin D, Small I, Defrance M-C, Pelletier G (1992) Sequence and transcript analysis of the *Nco*2.5 Ogura-specific fragment correlated with cytoplasmic male sterility in *Brassica* cybrids. Mol Gen Genet 235:340–348

Brown G-G, Formanová N, Jin H, Wargachuk R, Dendy C, Patil P, Laforest M, Zhang J, Cheung W-Y, Landry B-S (2003) The radish *Rfo* restorer gene of Ogura cytoplasmic male sterility encodes a protein with multiple pentatricopeptide repeats. Plant J 35:262–272

Chang S, Chen J, Wang Y, Gu B, He J, Chu P, Guan R (2013) The mitochondrial genome of *Raphanus sativus* and gene evolution of Cruciferous mitochondrial types. J Genet Genomics 40:117–126

Darling AE, Mau B, Perna NT (2010) progressiveMauve: Multiple genome alignment with gene gain, loss and rearrangement. PLoS ONE 5:e11147

Desloire S, Gherbi H, Laloui W, Marhadour S, Clouet V, Cattolico L, Falentin C, Giancola S, Renard M, Budar F, Small I, Caboche M, Delourme R, Bendahmane A (2003) Identification of the fertility restoration locus, *Rfo*, in radish, as a member of the pentatricopeptid-repeat protein family. EMBO Rep 4:588–594

Duroc Y, Gaillard C, Hiard S, Defrance M-C, Pelletier G, Budar F (2005) Biochemical and functional characterization of ORF138 a mitochondrial protein responsible for Ogura cytoplasmic male sterility in *Brassiceae*. Biochimie 87:1089–1100

Giancola S, Rao Y, Chaillou S, Hiard S, Martin-Canadell A, Pelletier G, Budar F (2007) Cytoplasmic suppression of Ogura cytoplasmic male sterility in European natural populations of *Raphanus raphanistrum*. Theor Appl Genet 114:1333–1343

Grelon M, Budar F, Bonhomme S, Pelletier G (1994) Ogura cytoplasmic male-sterility (CMS)-associated *orf138* is translated into a mitochondrial membrane polypeptide in male-sterile *Brassica* cybrids. Mol Gen Genet 243:540–547

Handa H (2003) The complete nucleotide sequence and RNA editing content of the mitochondrial genome of rapeseed (*Brassica napus L.*): comparative analysis of the mitochondrial genomes of rapeseed and *Arabidopsis thaliana*. Nucl Acids Res 31:5907–5916

Ikegaya Y (1986a) Practical cytoplasmic male sterile line "UK-1" obtained from Chinese radish. Jpn J Breed 36 (Sup 1):104–105

Ikegaya Y (1986b) Frequent appearance of cytoplasmic male sterile plants in a radish cultivar "Kosena". Jpn J Breed 36(Sup 2):106–107

Iwabuchi M, Koizuka N, Fujimoto H, Sakai T, Imamura J (1999) Identification and expression of the kosena radish (*Raphanus sativus* cv. Kosena) homologue of the ogura radish CMS-associated gene, *orf138*. Plant Mol Biol 39:183–188

Jeong Y-M, Chung W-H, Choi AY, Mun J-H, Kim N, Yu H-J (2016) The complete mitochondrial genome of cultivated radish WK10039 (*Raphanus sativus* L.). Mit DNA Part A 27:941–942

Jikuya M, Terachi T, Yamagishi H (2015) Variations of the Ogura cytoplasmic male-sterility gene (*orf138*) and its fertility restorer genes in cultivated radishes. Breed Res 17(Sup 2):220

Koizuka N, Imai R, Fujimoto H, Hayakawa T, Kimura Y, Kohno-Murase J, Sakai T, Kawasaki S, Imamura J (2003) Genetic characterization of a pentatricopeptide repeat protein gene, *orf687*, that restores fertility in the cytoplasmic male-sterile Kosena radish. Plant J 34:407–415

Krishnasamy S, Makaroff C-A (1993) Characterization of the radish mitochondrial *orfB* locus: possible relationship with male sterility in Ogura radish. Curr Genet 24:156–163

Krishnasamy S, Makaroff C-A (1994) Organ-specific reduction in the abundance of a mitochondrial protein accompanies fertility restoration in cytoplasmic male-sterile radish. Plant Mol Biol 26:935–946

Lee Y-P, Park S, Lim C, Kim H, Ahn Y, Sung S-K, Yoon M-K, Kim S (2008) Discovery of a novel cytoplasmic male-sterility and its restorer lines in radish (*Raphanus sativus* L.). Theor Appl Genet 117:905–913

Makaroff C-A, Palmer J-D (1988) Mitochondrial DNA rearrangements and transcriptional alterations in the male-sterile cytoplasm of Ogura radish. Mol Cell Biol 8:1474–1480

Makaroff C-A, Apel I-J, Palmer J-D (1989) The *atp6* coding region has been disrupted and a novel reading frame generated in the mitochondrial genome of cytoplasmic male-sterile radish. J Biol Chem 264:11706–11713

Makaroff C-A, Apel I-J, Palmer J-D (1990) Characterization of radish mitochondrial *atpA*: influence of nuclear background on transcription of *atpA*-associated sequences and relationship with male sterility. Plant Mol Biol 15:735–746

Makaroff C-A, Apel I-J, Palmer J-D (1991) The role of *coxI*-associated repeated sequences in plant mitochondrial DNA rearrangements and radish cytoplasmic male sterility. Curr Genet 19:183–190

Nahm S-H, Lee H-J, Lee S-W, Joo G-Y, Harn C-H, Yang S-G, Min B-W (2005) Development of a molecular marker specific to a novel CMS line in radish (*Raphanus sativus* L.). Theor Appl Genet 111:1191–1200

Ogura H (1968) Studies on the new male-sterility in Japanese radish, with special reference to the utilization of this sterility towards the practical raising of hybrid seeds. Mem Fac Agric Kagoshima Univ 6:39–78

Palmer J-D, Herbon L-A (1986) Tricircular mitochondrial genomes of *Brassica* and *Raphanus*: reversal of repeat configurations by inversion. Nucl Acids Res 14 (24):9755–9764

Park JY, Lee Y-P, Lee J, Choi B-S, Kim S, Yang T-J (2013) Complete mitochondrial genome sequence and identification of a candidate gene responsible for cytoplasmic male sterility in radish (*Raphanus sativus* L.) containing DCGMS cytoplasm. Theor Appl Genet 126:1763–1774

Paulmann W, Röbbelen G (1988) Effective transfer of cytoplasmic male sterility from radish (*Raphanus sativus* L.) to rape (*Brassica napus* L.). Plant Breed 100:299–309

Sakai T, Imamura J (1992) Alteration of mitochondrial genomes containing *atpA* genes in the sexual progeny of cybrids between *Raphanus sativus* cms line and *Brassica napus* cv. Westar. Theor Appl Genet 84:923–929

Sloan D-B (2013) One ring to rule them all? Genome sequencing provides new insights into the 'master circle' model of plant mitochondrial DNA structure. New Phytol 200:978–985

Small ID, Isaac PG, Leaver CJ (1987) Stoichiometric differences in DNA molecules containing the *atpA* gene suggest mechanisms for the generation of mitochondrial genome diversity in maize. EMBO J 6:865–869

Tanaka Y, Tsuda M, Yasumoto K, Yamagishi H, Terachi T (2012) A complete mitochondrial genome sequence of Ogura-type male-sterile cytoplasm and its comparative analysis with that of normal cytoplasm in radish (*Raphanus sativus* L.). BMC Genom 13:352

Tokumasu S (1951) Male-sterility in Japanese radish (*Raphanus sativus* L.). Sci Bull Fac Agric Kyushu Univ 13:83–89

Wolfe KH, Li WH, Sharp PM (1987) Rates of nucleotide substitution vary greatly among plant mitochondrial, chloroplast, and nuclear DNAs. Proc Natl Acad Sci USA 84:9054–9058

Yamagishi H, Terachi T (1994) Molecular and biological studies on male sterile cytoplasm in Cruciferae. II. The origin of Ogura male sterile cytoplasm inferred from the segregation pattern of male sterility in the F_1 progeny of wild and cultivated radishes (*Raphanus sativus* L.). Euphytica 80:201–206

Yamagishi H, Terachi T (1997) Molecular and biological studies on male-sterile cytoplasm in the Cruciferae. IV. Ogura-type cytoplasm found in the wild radish, Raphanus raphanistrum. Plant Breed 116:323–329

Yamagishi H, Terachi T (2001) Intra- and inter-specific variations in the mitochondrial gene *orf138* of Ogura-type male-sterile cytoplasm from *Raphanus sativus* and *Raphanus raphanistrum*. Theor Appl Genet 103:725–732

Yamagishi H, Terachi T (2003) Multiple origins of cultivated radishes as evidenced by a comparison of the structural variations in mitochondrial DNA of *Raphanus*. Genome 46:89–94

Yasumoto K, Terachi T, Yamagishi H (2009) A novel *Rf* gene controlling fertility restoration of Ogura male sterility by RNA processing of *orf138* found in Japanese wild radish and its STS markers. Genome 52:495–504

Gene Expression Profiles During Tuberous Root Development

8

Yuki Mitsui

Abstract

Great variation in root size and shape can be observed in local and commercial varieties of radish. At present, little is known about the underlying molecular mechanisms that determine the size and shape of thickening roots not only in radish but also in other root crops. This chapter first summarizes the properties of vegetative growth, developmental patterns, and anatomical structures of radish tuberous roots. Next, genes expressed in radish roots, particularly in the key developmental stages and proliferating tissues during tuberization, are presented based on several studies using next-generation sequencing data. Transcriptome analyses revealed that genes and pathways related to starch and sucrose metabolism are particularly activated in thickening roots and their cell proliferating tissues. The examinations of expression patterns of genes involved in the sucrose metabolism pathway during root development revealed that genes encoding sucrose synthase, a key enzyme involved in sucrose cleavage, shows consistently high expression patterns in radish thickening roots, whereas low expression was observed in young seedling roots, suggesting that this enzyme plays a critical role in radish root tuberization.

8.1 Introduction

Radish roots develop deep in the ground and grow remarkably larger than those of other plants. Its tuberous roots are used as vegetables around the world. Radish was domesticated in the eastern Mediterranean area and subsequently spread to Asia. Throughout its long history of domestication, a large number of local and commercial varieties of radish have been developed in each region. The most remarkable characteristic of radish is its tuberous roots, which vary greatly in shape. The root shape, that is, round, round-cylindrical, long-cylindrical, thick-cone, thin-cone, or long-cone, characterizes each cultivar (Fig. 8.1). A wide variation in root size is also

Y. Mitsui (✉)
Tokyo University of Agriculture, 1-1-1, Sakuragaoka, Setagaya-Ku, Tokyo 156-8502, Japan
e-mail: y3mitsui@nodai.ac.jp

T. Nishio and H. Kitashiba (eds.), *The Radish Genome*,
Compendium of Plant Genomes, DOI 10.1007/978-3-319-59253-4_8

Fig. 8.1 Diversity of radish cultivars

observed. For example, 'Sakurajima-daikon' is a Japanese local radish variety, the tuberous root of which reaches a diameter of more than 30 cm. Conversely, another Japanese local variety, 'Moriguchi-daikon,' has an extremely long tuber with a length of 1–2 m (with a width of approximately 2 cm). In Western countries, radish is commonly recognized as a small-rooted vegetable. In East Asia, where the biggest producers and consumers of radish are located, the large-rooted varieties are produced much more widely than the small-rooted ones.

For root vegetables, the root shape and size are important traits that directly influence the quality of the product, including the efficiency of transportation, processing, and consumer appeal. Thus far, little is known about the underlying molecular mechanisms that determine the size and shape of thickening roots, not only in radish but also in other root crops. Understanding the molecular mechanisms that control tuberous storage root formation and radish development will contribute to the fields of breeding, food production, medicinal care, and plant root organogenesis. Next-generation sequencing (NGS) technology has provided a novel method for transcriptome analysis (Schuster 2007; Shendure and Ji 2008; Wang et al. 2009). RNA sequencing (RNA-Seq) analysis, which is based on NGS, does not depend on prior information about the genomic sequence of the target species and is a powerful tool for analyzing comprehensive gene expression profiles. This chapter first explains the physiology and developmental process of the radish root. Next, the gene expression patterns in radish roots in the key developmental stages and proliferating tissues during tuberization based on several NGS studies are reviewed to identify genes and pathways that are involved in tuberous root formation and development.

8.2 Physiology of Radish Root Development

Radish prefers to grow in a cool climate; a suitable temperature is around 20 °C, although the range differs between cultivars. Tuberous root development is controlled by complex interactions involving genetic, environmental, and physiological factors. The effects of soil condition on root development have been well investigated. As the main root develops deep in the ground, the soil should be sufficiently soft. Radish varieties with various root forms have been bred to adapt to the soil environment in each region. In deeply cultivated soil with a water-holding capacity and good drainage, such as sandy loam, alluvial soil, and volcanic ash soil, long-rooted varieties have been preferably bred. In shallow soil, root development is suppressed and short-rooted varieties have been selected.

The main root (primary root) is the first of the radish roots to develop, followed by the lateral roots (secondary roots), which arise from the primary root. Tertiary roots also emerge from the secondary roots, resulting in a wide-ranging root system. Secondary and tertiary roots are very thin compared with the main root, which functions as a storage root (i.e., a sink organ). Soil with poor physical characteristics (shallowly cultivated, with large stones or lumps of soil beneath the main root, or containing immature compost and fertilizer) causes an abnormal, branched form of the root. The use of an old seed with low vitality and damage by nematodes also causes branched roots. Branched roots are caused by damage to the meristem on the main root. An extremely curved hypocotyl just after germination also causes branched roots.

Vegetative growth of radish is largely divided into four stages: germination, early seedling, rosette, and maturity (Fig. 8.2). First, radish seeds germinate approximately 3 days after absorbing water and the cotyledon spreads 7–8 days after germination (DAG). Thereafter, rosette leaves are differentiated one by one in approximately 2–3 days, spreading horizontally until approximately 10 DAG; this is referred to as the early seedling stage. During the early seedling stage, root growth mostly occurs through elongation. In a few DAG, the main root grows significantly, reaching tens of centimeters below the ground. Next, many thin lateral roots branch from the main root and the third and fourth roots are divided, forming a fairly deep spread root system. In the maturity stage, the main root reaches approximately 200 cm and

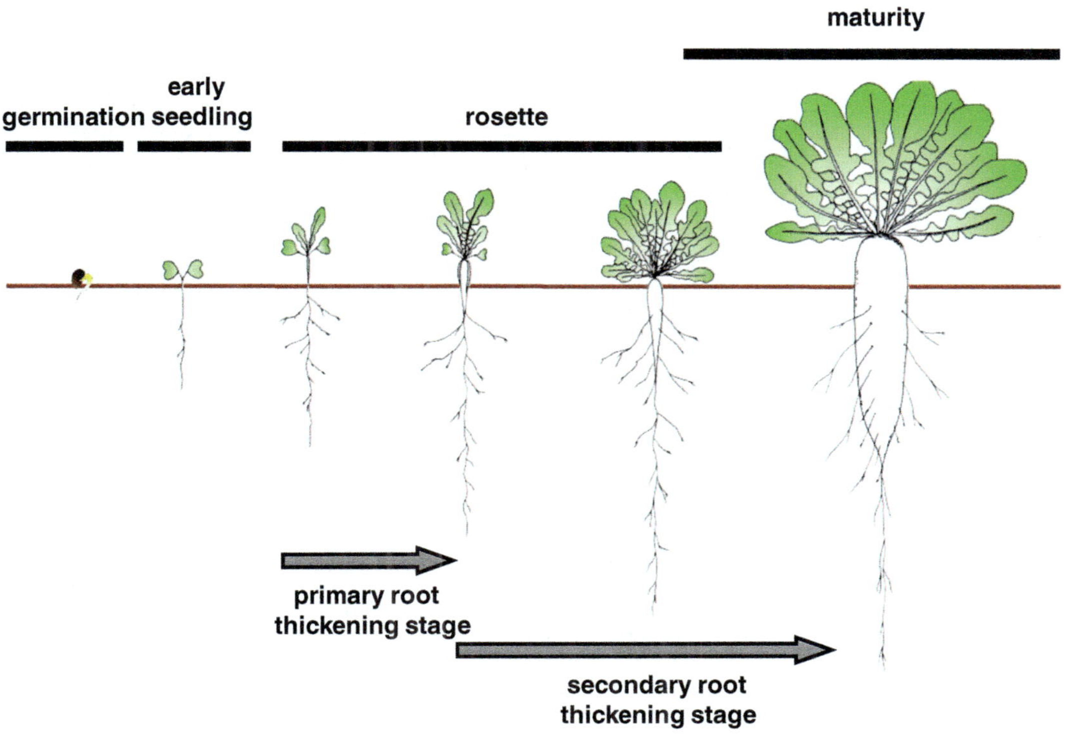

Fig. 8.2 Vegetative growth of radish (*Raphanus sativus* cv. 'Aokubi')

lateral roots reach approximately 100 cm in the large-rooted varieties.

Root tuberization of radish is largely divided into two stages: the primary thickening stage and the secondary thickening stage. At approximately 10 DAG, rosette leaves appear and root tuberization initiates, although the timing of tuber initiation differs slightly among cultivars; this is the beginning of the primary root thickening stage. By developing rosette leaves, photosynthetic activities gradually increase, leading to root enlargement. The radish root is comprised of two anatomically distinct parts. The upper part originates from the hypocotyl on which lateral roots are absent. The lower part consists of true root tissue, and lateral roots are developed in this part. Root tuberization is caused by the enlargement of both the hypocotyl and the root. The stem is barely present between the leaves and hypocotyl.

Anatomically, roots in the early seedling stage are mostly comprised of 6–7 layers of cortex cells; the endodermis and pericycle are located inside (Mitsui et al. 2015; Fig. 8.3a). The primary phloem and xylem are differentiated inside the pericycle. The tissue structures of the hypocotyl and root differ slightly, especially the upper part of the hypocotyl, which is structurally similar to the stem (Hayward 1938). At the root tip, cell division occurs in the cambium, which is located at the front end of the root crown, leading to root elongation by increasing the cell number. When two to three leaves appear, the root vascular cambium is formed between the primary xylem and phloem (Fig. 8.3b). After the cambium is formed, the secondary xylem and phloem are gradually differentiated inside and outside of the cambium, respectively. Simultaneously, primary root thickening begins. Cell division actively occurs in the xylem parenchyma and vessels are arranged radially. Enlargement of radish roots is mainly caused by the development of xylem tissue, in which proliferating cells are supplied inward from the vascular cambium layer. Conversely, carrot is an example in which root enlargement occurs due to the development of phloem tissues. The phloem tissues of radish roots are very thin.

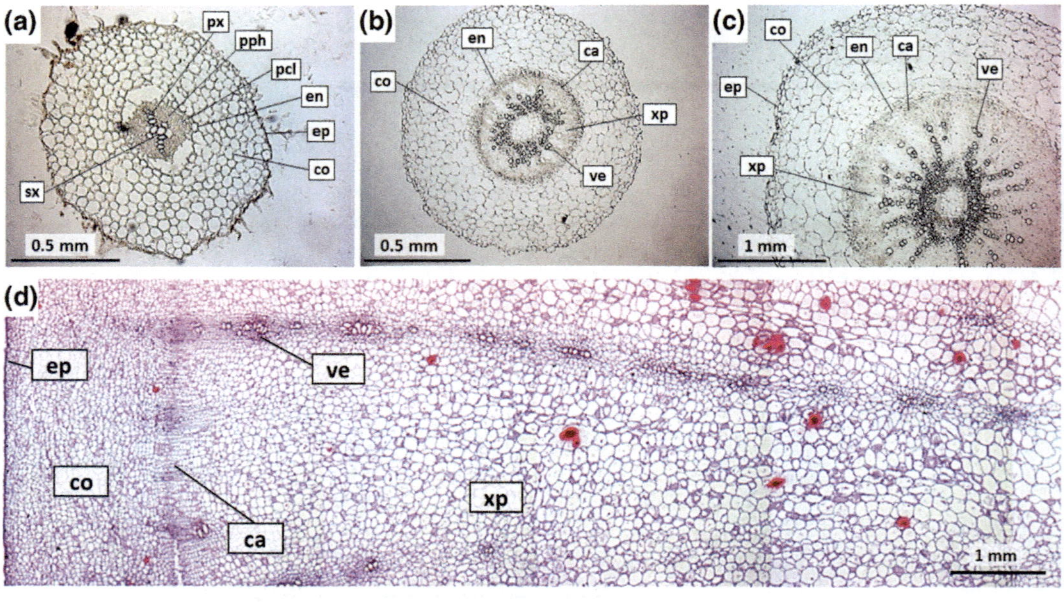

Fig. 8.3 Transverse section of thickening roots of radish (*Raphanus sativus* cv. 'Aokubi') modified from Mitsui et al. (2015). **a** 7 days after germination (DAG), **b** 14 DAG, **c** 20 DAG, **d** 40 DAG. *px* primary xylem, *pph* primary phloem, *pcl* pericycle, *en* endodermis, *ep* epidermis, *co* cortex, *sx* secondary xylem, *xp* xylem parenchyma, *ca* vascular cambium

The primary root thickening stage continues until approximately 3–4 weeks after germination; the root slowly thickens in this period. When three to four rosette leaves appear, the primary cortex splits due to the pressure of the developing stele because the primary cortex cells cannot divide and expand (Fig. 8.3c). The surface of the hypocotyl begins to tear and ultimately spreads throughout the entire root. The outer side of the pericycle develops to form a secondary epidermis and cortex tissues of the thickening root. Secondary cortex tissue, located outside of the vascular cambium, forms a thin layer compared to the xylem tissue. Inside the vascular cambium, vessels are arranged concentrically and are connected to the xylem parenchyma.

Cortex splitting is the sign of entry into the secondary root thickening stage. Until this point, there are only small differences in root morphology among cultivars. In the secondary thickening stage, root tuberization progresses more rapidly and the root weight continues to increase. The rapid tuberization is mainly caused by the development of the vascular cambial layer, which vigorously supplies parenchyma cells to xylem tissues. During the late thickening stage, parenchyma cells in the xylem vigorously differentiate and increase the cell size (Fig. 8.3d). The parenchyma cells in the middle part of the xylem are remarkably enlarged. In this way, xylem parenchyma cells are continuously differentiated and enlarged to promote root tuberization. Meanwhile, the thickness of the cortex increases slowly.

Furthermore, some parenchyma cells in the xylem and ray tissues proliferate in an indeterminate direction, forming groups of proliferating cells. These groups of cells create a new cambium that differentiates into the phloem inside and the xylem outside, forming concentric vascular bundles (Fig. 8.4). This secondary cambium emerges at the beginning of the vessel's concentric arrangement. Differentiation of the secondary cambium begins from the inner part of the xylem and extends to the outer part, reaching the cambial layer in thick-rooted varieties (Fig. 8.5). In thin-rooted varieties, however, only sieve elements are formed in the secondary cambium-like structure. The thick-rooted varieties reach maturity and can be harvested at approximately 60–80 DAG. In the fully matured root, the differentiation ability of parenchyma cells begins to decline. The cell membrane and vessels of parenchyma cells become woody and air holes emerge in the root tuber.

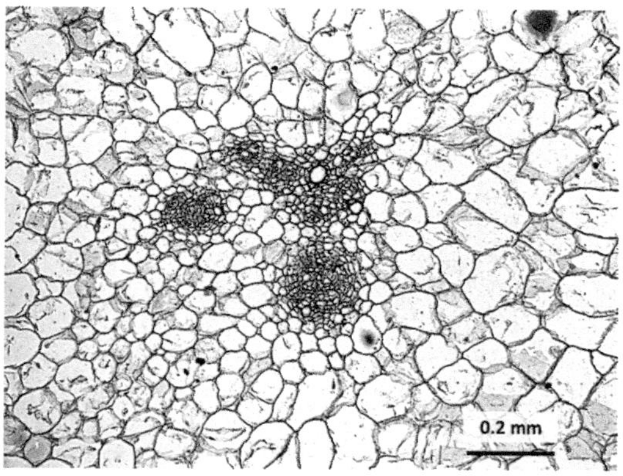

Fig. 8.4 Secondary cambium differentiated in 40 DAG roots of radish (*Raphanus sativus* cv. 'Hatsuka')

8.3 Differentially Expressed Genes Among Roots in Key Developmental Stages

In recent years, NGS technologies have been accelerating whole-genome and RNA sequencing of radish species. Kitashiba et al. (2014) first reported the draft genome sequence of *Raphanus sativus* cv. 'Aokubi-daikon', the most popular Japanese radish strain with a long and thick root, and predicted 61,572 gene models. The following year, the draft genome sequence of another strain of *R. sativus* cv. 'Aokubi-daikon' was reported by Mitsui et al. (2015); they constructed 64,657 gene models. Moghe et al. (2014) determined the genome sequence of wild radish *Raphanus raphanistrum* and the transcriptome of floral organ (flower buds) using NGS by investigating duplicate gene functions among Brassicaceae species. They also sequenced expressed sequence tags (ESTs) from three *R. sativus* cultivars (var. *sativus*, *caudatus*, and *oleifera*) and

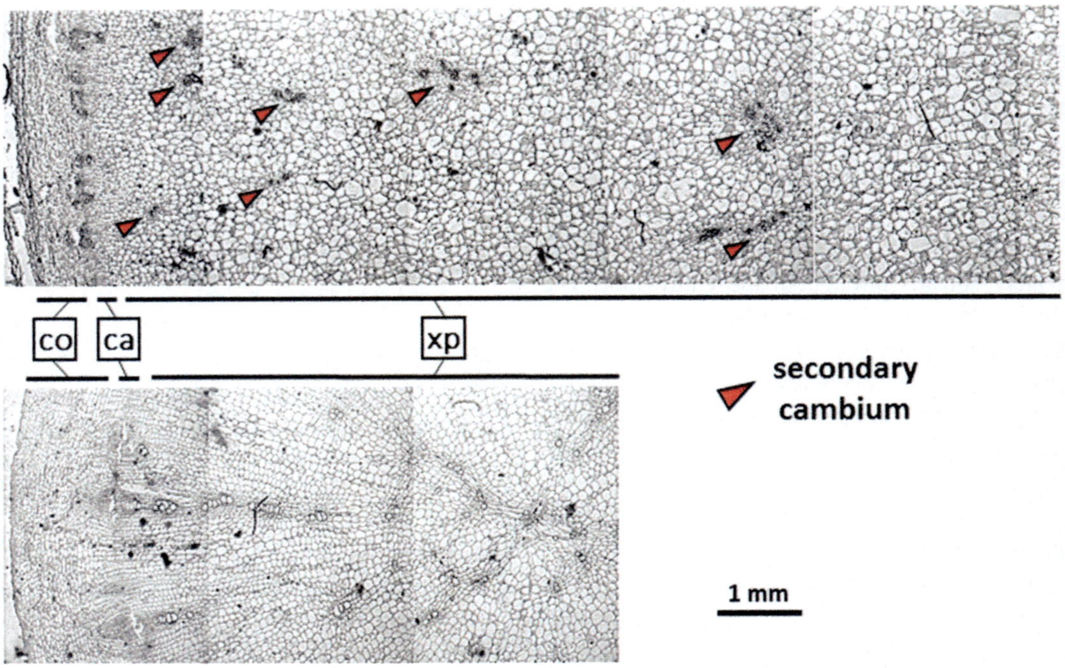

Fig. 8.5 Transverse section of roots of thick-rooted radish variety (cv. 'Hatsuka': *upper*) and thin-rooted variety (cv. 'Moriguchi': *lower*). The position of secondary cambium is indicated by *arrows*. *co* cortex, *ca* cambium, *xp* xylem parenchyma

four *R. raphanistrum* (wild radish) populations (subsp. *raphanistrum* from New York and Central Spain, subsp. *maritimus* from coastal Spain, and subsp. *landra* from France). A total of 185.4 Mb and 310,844 ESTs were generated. In another study, the transcriptome of radish roots at different developmental stages (seedling, thickening, and mature stages) was investigated by Wang et al. (2013a). They obtained 73,084 unigenes and found genes involved in glucosinolate metabolism during tuberous root formation and development. Differentially expressed genes (DEGs) under lead (Pb) stress in radish roots were investigated by Wang et al. (2013b); 33,337 unigenes were obtained. Xie et al. (2015) reported DEGs under chromium stress in radish roots; 30,676 unigenes were obtained. The transcriptome of radish leaves was investigated by Wu et al. (2015). Combined with the transcriptome data of roots from Wang et al. (2013a), they obtained 106,874 unigenes from leaf and root tissues and 7,453 and 15,043 unigenes were specifically identified in the leaf and root tissues, respectively. They reported that 89% of the leaf unigenes matched perfectly to root unigenes, indicating that the transcripts of root and leaf tissues of *R. sativus* are very similar.

Using the *R. sativus* thick-root variety, Wang et al. (2012) compared the gene expression patterns between roots in the early seedling stage (7 DAG) and the late seedling (20 DAG) stage based on RNA-Seq data to detect DEGs between non-thickening and thickening stages; 61,884 and 64,572 unigene contigs were obtained from the early and late seedling roots, respectively. They reported that 21,842 unigenes were identified as DEGs (false discovery rate [FDR]-adjusted significance level of <0.001, an absolute value of $\log_2\text{Ratio} \geq 1$ and twofold or more change) between roots of the early seedling stage and the primary thickening stage. Among them, 13,453 unigenes were up-regulated and 8389 unigenes were down-regulated in late seedling stage roots. Mitsui et al. (2015) presented comprehensive gene expression profiles in the key developmental stages and proliferating tissues of the root of *R. sativus* cv. 'Aokubi'. RNA-Seq analysis revealed DEGs between roots at the early seedling stage (7 DAG roots) and at the primary thickening stage (14 and 20 DAG roots) by analysis of variance (ANOVA) (false discovery rate [FDR]-adjusted significance level of <0.05, fold change >2, and minimum reads per kilobase of transcript per million mapped reads [RPKM] difference of >10). A total of 1352 DEGs were found between 7 DAG roots and 14 DAG roots, and 3032 DEGs were found between 7 DAG roots and 20 DAG roots. In contrast, only 73 DEGs were found between 14 DAG roots and 20 DAG roots. Thus, the DEGs were mostly found between roots at the early seedling stage and those at the primary thickening stage. Further cluster analysis revealed that 1538 genes in 7 DAG roots, 1035 genes in 14 and 20 DAG roots, and 1539 genes in 20 DAG roots were clustered as overexpressed genes. The different criteria for detecting DEGs possibly resulted in the difference in the number of DEGs in the reports by Wang et al. (2012) and Mitsui et al. (2015).

Root tuberization of radish mostly occurs by the supplementation of proliferating cells from the root vascular cambium to xylem parenchyma tissues. The proliferation and enlargement of parenchyma cells in the xylem also contribute to tuberization. Thus, DEGs were detected between these proliferating tissues and cortex tissues in which cell proliferation is limited (Mitsui et al. 2015). As expected, most DEGs were found between proliferating tissues and cortex tissues, suggesting that similar genes can be involved in cell proliferation in cambium and xylem tissues.

8.4 Genes and Pathways Involved in Tuberous Root Formation and Development

The Gene Ontology (GO) database describes gene functional properties and includes information on biological processes (BP), molecular functions (MF), and cellular components (CC) in any organism. GO is an international standardized gene functional classification system that is frequently updated. In the radish genome, 40,705 of 65,457 *R. sativus* gene models were assigned GO terms (Mitsui et al. 2015). GO enrichment analysis can detect significantly enriched GO classification in a large number of expressed genes. Mitsui et al. (2015) showed that the predominant GO terms enriched in genes expressed in the roots of the primary thickening stage were related to ribosomal activity, structural molecule activity, and translation. In proliferating tissues (vascular cambium and xylem parenchyma) of the secondary thickening stage, the predominantly enriched GO terms were membrane activities, transcription, and cell development. In contrast, GO terms involved in stress and stimulus responses, transport, and membrane activities were predominant in the early seedling stage of the root and leaf. Transport- and signaling-related terms were also activated in cortex tissues of thickening roots, suggesting that cortex tissues play a central role in transport and signaling activities because early seedling roots largely comprise cortex tissues.

Pathway-based analysis is useful to obtain a deeper understanding of the biological functions and interactions between genes. Wang et al. (2012) identified 29 significantly enriched pathways for the 21,842 DEGs between early and late seedling roots. Most included metabolic pathways such as carbohydrate, energy, lipid, amino acid, terpenoid, and polyketide metabolism, and the biosynthesis of other secondary metabolites. Among them, the starch and sucrose metabolism pathway and the phenylpropanoid biosynthesis pathway were significantly enriched, as indicated by their p values. Similarly, Mitsui et al. (2015) reported that the starch and sucrose metabolism pathway was the most significantly enriched in thickening roots and proliferating tissues. In addition, Wang et al. (2013a, b) conducted a transcriptome analysis under heavy metal (Pb) stress. They showed that carbohydrate metabolism pathways, including pentose and glucuronate interconversions, and starch and sucrose metabolism, were most significantly down-regulated in the roots under Pb stress. Heavy metal stress usually restricts root tuberization. These results strongly suggest that the carbohydrate metabolism pathway plays a critical role in regulating radish tuberous root formation. In young seedling roots in which tuberization is not occurring, the phenylpropanoid biosynthetic pathway, which synthesizes lignin, is activated (Wang et al. 2012). Mitsui et al. (2015) also reported that the phenylpropanoid biosynthetic pathway is significantly enriched in root cortex tissues. The early seedling roots are largely occupied by cortex tissue. Therefore, the phenylpropanoid biosynthetic pathway is activated in cortex tissue and accumulates lignin, which may help maintain the strength of radish roots.

8.5 Roles of Sucrose Metabolism in Radish Thick-Root Development

Several transcriptome analyses revealed that the starch and sucrose metabolism pathway is significantly activated, indicating that this pathway plays a central role in radish root development. In most plants, assimilated carbon in source leaves is transported as sucrose into sink organs, including roots, tubers, fruit, and seeds. Sucrose is a disaccharide in which glucose and fructose are linked via an *O*-glycosidic bond. To utilize sucrose, this bond should be cleaved to generate the two monosaccharides. This reaction is catalyzed by two enzymes: invertase and sucrose synthase. Invertase catalyzes the cleavage of an alpha-1,4-glycosidic bond in a hydrolysis reaction and generates glucose and fructose. Radish has three types of invertase genes: cell wall invertases (CWIs), cytoplasmic invertases (CINs), and vacuolar invertases (VINs). Mitsui et al. (2015) reported comprehensive expression patterns of genes involved in sucrose metabolism during radish root development (Fig. 8.6). The expression of CINs increased only in the young roots whereas VINs exhibited relatively low expression during all developmental stages. CWIs showed almost no expression. In radish, CINs may play a role in sink initiation because they are involved in the control of cell differentiation and the initiation of new sink structures (Sturm and Tang 1999).

Sucrose synthase is the other key enzyme that catalyzes both the synthesis and the cleavage of sucrose. Unlike invertase, sucrose synthase is a glycosyl transferase, which converts sucrose into UDP-glucose and fructose in the presence of uridine diphosphate (UDP). *R. sativus* has five homologs of sucrose synthase genes (SUSs; Mitsui et al. 2015). Among them, SUS1 shows consistently high expression patterns in radish thickening roots, whereas low expression was observed in early seedling roots (Fig. 8.6). Another study by Rouhier and Usuda (2001) indicated that sucrose synthase is a major enzyme involved in the storage root development of radish. They investigated the localization of sucrose synthase in the root tissues of two radish cultivars with different sink capacities using immunogold labeling. The results revealed a very low level of SuSy (a synonym of SUS) in early seedling roots and the highest level was observed after the entry into the primary root developing stage. A high level of sucrose synthase was found mostly in the companion cells in close proximity to the sieve element. Their observation confirmed that the cleavage of sucrose by sucrose synthase occurs in the companion cells and the products of fructose and UDP-glucose are utilized for sink development. Sucrose synthase was also found in the xylem parenchyma and in some cortical cells. In other plants, such as barley and *Vicia*, similar expressional shifts from invertase to sucrose synthase at the later developmental stages of sink organs have been observed (Borisjuk et al. 2002; Weschke et al. 2003). These results suggest that sucrose synthase specifically facilitates the storage and maturation of sinks. Additionally, several studies have reported that SUS plays a crucial role in the development of sink organs in sweet potato (Martin et al. 1993; Zrenner et al. 1995; Li and Zhang 2003; Baroja-Fernandez et al. 2009) and potato tubers (Martin et al. 1993; Zrenner et al. 1995; Baroja-Fernandez et al. 2009), as well as tomato fruit (Wang et al. 1993). These results confirm that plant sink development is commonly regulated in sucrose metabolism and that sucrose synthase is critically involved.

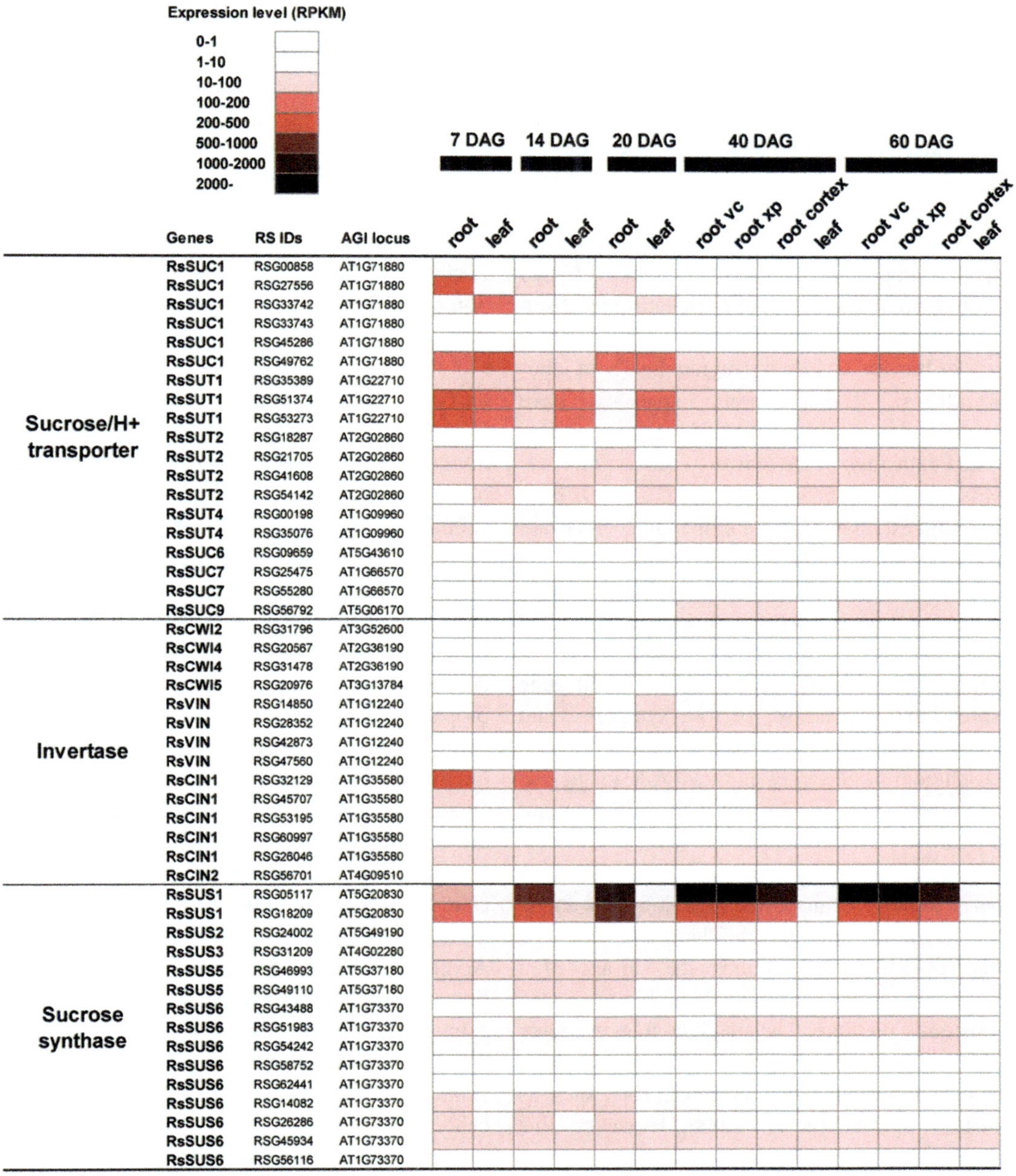

Fig. 8.6 Expression patterns of genes encoding sucrose metabolism in multiple developmental stages and tissues (modified from Mitsui et al. 2015)

For the utilization or storage of sucrose in sink organs, long-distance transport of sucrose from the source to the sink is essential. Sucrose transporters play a central role in sucrose allocation both intracellularly and at the whole-plant level. Membrane-localized, energy-dependent, sucrose/H^+ transporters (typically named SUT or SUC proteins) mediate sucrose partitioning between the source and sink tissues in the phloem. Radish SUT or SUC genes were mostly detected in the roots and leaves of young seedlings, whereas low expression was observed in tuberizing root tissues (Fig. 8.6). In young seedlings, sucrose transport is believed to be mediated mostly by sucrose transporters. After entry into root tuberization, the cleavage of sucrose in sink tissues is accelerated by sucrose synthase, lowering the local concentration of sucrose. This concentration gradient enables sucrose transport from the sink to source tissues; sucrose transporters may therefore not be needed. Simultaneously, up-regulation of photosynthetic activity in leaves is important for sucrose transport. Usuda (2006) demonstrated that CO_2 elevation results in an enhanced sink capacity of radish, which is responsible for absorbing higher levels of photosynthates. Transcriptomic analyses have revealed significantly activated photosynthesis-related pathways in the leaves of tuberizing roots (Mitsui et al. 2015; Wu et al. 2015). These findings suggest that high amounts of sucrose are produced following an increase in photosynthetic activity and accumulate in the phloem, generating a concentration gradient of sucrose between the source and sink tissues that drive the mass flow of sucrose. Overall, the spatial and temporal gene expression of invertase, sucrose synthases, and sucrose transporters in source leaves, sink roots, and phloem tissues play a crucial role in sucrose metabolism, resulting in root tuber initiation and subsequent root development.

Sucrose generated from photosynthates in source organs is transported to sink organs and is then synthesized into starch. Plants store sugar as polymerized starch, enabling the storage of a larger amount of sugar without problems caused by osmotic pressure. The starch content accounts for 0.5–1% of fresh weight of radish roots, which is much lower than that in other crops such as potato (Sugimoto et al. 1988). However, this amount is not so small because the water content in radish roots reaches 90%. The radish root stores photosynthetic products as starch during vegetative growth (usually fall to winter) and uses the starch as an important energy source for reproductive growth in the spring. The thickening root becomes skinny when entering the flowering period. A digestive enzyme of amylase stored in vacuoles can contact with starch grains in the protoplasm in aging roots. Moreover, other hydrolytic enzymes, except for amylase, can also access and digest starch.

8.6 Genes Involved in Root Development in Different Cultivars

Radish cultivars exhibit great variation in root shape and size. Elucidating the molecular mechanisms underlying the variation in root traits in radish cultivars is another important task. Zaki et al. (2010) identified DEGs during root formation and development using two radish cultivars with different root shapes: thick and skinny. The root thickness was significantly different between the two cultivars from 4 weeks after sowing. Hence, they isolated differentially expressed non-redundant gene sequences between the roots of the two cultivars at this stage using suppression subtractive hybridization (SSH). They identified 140 and 70 genes that were specifically expressed in skinny-type roots and thick-type roots, which are considered to be candidates of the genes promoting or repressing root enlargement. Among them, some genes involved in phenylpropanoid metabolism were overexpressed in the skinny-root-type cultivar. As mentioned above, phenylpropanoids are the precursors in lignin synthesis. The lignin amount is much higher in the thin-rooted variety of radish and their roots are relatively woody. Therefore, a functional change in the lignin biosynthetic pathway may be involved in the variation in radish root thickness. Additionally, genes related to ethylene production and root hair elongation are overexpressed in skinny-type roots. They discussed that the enhanced ability of ethylene production explains skinny-root formation because ethylene inhibits cell expansion and elongation throughout development in most tissues in *Arabidopsis* (Kieber et al. 1993). Many skinny-root cultivars, as well as wild radish, produce well-developed lateral roots compared to thick-root cultivars. Because lateral root development may inhibit main and thick-root formation, the differential expression of root hair elongation genes between skinny- and thick-root cultivars is probably involved in root type variation. In the future, a transcriptomic analysis of radish cultivars with different root types will identify genes that promote and/or restrict root tuberization.

References

Baroja-Fernandez E, Munoz FJ, Montero M, Etxeberria E, Sesma MT, Ovecka M, Bahaji A, Ezquer I, Li J, Prat S, Pozueta-Romero J (2009) Enhancing sucrose synthase activity in transgenic potato (*Solanum tuberosum* L.) tubers results in increased levels of starch, ADPglucose and UDPglucose and total yield. Plant Cell Physiol 50:1651–1662

Borisjuk L, Walenta S, Rolletschek H, Mueller-Klieser M, Wobus U, Weber H (2002) Spatial analysis of plant metabolism: sucrose imaging within *Vicia faba* cotyledons reveals specific developmental patterns. Plant J 29:521–530

Hayward H (1938) The structure of economic plants. The Macmillan Company, New York

Kieber JJ, Rothenberg M, Roman G, Feldmann KA, Ecker JR (1993) CTR1, a negative regulator of the ethylene response pathway in *Arabidopsis*, encodes a member of the raf family of protein kinases. Cell 72:427–441

Kitashiba H, Li F, Hirakawa H, Kawanabe T, Zou Z, Hasegawa Y, Tonosaki K, Shirasawa S, Fukushima A, Yokoi S, Takahata Y, Kakizaki T, Ishida M, Okamoto S, Sakamoto K, Shirasawa K, Tabata S, Nishio T (2014) Draft sequences of the radish (*Raphanus sativus* L.) genome. DNA Res 21:481–490

Li XQ, Zhang D (2003) Gene expression activity and pathway selection for sucrose metabolism in developing storage root of sweet potato. Plant Cell Physiol 44:630–636

Martin T, Frommer WB, Salanoubat M, Willmitzer L (1993) Expression of an *Arabidopsis* sucrose synthase gene indicates a role in metabolization of sucrose both during phloem loading and in sink organs. Plant J 4:367–377

Mitsui Y, Shimomura M, Komatsu K, Namiki N, Shibata-Hatta M, Imai M, Katayose Y, Mukai Y, Kanamori H, Kurita K, Kagami T, Wakatsuki A, Ohyanagi H, Ikawa H, Minaka N, Nakagawa K, Shiwa Y, Sasaki T (2015) The radish genome and comprehensive gene expression profile of tuberous root formation and development. Sci Rep 5:10835

Moghe GD, Hufnagel DE, Tang H, Xiao Y, Dworkin I, Town CD, Conner JK, Shiu SH (2014) Consequences of whole-genome triplication as revealed by comparative genomic analyses of the wild radish *Raphanus raphanistrum* and three other Brassicaceae species. Plant Cell 26:1925–1937

Rouhier H, Usuda H (2001) Spatial and temporal distribution of sucrose synthase in the radish hypocotyl in relation to thickening growth. Plant Cell Physiol 42:583–593

Schuster SC (2007) Next-generation sequencing transforms today's biology. Nature 200:16–18

Shendure J, Ji H (2008) Next-generation DNA sequencing. Nat Biotechnol 26:1135–1145

Sturm A, Tang GQ (1999) The sucrose-cleaving enzymes of plants are crucial for development, growth and carbon partitioning. Trends Plant Sci 4:401–407

Sugimoto Y, Yamamoto M, Abe K, Fuwa H (1988) Developmental changes in the properties of the starch of two varieties of Chinese radish (*Raphanus sativa* L.). J Jpn Soc Starch Sci 35:19–27

Usuda H (2006) Effects of elevated CO_2 on the capacity for photosynthesis of a single leaf and a whole plant, and on growth in a radish. Plant Cell Physiol 47:262–269

Wang F, Sanz A, Brenner ML, Smith A (1993) Sucrose synthase, starch accumulation, and tomato fruit sink strength. Plant Physiol 101:321–327

Wang S, Wang X, He Q, Liu X, Xu W, Li L, Gao J, Wang F (2012) Transcriptome analysis of the roots at early and late seedling stages using Illumina paired-end sequencing and development of EST-SSR markers in radish. Plant Cell Rep 31:1437–1447

Wang Y, Pan Y, Liu Z, Zhu X, Zhai L, Xu L, Yu R, Gong Y, Liu L (2013a) *De novo* transcriptome sequencing of radish (*Raphanus sativus* L.) and analysis of major genes involved in glucosinolate metabolism. BMC Genomics 14:836

Wang Y, Xu L, Chen Y, Shen H, Gong Y, Limera C, Liu L (2013b) Transcriptome profiling of Radish (*Raphanus sativus* L.) root and identification of genes involved in response to lead (Pb) stress with next generation sequencing. PLoS ONE 8:e66539

Wang Z, Gerstein M, Snyder M (2009) RNA-seq: a revolutionary tool for transcriptomics. Nat Rev Genet 10:57–63

Weschke W, Panitz R, Gubatz S, Wang Q, Radchuk R, Weber H, Wobus U (2003) The role of invertases and hexose transporters in controlling sugar ratios in maternal and filial tissues of barley caryopses during early development. Plant J 33:395–411

Wu G, Zhang L, Yin Y, Wu J, Yu L, Zhou Y, Li M (2015) Sequencing, *de novo* assembly and comparative analysis of *Raphanus sativus* transcriptome. Front Plant Sci 6:198

Xie Y, Ye S, Wang Y, Xu L, Zhu X, Yang J, Feng H, Yu R, Karanja B, Gong Y (2015) Transcriptome-based gene profiling provides novel insights into the characteristics of radish root response to Cr stress with next-generation sequencing. Front Plant Sci 6:202

Zaki HE, Yokoi S, Takahata Y (2010) Identification of genes related to root shape in radish (*Raphanus sativus*) using suppression subtractive hybridization. Breed Sci 60:130–138

Zrenner R, Salanoubat M, Willmitzer L, Sonnewald U (1995) Evidence of the crucial role of sucrose synthase for sink strength using transgenic potato plants (*Solanum tuberosum* L.). Plant J 7:97–107

9 MicroRNAs Associated with Tuberous Root Development

Yuyan Sun and Xixiang Li

Abstract

MicroRNAs (miRNAs) are a class of small noncoding RNAs that perform important regulatory roles in plant growth and development. In this chapter, we introduced the miRNAs associated with radish tuberous root development at the whole genome level. Five small RNA (sRNA) libraries were constructed from roots of Asian big radish at different developmental stages. A total of 147.99 M clean reads were generated, from which 494 known miRNAs belonging to 434 families, and 220 putative novel miRNAs were obtained. Combined with target prediction and annotation, 77 (71 known and 6 novel miRNAs) differentially expressed miRNAs were potentially associated with tuberous root development. Target transcripts of these tuberous root associated miRNAs were involved in various biological processes, including development process, transcription factor regulation, plant hormone signaling, cell cycle regulation, and carbohydrate metabolism.

9.1 Introduction

Radish (*Raphanus sativus* L, $2n = 2x = 18$), as a major member of the Brassicaceae family, is an important annual or biennial root vegetable crop worldwide. The species was domesticated with the development of the tuberous root in the eastern Mediterranean and the Middle East, and subsequently spread to China, Korea, and Japan (Rubatzky and Yamaguchi 1997). The radish tuberous root can be used in making salad, cooking, pickling or drying. In addition, radish has been suggested as an alternative treatment for various ailments, including hyperlipidemia, coronary heart diseases, and cancer (Curtis 2003). The radish tuberous root is composed of two anatomically distinct parts, the upper part originating from the hypocotyls and the lower part consisting of true root tissue (Tsuro et al.

Y. Sun · X. Li (✉)
Institute of Vegetables and Flowers, Chinese Academy of Agricultural Sciences, Beijing 100081, China
e-mail: Lixx0612@163.com

Y. Sun
e-mail: syy1111@126.com

T. Nishio and H. Kitashiba (eds.), *The Radish Genome*, Compendium of Plant Genomes, DOI 10.1007/978-3-319-59253-4_9

2008). These two tuberous root parts can thicken and form succulent tissues, serving as a vegetable (Tsuro et al. 2008). Radish tuberous root development is controlled by complex interactions among genetic, environmental, and physiological factors (Zaki et al. 2010). Currently, the molecular basis and inheritance of radish tuberous root development are not well understood (Iwata et al. 2004).

miRNAs are a class of 19–25 nucleotide endogenous small sRNA (sRNAs) that regulate gene expression. In plants, the biogenesis of miRNAs requires multiple biological processes, recruiting several evolutionary conserved protein families, to generate fully functional mature miRNAs (Rogers and Chen 2013). At first, plant miRNA genes are transcribed to primary miRNAs (pri-miRNAs) by RNA polymerase II (Lee et al. 2004), then these long pri-miRNAs are cleaved to hairpin-like miRNAs precursor (pre-miRNAs) (He and Hannon 2004), and the loop regions of the hairpin are excised by the RNAse III enzyme DICER-LIKE1 (DCL1) with the aid of HYL1 and SERRATE (Kurihara and Watanabe 2004; Reinhart et al. 2002). Eventually, released mature miRNAs are incorporated into a ribonucleoprotein complex known as the RNA-induced silencing complex (RISC), which inhibits translation elongation or triggers the degradation of target mRNAs (Hutvagne 2005; Kidner and Martienssen 2005).

Previous studies foreboded that miRNAs-mediated regulatory mechanism is important in plant root development. For instance, miR166 is involved in root and nodule development in *Medicago truncatula* via affecting the stability of HD-ZIP III transcripts (Boualem et al. 2008). Auxin response factors (ARFs) are essential for plant development. Several miRNA families, including miR160, miR164, miR167, and miR390, were involved in lateral root development and adventitious rooting through the auxin signaling in *Arabidopsis* and maize (Li et al. 2012; Gutierrez et al. 2009; Yoon et al. 2010). miR395 and miR399 define sulfate metabolism and phosphate signaling, which guarantee the nutrition balance for plant root development (Bari et al. 2006; Kawashima et al. 2009).

High-throughput sequencing technology has been successfully employed for miRNA discovery in some important vegetable crops, including cucumber (Mao et al. 2012), Chinese cabbage (Wang et al. 2012b), and tomato (Moxon et al. 2008). The tuberous root of radish, as an important vegetable worldwide, is a specialized product organ. The association between the tuberous root development and miRNA remains poorly understood. For this chapter, we sequenced the sRNAs of five libraries constructed from radish tuberous roots at different developmental stages and identified a set of tuberous root development-associated miRNAs.

9.2 MiRNAs Associated with Tuberous Root Development

9.2.1 Overview of Small RNA Sequencing Results

The HiSeq 2000 sequencing platform was used to investigate the genome-wide expression profiles of miRNAs in radish tuberous roots. A total of 149.69 M raw reads were produced. After data filtering, a total of 147.99 M clean reads were generated, with 31.25, 24.60, 29.05, 33.57, and 29.50 M clean reads distributed in libraries of early seeding stage (ESS), splitting stage (SS), early expanding stage (EES), rapid expanding stage (RES), and mature stage (MS), respectively. Moreover, 7.94, 7.40, 7.79, 10.15, and 9.22 M unique sequences were obtained from the five libraries, respectively, and over 65% unique sRNAs were mapping to the radish genome (Table 9.1).

The length of the sRNAs varied from 15 to 30 nt, with 21 and 24 nt sRNAs representing the most frequent lengths (Fig. 9.1). The length distribution of the sRNA in this study was consistent with previous reports on radish (Xu et al. 2013a, b), but is a little different from reports on other Brassicaceae plants, such as *Arabidopsis* and Chinese cabbage, in which only the 24 nt length is dominant (Fahlgren et al. 2007; Wang et al. 2012b).

Table 9.1 Statistics of small RNA sequences and genome mapping of five tuberous root libraries in Asian big radish

	ESS	SS	EES	RES	MS
Total reads	31,594,300	24,748,020	29,380,632	34,079,701	29,897,119
High quality reads	31,370,085	24,682,596	29,149,842	33,799,454	29,654,883
3′ adapter null	25,421	3667	26,322	29,957	27,835
Insert null	2356	1776	1759	2100	2049
5′ contaminants	63,374	51,363	50,609	72,600	74,025
Smaller than 18 nt	17,330	18,697	12,369	116,770	43,866
Poly A	3357	3903	6120	5544	4497
Clean reads	31,258,247	24,603,190	29,052,663	33,572,483	29,502,611
Total sRNAs mapped to genome	25,699,550 (82.22%)	20,178,861 (82.02%)	23,834,130 (82.04%)	27,372,335 (81.53%)	24,064,622 (81.57%)
Unique sRNAs	7,943,163	7,402,068	7,789,604	10,155,528	9219,003
Unique sRNAs mapped to genome	5,268,958 (66.33%)	4,997,548 (67.52%)	5,339,266 (68.54%)	6,965,312 (68.59%)	6,398,843 (69.41%)

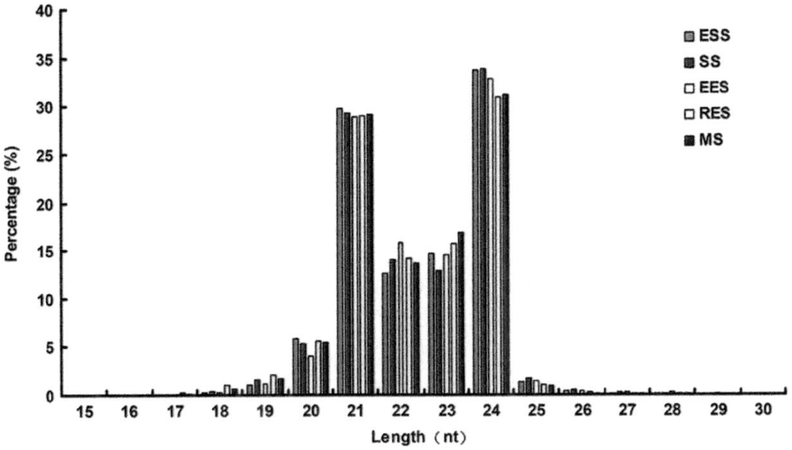

Fig. 9.1 Length distribution of small RNAs (sRNAs) in the libraries of the radish tuberous roots. The *y-axis* indicates the percentage of sRNA reads in each library, whereas the *x-axis* corresponds to the nucleotide (nt) lengths of sRNAs. ESS, SS, EES, RES, and MS represent libraries of early seedling, splitting, early expanding, rapid expanding, and mature stages, respectively

The sRNA sequences in each library were subjected to a BLAST search against the radish genome (Zhang et al. 2015), NCBI GenBank (Benson et al. 2011), Rfam10.1 (Griffiths-Jones et al. 2005), and miRBase20.0 databases (Griffiths-Jones et al. 2008) for annotation. They were classified into eleven categories (Fig. 9.2). Of these, only a few unique sRNAs were annotated as miRNAs in five libraries. Additionally, over 60% unique sequences were unannotated sRNAs, suggesting a large number of novel miRNAs are found in radish tuberous roots.

9.2.2 Identification of Known and Novel MiRNAs from Radish Tuberous Root

Using SOAP2 software (Li et al. 2009), the clean reads were aligned against the radish genome (Zhang et al. 2015) and Rfam databases (Griffiths-Jones et al. 2005). Reads annotated into the appropriate noncoding RNA categories or considered to be degradation fragments of mRNAs were removed. The remaining sRNA

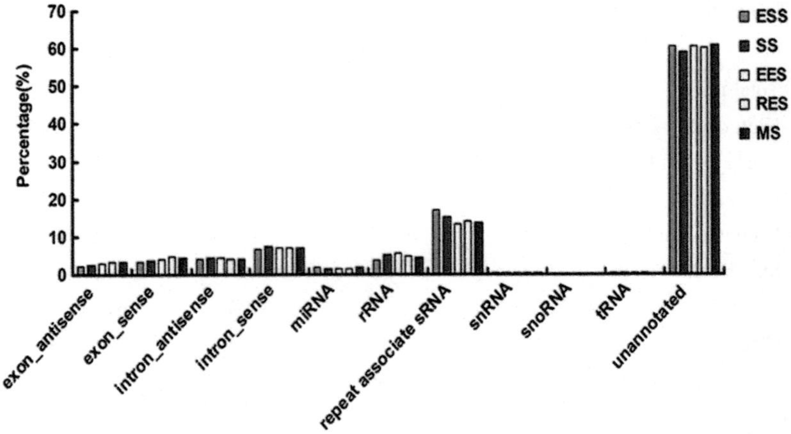

Fig. 9.2 Classification of unique sRNAs from the radish tuberous roots. The reads were categorized into 11 different classes of sRNAs. Their distributions are shown as percentages

sequences were subjected to BLAST against the miRBase20.0 (Griffiths-Jones et al. 2008). The nearly perfectly matched (less than two mismatches) sequences were considered to be known miRNAs (Meyers et al. 2008). A homology search using miRBase20.0 revealed that a total of 494 known miRNAs representing 434 families were identified from the five libraries, with 294, 252, 283, 326, and 319 miRNAs distributed in ESS, SS, EES, RES, and MS libraries, respectively. The majority of these miRNA families were composed of one member, whereas miR160 and miR169 contained four members. miR156, miR159, miR167, and miR172 had three members each. The majority of these miRNAs, including 205 (69.73%) miRNAs for ESS, 164 (65.08%) miRNAs for SS, 182 (64.31%) miRNAs for EES, 204 (62.57%) miRNAs for RES, and 202 (63.32%) miRNAs for MS, had more than 100 reads. Among these miRNAs, miR157a, miR166a, and miR168a were highly expressed in five libraries.

The remaining unannotated sRNA sequences were used to identify potential novel miRNA. Thus, a total of 220 unique miRNA sequences were identified from five libraries, with 84, 56, 57, 75, and 84 novel miRNA precursor candidates identified in ESS, SS, EES, RES, and MS libraries, respectively. The length of the novel miRNA precursors varied greatly from 69 to 357 nt, with an average of 150 nt. We also calculated the minimum folding free energies of putative radish miRNA precursors for the libraries, which ranged from −215.8 to −18.0 kcal/mol, with an average of −55.4 kcal/mol. The minimum free energy (MFE) in radish is similar to that in *Vitis amurensis* Rupr. (Wang et al. 2012a) and *Triticuma estivum* L (Yao et al. 2007). The length of the mature miRNAs ranged from 19 to 27 nt, with the majority being 23 nt (90; 40.9%). In contrast with the known miRNAs, the novel miRNAs often expressed at a relative low level. Only 22, 22, 13, 15, and 16 novel miRNAs were detected as greater than 100 reads in the ESS, SS, EES, RES, and MS, respectively.

9.2.3 Differential Expression Analysis of miRNAs

Differential expression was analyzed to identify miRNAs involved in tuberous root development. miRNAs with a *P* value of <0.01 and an absolute value of $\log_2$ greater than 1 were considered to be differentially expressed miRNAs (Eldem et al. 2012). In all, 429 differentially expressed known miRNAs were identified, including 206 (85 up-regulated and 121 down-regulated), 226 (108 up- and 118 down-regulated), 276 (151 up- and 125 down-regulated), and 294 (159 up- and 135 down-regulated) miRNAs that were differentially expressed in SS/ESS, EES/ESS, RES/ESS, and MS/ESS libraries, respectively. Among which, 145 highly differentially expressed known miRNAs (transcripts per million, TPM $\geq$ 30) were performed on hierarchical clustering. As shown in Fig. 9.3, 59 miRNAs showed an up-regulated pattern and 46 miRNAs were down-regulated with tuberous root development. Moreover, 40 miRNAs showed fluctuated expression overtime. The results implied that these differentially expressed miRNAs might play essential regulatory roles during tuberous root development.

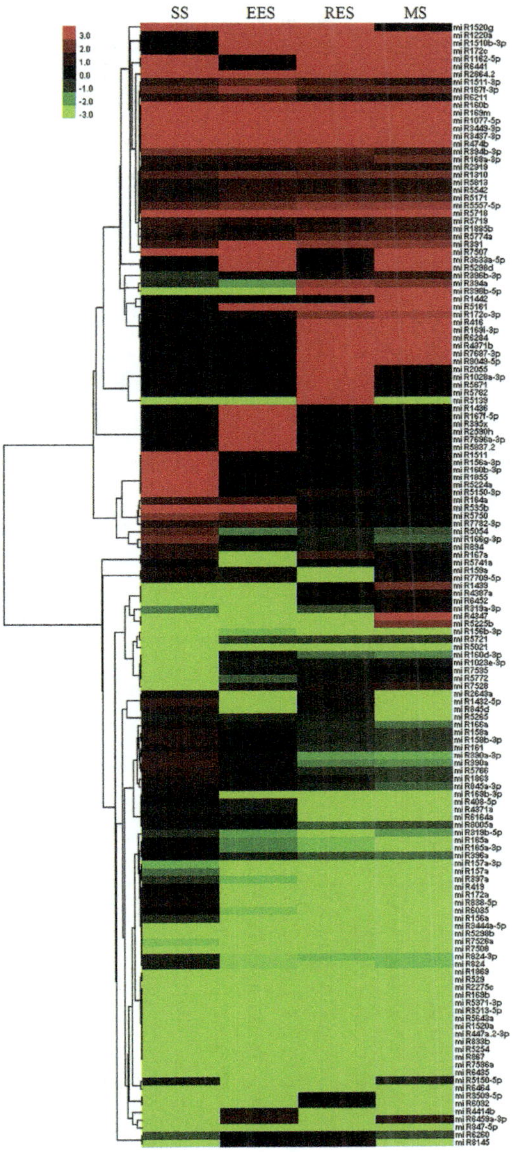

Fig. 9.3 Heatmap of highly expressed known miRNAs in developing radish tuberous roots. The *color scale* represents the log_2-transcripts per million ratios of SS, EES, RES, and MS compared with ESS. The *upper left* shows the *color bar*. The trees on the left were generated using Cluster 3.0

In addition, a total of 29 highly differentially expressed (TPM $\geq$ 5) novel miRNAs were also identified (Table 9.2). Five novel miRNAs, novel_mir_9, novel_mir_17, novel_mir_18, novel_mir_24, and novel_mir_29, were expressed in all five libraries. However, ten novel miRNAs were only detected in certain stages. For instance, novel_mir_58, novel_mir_63, novel_mir_70, and novel_mir_105 were expressed only in the ESS library. Novel_mir_40 was detected in the EES library. Novel_mir_176 and novel_mir_183 were present only in the RES library. Novel_mir_129, novel_mir_151, and novel_mir_161 were detected just in the MS library. These miRNAs might be induced or repressed during radish tuberous root development.

9.2.4 Target Prediction for Tuberous Root Development-Associated miRNAs

Target genes of the 145 highly and differentially expressed miRNAs were predicted according to previous criteria (Allen et al. 2005; Schwab et al. 2005). A total of 77 miRNAs including 71 known and 6 novel miRNAs targeted the transcripts. A large number of the targets for miRNAs encode transcription factors, such as squamosa promoter binding (SPB), MYBs/TCPs, ARF, no apical meristem (NAM), HD-ZIPs, APETALA2 (AP2), helix-loop-helix (HLH), F-box domain, zinc finger, and MADS-box, which might regulate corresponding transcriptional process in radish tuberous root development. Moreover, a few target transcripts which encode functional enzymes, including pentatricopeptide repeat, protein kinase, multicopper oxidase, xyloglucan fucosyltransferase, trehalose-phosphatase, ABC transporter protein, and sugar/inositol transporter, were also found to be involved in tuberous root development (Table 9.3).

Table 9.2 Novel miRNAs highly differentially expressed in developing radish tuberous roots

miRNA	Sequence	Length (nt)	Abundance (TPM)				
			ESS	SS	EES	RES	MS
novel_mir_1	TGAGCTCAGAATGACACTTATTA	23	0.01	13.35	20.66	21.17	24.68
novel_mir_7	GTCGGCGTCGGCGTCAGCGGCTG	23	0.01	84.96	61.88	0.01	0.01
novel_mir_9	TATGAATGATGCGGGAGATGT	21	456.41	283.53	227.60	82.40	43.64
novel_mir_12	AGCTAGATGGACTTTACACTTAA	23	10.36	0.01	12.63	20.57	0.01
novel_mir_15	CTAGGCGATACTAGATCTCATCG	23	10.59	11.32	9.97	11.84	0.01
novel_mir_17	AGCGAAAGATGAATTTGTGCC	21	190.15	325.66	298.71	373.44	398.01
novel_mir_18	AGCTCCAGACTCATTCATTCA	21	308.39	518.67	3731.75	5191.22	10,843.02
novel_mir_24	TTCATCAAAGTTGGATTTCCT	21	29.05	27.27	20.14	14.13	18.74
novel_mir_27	TAGCCAAGGATGACTTGCCTG	21	0.01	7.91	0.78	0.01	0.01
novel_mir_29	GGCAGTCTCCTTGGCTATT	19	13.67	10.88	1.41	1.07	0.92
novel_mir_36	CGAAGTGACTTATAATGATCT	21	0.01	5.38	4.54	0.01	0.01
novel_mir_40	ATGCGGGTTTGTAATAGTGTATA	23	0.01	0.01	5.27	0.01	0.01
novel_mir_47	AATGAATGATGCGGTAGACAAAT	23	60.64	62.69	103.10	0.01	0.01
novel_mir_51	TCAATGAAAGGTATGATTCCC	21	0.01	0.01	23.22	37.07	34.73
novel_mir_53	TCAAGCTTGAGGTAGTCTGGGTA	23	5.85	0.01	6.57	6.86	5.62
novel_mir_58	AGGACGTAGATCCACATGATGG	22	6.54	0.01	0.01	0.01	0.01
novel_mir_63	GCAAGTCGACTTCGGCTCTAT	21	6.58	0.01	0.01	0.01	0.01
novel_mir_70	TGTAGTAAGGATGTATATGAAG	22	17.59	0.01	0.01	0.01	0.01
novel_mir_76	GGCAGTCTCCTTGGCTATC	19	13.63	3.92	0.01	0.01	0.70
novel_mir_94	TTTAGGTCTGAGTATCTGGATCCGG	25	13.03	13.60	0.01	0.01	0.01
novel_mir_95	ATGACTTGCCGGAATGCGTTCAT	23	24.26	18.28	0.01	0.01	0.01
novel_mir_100	TCGAAACGTGGCTGGCTGGAGA	22	13.26	0.01	0.01	8.86	9.99
novel_mir_105	TAGCCAAGGATGACTTGCCTA	21	6.49	0.01	0.01	0.01	0.01
novel_mir_107	CATCAATGAAAGGTATGATTCC	22	22.93	32.90	0.01	0.01	0.01
novel_mir_129	TAAAGGTGGACTTCATATATTTA	23	0.01	0.01	0.01	0.01	8.80
novel_mir_151	CCCAGCACTCGTGGATCGGACGTA	24	0.01	0.01	0.01	0.01	5.13

(continued)

Table 9.2 (continued)

miRNA	Sequence	Length (nt)	Abundance (TPM)				
			ESS	SS	EES	RES	MS
novel_mir_161	TTCTCGTAGACGACTTATATATA	23	0.01	0.01	0.01	0.01	116.45
novel_mir_176	ACCCAGTTGGACTGTCCATGGA	22	0.01	0.01	0.01	5.60	0.01
novel_mir_183	TAGCTTCCGACTCATTCATCCA	22	0.01	0.01	0.01	12.12	0.01

Table 9.3 miRNAs and their targets potentially associated with tuberous root development

miRNA member	ESS-std	SS-std	EES-std	RES-std	MS-std	Description of targets
miR1023e-3p	122.10	0.01	82.91	91.73	94.41	Acylphosphatase-like
miR1077-5p	0.01	6770.88	3740.88	8825.81	9124.70	Amino acid-binding ACT
miR1436	0.01	0.01	98.35	0.01	0.01	Proteasome inhibitor PI31; lipase
miR1439	7.97	0.01	0.01	7.60	31.81	Lipase
miR1442	0.01	0.01	0.01	0.01	511.16	Pectinesterase inhibitor; F-box domain
miR1510b-3p	0.01	0.01	309.72	816.65	659.48	Leucine-rich repeat; glycoside hydrolase
miR1511	3.32	385.64	4.07	3.59	3.89	RNA-binding S4
miR156a	32,383.76	17,541.59	3002.26	2037.38	2173.31	SPB-box
miR157a	77,852.01	25,697.75	1838.25	353.76	519.91	SPB-box
miR158a	5754.05	10,991.36	5435.69	3418.49	2823.72	Xyloglucan fucosyltransferase; pentatricopeptide repeat
miR158b-3p	5811.56	11,085.11	5477.38	3454.30	2848.62	Xyloglucan fucosyltransferase; pentatricopeptide repeat
miR159a	20.72	35.93	31.72	0.01	25.98	Myb
miR160b	0.01	44.60	7.36	8.21	7.29	Auxin response factor
miR160d-3p	479.39	0.01	391.07	164.43	135.84	Pirin
miR161	24.08	37.70	22.38	12.78	11.40	Pentatricopeptide repeat
miR164a	1913.78	4759.08	5055.01	2363.02	2266.37	No apical meristem (NAM) protein
miR165a	2129.95	2547.56	573.22	377.64	208.75	Homeobox-leucine zipper protein
miR165a-3p	2028.33	2416.55	549.68	350.69	197.03	Homeobox-leucine zipper protein
miR166a	24,189.07	34,396.56	20,176.42	13,497.76	8363.78	Homeobox-leucine zipper protein
miR166 g-3p	301.76	1256.86	316.61	196.65	105.92	Homeobox-leucine zipper protein
miR167a	549.60	1163.61	0.01	1330.46	633.07	Auxin response factor
miR167f-5p	0.01	0.01	678.41	0.01	0.01	Auxin response factor
miR169i-3p	0.01	0.01	0.01	2162.36	2179.47	ZF-HD
miR172a	379.34	528.79	0.01	0.01	0.01	AP2
miR172c	0.01	0.01	1941.97	2291.95	3266.13	AP2; trehalose-phosphatase
miR1855	0.01	30.05	0.01	0.01	0.01	Helix-loop-helix; zinc finger
miR1885b	228.74	503.74	332.00	494.32	555.02	Toll-Interleukin receptor
miR2643a	160.69	132.53	0.01	125.96	0.01	ABC transporter
miR2864.2	0.01	307.38	1.30	1.07	2.32	Oxoglutarate/iron-dependent oxygenase
miR2919	21.96	36.88	34.59	48.83	40.83	Extensin repeat
miR319a-3p	28.09	8.67	2.30	12.17	39.10	MYB, TCP
miR3437-3p	0.01	1978.22	4012.93	3485.78	2837.71	Mitochodrial transcription termination factor
miR3633a-5p	0.01	0.01	158.10	0.01	177.21	Rop nucleotide exchanger
miR390a	664.66	1562.03	820.59	194.14	174.62	Protein kinase
miR391	132.05	327.88	1307.97	823.92	726.99	RNA binding

(continued)

Table 9.3 (continued)

miRNA member	ESS-std	SS-std	EES-std	RES-std	MS-std	Description of targets
miR394a	6.31	3.67	1.51	38.24	30.73	F-box domain cyclin
miR395x	0.01	0.01	4458.89	0.01	0.01	Protein kinase
miR396a	95.17	90.53	58.80	37.63	31.38	Protein kinase; HLH; WD40
miR397a	135.92	70.03	25.20	6.16	10.75	Multicopper oxidase
miR416	0.01	0.01	0.01	25.09	34.51	Protein kinase
miR419	186.47	214.39	3.18	5.50	3.46	Zinc finger; WRKY; MYB, CYC
miR4371b	0.01	0.01	0.01	59.04	62.87	Peptidase S8/S53
miR5021	107.05	0.01	0.26	7.04	10.59	Protein kinase; sugar/inositol transporter; pectinesterase inhibitor
miR5054	39.23	120.01	12.05	23.69	16.04	FAD-dependent oxidoreductase
miR5150-3p	0.83	80.21	0.89	1.49	0.92	Ribonuclease II/R
miR5150-5p	33.15	22.90	0.01	0.01	19.93	Remorin
miR5161	0.01	0.01	0.26	0.28	1449.45	BEACH domain
miR5254	1282.68	0.01	0.01	0.01	0.01	RNA recognition motif domain
miR5265	121.69	83.38	79.83	84.83	8.64	Methyltransferase type 11
miR529	48.16	0.38	0.68	2.56	3.24	FMN-dependent dehydrogenase; tetratricopeptide repeat
miR5298b	344.21	39.85	0.01	0.01	0.01	3′–5′ exonuclease
miR5298d	0.01	0.01	31.05	0.01	31.16	Kelch repeat type 1
miR535b	0.01	161.76	143.02	0.01	0.01	Pentatricopeptide repeat
miR5371-3p	34.58	0.01	0.01	0.01	0.01	Abortive infection protein
miR5542	44.20	67.94	99.97	89.21	102.14	ATPase
miR5671	0.01	0.01	0.01	419.61	0.01	Myb, DNA binding
miR5782	0.01	0.01	0.01	4268.43	0.01	Cyclin-dependent kinase inhibitor
miR5837.2	0.01	0.01	109.68	0.01	0.01	MATH; F-box domain
miR6164a	27.03	31.82	24.78	0.01	0.01	Lipase
miR6435	60.36	0.01	0.01	0.01	0.01	Shikimate kinase
miR6441	0.01	50.36	0.01	71.68	67.41	Protein kinase; ABC transporter; Hsp
miR7508	72.56	4.49	0.01	0.01	0.01	Lipoxygenase
miR7526a	446.38	75.03	0.01	0.01	0.01	Protein of unknown function
miR7535	176.66	0.01	142.44	143.03	166.63	F-box domain; Leucine-rich repeat
miR7536a	40.24	0.01	0.01	0.01	0.01	Photosystem antenna protein-like
miR7696a-3p	0.01	0.01	332.99	0.01	0.01	No apical meristem (NAM) protein
miR8049-5p	0.01	0.01	0.01	499.31	469.68	MATH
miR824	186.84	118.23	27.91	25.04	34.78	MADS-box; ATPase
miR824-3p	57.97	35.49	8.45	10.59	10.59	Protein kinase; Kinesin
miR838-5p	35.96	53.96	0.01	0.01	0.01	Amidase
miR894	207.33	419.86	168.22	147.27	130.11	Methylenetetrahydrofolate reductase
novel_mir_18	308.39	518.67	3731.75	5191.22	10,843.02	

(continued)

Table 9.3 (continued)

miRNA member	ESS-std	SS-std	EES-std	RES-std	MS-std	Description of targets
						Cellular retinaldehyde binding; proteasome
novel_mir_29	13.67	10.88	1.41	1.07	0.92	Ribosomal protein; proteinase inhibitor
novel_mir_100	13.26	0.01	0.01	8.86	9.99	Arabidopsis retrotransposon ORF-1 protein
novel_mir_76	13.63	3.92	0.01	0.01	0.70	Ribosomal protein
novel_mir_7	0.01	84.96	61.88	0.01	0.01	Calcium-binding EF-hand
novel_mir_58	6.54	0.01	0.01	0.01	0.01	Transmembrane receptor

9.2.5 GO Analysis of Target Transcripts During Tuberous Root Development

To provide further insight into the mechanisms underlying radish tuberous root development, GO analysis was performed on the targets. The target transcripts with a known function were categorized into biological process, cellular component, and molecular function according to the ontological definitions. The significantly enriched GO terms (*P* value <0.05) of the five libraries were identified and are summed up in Table 9.4. Only one significantly enriched term, necleoid, was identified in the cellular component category. Eight significantly enriched terms, including nucleic acid binding transcription factor activity, nucleic acid binding, heterocyclic compound binding, organic cyclic compound binding, hormone binding, alcohol binding, DNA binding, and antiporter activity, were identified in the molecular function category.

A total of 80 significantly enriched terms were determined in the biological process category. Most terms were associated with metabolic processes, developmental processes, biological regulation, and response to stimulus. Additionally, a few enriched terms in biological process had stage-specific features. For example, axis specification, cellular response to chemical stimulus, organ morphogenesis, and response to hormone stimulus were significantly enriched in the ESS, while response to abscisic acid stimulus and response to alcohol were significantly enriched in the EES. Moreover, some terms, including regulation of cell differentiation and signaling, were only significantly enriched in the MS.

9.2.6 miRNA-Mediated Regulatory Roles in Tuberous Root Development

Target predictions indicated that the tuberous root development-associated miRNAs potentially targeted a set of radish genes that were involved in various biological processes of radish tuberous root development. A set of the differentially expressed miRNA with their target annotations listed in Table 9.3 and Fig. 9.4 were discussed. For example, the high abundance miR156a and miR157a, ubiquitously expressed in five libraries and potentially targeting the *SPB* gene, were detected with significant down-regulation and acted as negative regulators of tuberous root development. The result was opposite to the process of wheat grain-filling with the up-regulated expression level (Meng et al. 2013). Cell walls are crucial for development, signal transduction, and disease resistance in plants (Perrin et al. 1999). The xyloglucan fucosyltransferase, essential for cell wall synthesis, was targeted by miR158a. miR158a showed slightly

Table 9.4 GO enrichment terms for miRNA target transcripts

	Gene ontology term	ESS *P* value	SS *P* value	EES *P* value	RES *P* value	MS *P* value
Cellular component	Nucleoid	–	0.02976	–	–	–
Molecular function	Nucleic acid binding transcription factor activity	1.83E−26	4.09E−15	4.11E−25	5.78E−18	6.26E−20
	Nucleic acid binding	4.08E−11	0.0032	9.28E−10	0.00821	0.00682
	Heterocyclic compound binding	9.74E−06	–	0.00045	–	–
	Organic cyclic compound binding	1.07E−05	–	0.00049	–	–
	Hormone binding	1.71E−05	8.97E−05	–	0.00129	0.00185
	Alcohol binding	3.67E−05	0.00019	–	0.00272	0.00389
	DNA binding	1.81E−02	–	0.00248	–	–
	Antiporter activity	–	–	–	–	0.03693
Biological process	Adaxial/abaxial axis specification	0.00099	0.00598	0.003	0.00953	–
	Adaxial/abaxial pattern specification	0.00156	0.00929	0.00467	0.01581	–
	Anatomical structure arrangement	1.68E−14	1.24E−12	2.18E−13	7.12E−10	1.53E−09
	Anatomical structure development	1.01E−08	2.80E−06	1.68E−07	0.00383	0.00386
	Anatomical structure morphogenesis	5.32E−06	0.00049	5.73E−05	–	–
	Aromatic compound biosynthetic process	3.97E−17	8.03E−10	9.61E−19	5.04E−12	5.48E−15
	Axis specification	0.02024	–	–	–	–
	Biological regulation	0.00193	0.00663	0.03174	0.00034	0.00016
	Biosynthetic process	5.01E−12	5.50E−05	1.48E−11	1.13E−06	7.31E−09
	Cell proliferation	0.00625	0.0055	0.02107	0.0274	0.04214
	Cellular aromatic compound metabolic process	2.25E−13	4.86E−06	2.85E−13	6.88E−06	5.44E−07
	Cellular biosynthetic process	8.26E−13	1.34E−05	2.16E−12	5.67E−07	1.45E−09
	Cellular macromolecule biosynthetic process	3.79E−19	2.70E−11	6.20E−22	7.99E−14	3.20E−18
	Cellular macromolecule metabolic process	2.93E−06	0.00011	1.94E−07	0.0007	6.01E−06
	Cellular nitrogen compound biosynthetic process	1.20E−18	8.66E−11	5.66E−20	9.10E−14	4.71E−17
	Cellular nitrogen compound metabolic process	1.24E−11	1.57E−06	1.45E−11	5.58E−06	5.78E−07
	Cellular response to chemical stimulus	0.00267	–	–	–	–
	Cellular response to endogenous stimulus	0.00057	0.00291	–	0.00197	0.01642
	Cellular response to hormone stimulus	0.00055	0.00283	–	0.00191	0.01594
	Cellular response to organic substance	0.00525	0.02923	–	0.03496	0.04848
	Developmental process	3.76E−07	2.50E−05	5.78E−06	0.04787	0.00535

(continued)

Table 9.4 (continued)

Gene ontology term	ESS *P* value	SS *P* value	EES *P* value	RES *P* value	MS *P* value
Developmental process involved in reproduction	1.27E−08	1.18E−06	9.07E−08	0.01494	0.00045
Gene expression	3.85E−13	5.38E−06	1.12E−14	2.78E−06	6.84E−08
Heterocycle biosynthetic process	4.98E−18	2.94E−10	2.78E−19	1.25E−12	8.83E−16
Heterocycle metabolic process	3.78E−12	3.70E−06	5.71E−12	1.55E−05	3.23E−06
Hormone-mediated signaling pathway	0.00038	0.00191	–	0.00116	0.01013
Macromolecule biosynthetic process	1.35E−18	8.21E−11	2.73E−21	3.63E−13	1.92E−17
Macromolecule metabolic process	0.00084	0.0394	0.00049	–	0.03548
Meristem development	5.57E−07	7.08E−07	1.04E−07	1.19E−07	1.09E−05
Meristem structural organization	1.68E−14	1.24E−12	2.18E−13	7.12E−10	1.53E−09
Multicellular organismal development	9.62E−07	4.57E−07	1.26E−05	0.00019	0.00046
Multicellular organismal process	1.43E−06	6.60E−07	3.14E−06	0.00067	0.00037
Nitrogen compound metabolic process	1.72E−11	8.40E−06	4.45E−10	5.59E−05	5.96E−06
Nucleic acid metabolic process	1.96E−15	1.69E−08	4.55E−17	3.49E−09	1.23E−11
Nucleobase-containing compound biosynthetic process	2.95E−20	2.74E−12	1.56E−22	9.11E−15	2.40E−18
Nucleobase-containing compound metabolic process	8.98E−13	3.96E−07	1.63E−13	1.44E−06	9.31E−08
Organ development	2.77E−07	5.20E−06	3.82E−05	0.00333	0.00753
Organ morphogenesis	0.03694	–	–	–	–
Organic cyclic compound biosynthetic process	1.10E−16	1.95E−09	3.08E−18	1.68E−11	2.19E−14
Organic cyclic compound metabolic process	8.79E−13	3.49E−06	4.48E−13	1.48E−05	1.32E−06
Organic substance biosynthetic process	8.54E−13	2.37E−05	5.27E−12	6.27E−07	1.47E−09
Pattern specification process	0.00244	0.00016	0.01725	0.0001	0.04421
Phenylpropanoid metabolic process	0.00108	–	0.0109	–	–
Phyllome development	0.00023	0.00183	1.20E−05	0.00278	0.00626
Post-embryonic development	0.00745	0.02121	0.02883	–	–
Post-embryonic organ development	0.00104	0.02754	–	–	–
Post-embryonic root development	0.00017	0.00421	–	–	–
Programmed cell death	0.03267	–	0.00464	–	0.02983
Regionalization	0.00087	4.22E−05	0.00648	2.04E−05	0.01301
Regulation of anatomical structure morphogenesis	3.02E−05	0.00014	7.95E−05	0.00173	0.00226
Regulation of biological process	2.72E−05	0.0001	0.00016	6.18E−06	1.10E−05
Regulation of cell differentiation	–	–	–	–	0.03439

(continued)

Table 9.4 (continued)

Gene ontology term	ESS *P* value	SS *P* value	EES *P* value	RES *P* value	MS *P* value
Regulation of cellular process	–	–	–	0.01847	0.0279
Regulation of development, heterochronic	0.00055	0.0055	0.00021	0.0274	0.04214
Regulation of developmental process	3.37E−07	8.53E−08	1.23E−11	2.01E−07	1.10E−07
Regulation of multicellular organismal development	0.00108	0.01301	6.94E−05	–	–
Regulation of multicellular organismal process	0.00154	0.01822	0.0001	–	–
Regulation of organ formation	4.05E−08	1.97E−07	1.07E−07	2.48E−06	3.26E−06
Regulation of organ morphogenesis	4.05E−08	1.97E−07	1.07E−07	2.48E−06	3.26E−06
Reproduction	1.13E−06	0.00012	3.84E−06	–	0.01967
Reproductive process	1.76E−07	2.04E−05	5.48E−07	–	0.00305
Response to endogenous stimulus	4.22E−07	0.03732	0.02989	0.00626	–
Response to hormone stimulus	0.00126	–	–	–	–
Response to organic substance	0.00047	–	–	0.00761	0.01822
RNA biosynthetic process	6.61E−22	1.94E−13	2.36E−25	1.67E−17	4.60E−21
RNA metabolic process	4.46E−17	2.13E−10	4.47E−20	1.28E−12	1.72E−14
Root development	0.00021	0.00216	–	–	–
Root system development	0.00021	0.00216	–	–	–
Shoot axis formation	0.00055	0.00258	0.00143	0.00158	0.00218
Shoot system development	3.08E−06	5.25E−05	2.00E−07	4.77E−05	0.00013
Shoot system morphogenesis	8.02E−05	0.00065	0.00028	0.00148	0.00224
Single-multicellular organism process	6.85E−06	1.96E−06	3.91E−05	0.00032	0.00025
Single-organism developmental process	2.46E−08	3.93E−07	1.73E−06	0.00154	0.00039
Specification of symmetry	1.76E−12	2.17E−11	8.04E−12	1.03E−09	1.62E−09
System development	4.92E−08	4.84E−07	3.11E−07	0.00113	0.00404
Tissue development	0.00312	0.03885	0.00213	0.0103	
Transcription, DNA dependent	6.61E−22	1.94E−13	2.36E−25	1.67E−17	4.60E−21
Response to abscisic acid stimulus	–	–	0.03725	–	–
Response to alcohol	–	–	0.03957	–	–
Signaling	–	–	–	–	0.04991

up-regulated at the early stage, then down-regulated during the tuberous root development. The NAM protein, a potential target of miR164a, controls leaf, root, and fruit organs morphogenesis (Guo et al. 2005; Mallory et al. 2004). miR164a was up-regulated during the early to middle tuberous root growth stages in the present study. It has been shown that the

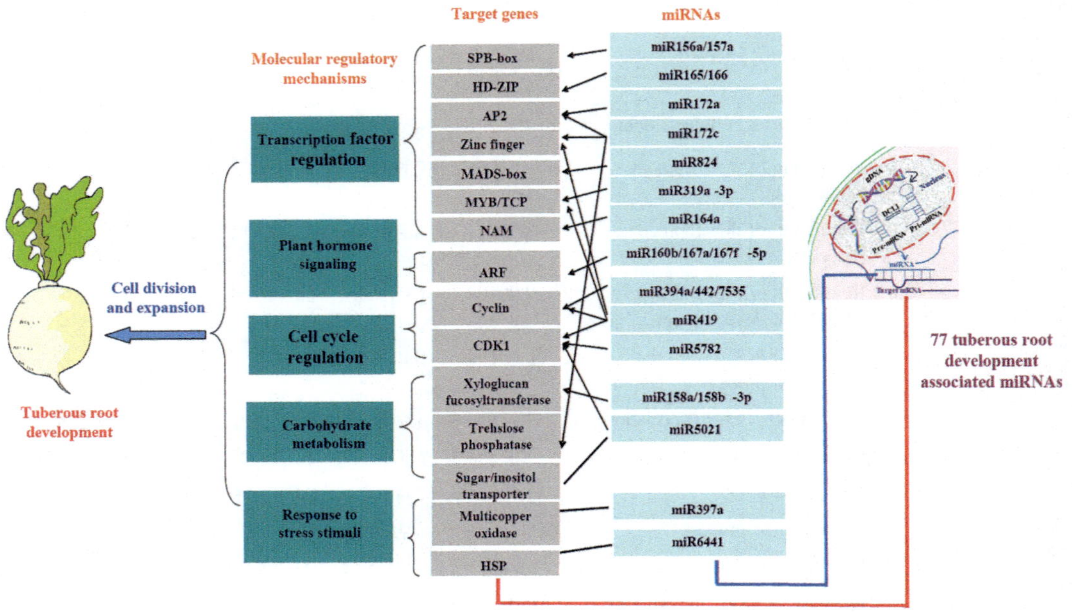

Fig. 9.4 Proposed miRNA-mediated regulatory network associated with tuberous root development in radish

miR165/166 group and its target genes regulate diverse aspects of plant development, including apical and lateral meristem formation of shoots, leaf polarity, floral and root development, and vascular development (Williams et al. 2005). HD-ZIP genes were targeted by miR165/166. The expression level of miR165a was slightly up-regulated at early stage, then down-regulated during the middle and late stage of tuberous root development. Another two miRNAs, miR160, and miR167 are known to target *ARFs* (Khraiwesh et al. 2012). It was reported that ARFs are transcription factors binding to auxin-responsive promoter elements to induce or repress auxin-regulated transcription during the plant development, such as root development and branching (Yoon et al. 2010). miR167a was up-regulated at SS and RES and down-regulated in EES, while miR160 was highly repressed during tuberous root development in the present study. miR172a, targeting the AP2 transcription factor, was slightly up-regulated at early stage and highly repressed later, acting as a crucial regulator of radish tuberous root development in the present study. The TCP transcription factors, regulate a series of plant growth and development process including shoot lateral organ morphogenesis (Tomotsugu et al. 2007), fiber elongation, and root hair development (Hao et al. 2012). miR319a-3p, targeting the TCP transcription factor, was down-regulated at the early to middle stages, especially the EES, and then slightly up-regulated at MS. miR824, potentially targeted the MADS-box genes, was gradually down-regulated in the SS and EES, and then maintained similar expression levels during the later developmental stages. F-box proteins, which are associated with cellular functions, such as signal transduction and regulation of the cell cycle, are broadly regulated by miRNA-mediated gene silencing (Jones-Rhoades et al. 2006). *F-box* genes were targeted by miR394a. miR394a was down-regulated during the early stage and up-regulated later with tuberous root development, implying that levels of F-box proteins are a possible factor in regulating the cell cycle of the tuberous root development. Overall, our results demonstrate that miRNA-mediated mechanisms are involved in the regulation of radish tuberous root development.

References

Allen E, Xie Z, Gustafson AM, Carrington JC (2005) microRNA-directed phasing during trans-acting siRNA biogenesis in plants. Cell 121:207–221

Bari R, Pant BD, Stitt M, Scheible WR (2006) PHO2, microRNA399, and PHR1 define a phosphate-signaling pathway in plants. Plant Physiol 141:988–999

Benson DA, Karsch-Mizrachi I, Lipman DJ, Ostell J, Sayers EW (2011) GenBank. Nucleic Acids Res 39: D32–D37

Boualem A, Laporte P, Jovanovic M, Laffont C, Plet J, Combier J, Niebel A, Crespi M, Frugier F (2008) MicroRNA166 controls root and nodule development in Medicago truncatula. Plant J 54:876–887

Curtis IS (2003) The noble radish: past, present and future. Trends Plant Sci 8:305–307

Eldem V, Akcay UC, Ozhuner E, Bakır Y, Uranbey S, Unver T (2012) Genome-wide identification of miRNAs responsive to drought in peach (*Prunus persica*) by high-throughput deep sequencing. PLoS ONE 7(12):e50298

Fahlgren N, Howell MD, Kasschau KD, Chapman EJ, Sullivan CM, Cumbie JS, Givan SA, Law TF, Grant SR, Dangl JL, Carrington JC (2007) High-throughput sequencing of Arabidopsis microRNAs: evidence for frequent birth and death of MIRNA genes. PLoS ONE 2:e219

Griffiths-Jones S, Moxon S, Marshall M, Khanna A, Eddy SR, Bateman A (2005) Rfam: annotating non-coding RNAs in complete genomes. Nucleic Acids Res 33:121–124

Griffiths-Jones S, Saini HK, van Dongen S, Enright AJ (2008) miRBase: tools for microRNA genomics. Nucleic Acids Res 36:154–158

Guo HS, Xie Q, Fei JF, Chua NH (2005) MicroRNA directs mRNA cleavage of the transcription factor NAC1 to down regulate auxin signals for Arabidopsis lateral root development. Plant Cell 17:1376–1386

Gutierrez L, Bussell DI, Pacurar DI, Schwambach J, Pacurar M, Bellini C (2009) Phenotypic plasticity of adventitious rooting in Arabidopsis is controlled by complex regulation of auxin response factor transcripts and microRNA abundance. Plant Cell 21 (10):3119–3132

Hao J, Tu L, Hu H, Tan J, Deng F, Tang W, Nie Y, Zhang X (2012) GbTCP, a cotton TCP transcription factor, confers fibre elongation and root hair development by a complex regulating system. J Exp Bot 63:6267–6281

He L, Hannon GJ (2004) MicroRNAs: small RNAs with a big role in gene regulation. Nat Rev Genet 5:522–531

Hutvagne G (2005) Small RNA asymmetry in RNAi: function in RISC assembly and gene regulation. FEBS Lett 579(26):5850–5857

Iwata H, Niikura S, Matsuura SY, Takano Y, Ukai Y (2004) Genetic control of root shape at different growth stages in radish (*Raphanus sativus* L.). Breed Sci 54:117–124

Jones-Rhoades MW, Bartel DP, Bartel B (2006) MicroRNAs and their regulatory roles in plants. Annu Rev Plant Biol 57:19–53

Kawashima CG, Yoshimoto N, Maruyama-Nakashita A, Tsuchiya YN, Saito K, Takahashi H, Dalmay T (2009) Sulphur starvation induces the expression of microRNA-395 and one of its target genes but in different cell types. Plant J 57:313–321

Kidner CA, Martienssen RA (2005) The developmental role of microRNA in plants. Curr Opin Plant Biol 8:38–44

Khraiwesh B, Zhu J-K, Zhu J (2012) Role of miRNAs and siRNAs in biotic and abiotic stress responses of plants. Biochim Biophys Acta 1819:137–148

Kurihara Y, Watanabe Y (2004) Arabidopsis micro-RNA biogenesis through Dicer-like 1 protein functions. PNAS 101:12753–12758

Lee Y, Kim M, Han J, Yeom K, Lee S, Baek SH, Kim V (2004) MicroRNA genes are transcribed by RNA polymerase II. EMBO J 23:4051–4060

Li R, Yu C, Li Y, Lam TW, Yiu SM, Kristiansen K, Wang J (2009) SOAP2: an improved ultrafast tool for short read alignment. Bioinformatics 25(15):1966–1967

Li J, Guo G, Guo W, Guo G, Tong D, Ni Z, Sun Q, Yao Y (2012) miRNA164-directed cleavage of ZmNAC1 confers lateral root development in maize (*Zea mays* L.). BMC Plant Boil 12:220

Mallory AC, Dugas DV, Bartel B (2004) MicroRNA regulation of NAC-domain targets is required for proper formation and separation of adjacent embryonic, vegetative, and floral organs. Curr Biol 14:1035–1046

Mao WH, Li ZY, Xia XJ, Li YD, Yu QJ (2012) A combined approach of high-throughput sequencing and degradome analysis reveals tissue specific 463 expression of microRNAs and their targets in cucumber. PLoS ONE 7:e33040

Meng F, Liu H, Wang K, Liu L, Wang S, Zhao Y, Yin J, Li Y (2013) Development-associated microRNAs in grains of wheat (*Triticum aestivum* L.). BMC Plant Biol 13:140

Meyers BC, Axtell MJ, Bartel B, Bartel DP, Baulcombe D, Bowman JL, Cao X, Carrington JC, Chen X, Green PJ, Griffiths-Jones S, Jacobsen SE, Mallory AC, Martienssen RA, Poethig RS, Qi Y, Vaucheret H, Voinnet O, Watanabe Y, Weigel D, Zhu JK (2008) Criteria for annotation of plant microRNAs. Plant Cell 20:3186–3190

Moxon S, Jing R, Szittya G, Schwach F, Rusholme-Pilcher RL, Moulton V, Dalmay T (2008) Deep sequencing of tomato short RNAs identifies microRNAs targeting genes involved in fruit ripening. Genome Res 18:1602–1609

Perrin RM, DeRocher AE, Bar-Peled M, Zeng W, Norambuena L, Orellana A, Raikhel NV, Keegstra K (1999) Xyloglucan fucosyltransferase, an enzyme involved in plant cell wall biosynthesis. Science 284:1976–1979

Reinhart BJ, Weinstein EG, Rhoades MW, Bartel B, Bartel DP (2002) MicroRNAs in plants. Genes Dev 16:1616–1626

Rogers K, Chen X (2013) Biogenesis, turnover, and mode of action of plant microRNAs. Plant Cell 25:2383–2399

Rubatzky VE, Yamaguchi M (1997) World vegetables: principles, production, and nutritive values, 2nd edn. Chapman & Hall, New York

Schwab R, Palatnik JF, Riester M, Schommer C, Schmid M, Weigel D (2005) Specific effects of microRNAs on the plant transcriptome. Dev Cell 8:517–527

Tomotsugu K, Masahiko F, Masao T, Masaru O (2007) TCP transcription factors control the morphology of shoot lateral organs via negative regulation of the expression of boundary-specific genes in *Arabidopsis*. Plant Cell 19:473–484

Tsuro M, Suwabe K, Kubo N, Matsumoto S, Hirai M (2008) Mapping of QTLs controlling root shape and red pigmentation in radish, *Raphanus sativus* L. Breed Sci 58:55–61

Wang C, Han J, Liu C, Kibet KN, Kayesh E, Shangguan LF, Li X, Fang J (2012a) Identification of microRNAs from Amur grape (*vitis amurensis* Rupr.) by deep sequencing and analysis of microRNA variations with bioinformatics. BMC Genom 13:122

Wang FD, Li LB, Liu LF, Li HY, Zhang YH, Yao YY, Ni ZF, Gao JW (2012b) High-throughput sequencing discovery of conserved and novel microRNAs in Chinese cabbage (*Brassica rapa* L. ssp. pekinensis). Mol Genet Genomics 287(7):555–563

Williams L, Grigg SP, Xie MT, Christensen S, Fletcher JC (2005) Regulation of *Arabidopsis* shoot apical meristem and lateral organ formation by microRNA miR166 g and its AtHD-ZIP target genes. Development 132:3657–3668

Xu L, Wang Y, Zhai L, Xu Y, Wang L, Zhu X, Gong Y, Yu R, Limera C, Liu L (2013a) Genome-wide identification and characterization of cadmium-responsive microRNAs and their target genes in radish (*Raphanus sativus* L.) roots. J Exl Bot 64(14):4271–4287

Xu L, Wang Y, Xu Y, Wang L, Zhai L, Zhu X, Gong Y, Ye Q, Liu L (2013b) Identification and characterization of novel and conserved microRNAs in radish (*Raphanus sativus* L.) using high-throughput sequencing. Plant Sci 201–202:108–114

Yao YY, Guo GG, Ni ZF, Sunkar R, Zu JK, Sun QX (2007) Cloning and characterization of microRNAs from wheat (*Triticuma estivum* L.). Genome Biol 8: R96

Yoon EK, Yang JH, Lim J, Kim SH, Kim SK, Lee WS (2010) Auxin regulation of the microRNA390-dependent trans-acting small interfering RNA pathway in *Arabidopsis* lateral root development. Nucleic Acids Res 38(4):1382–1391

Zaki HEM, Yokoi S, Takahata Y (2010) Identification of genes related to root shape in radish (*Raphanus sativus*) using suppression subtractive hybridization. Breed Sci 60:130–138

Zhang X, Yue Z, Mei S, Qiu Y, Yang X, Chen X, Chen F, Wu Z, Sun Y, Jing Y, Liu B, Shen D, Wang H, Cui N, Duan Y, Wu J, Wang J, Gan C, Wang J, Wang X, Li X (2015) A de novo genome of a Chinese radish cultivar. Hortic Plant J 1(3):155–164

Genetic Profile of Glucosinolate Biosynthesis

10

Tomohiro Kakizaki and Masahiko Ishida

Abstract

Recent advances in science have clarified the biosynthesis pathway and functional role of secondary metabolites. They play a major role not only for completion of the plant life cycle but also for communication with other organisms. In Brassicaceae, including radish, the most well-characterized secondary metabolite is glucosinolate. Glucosinolates are sulfur-containing metabolite and their associated degradation products have distinctive benefits for human diet and defense against pests. Plants produce approximately 200 types of different glucosinolates and those from different species show great diversity, with their contents being affected by the environment, cultivation conditions, and genetic background. The profile of glucosinolates in radish is attractive, but its biosynthesis pathway remains unclear. Here, we highlight recent progress in glucosinolate research of model plant *Arabidopsis thaliana*. To compare researches on glucosinolate between radish and *A. thaliana*, we further discuss with specificity the nature of glucosinolate in radish.

10.1 Introduction

Glucosinolates are well-characterized plant secondary metabolites found in Brassicaceae (e.g., radish, cabbage, rapeseed, and *Arabidopsis*) and related plant families. An overview of the biosynthesis pathway of *Arabidopsis thaliana* was presented based on the observed natural variation in glucosinolate composition and genetic analysis. Furthermore, the advent of next-generation sequencing techniques allows for transcriptome analysis of Brassicaceae plants, which reveal the similarities and differences in

T. Kakizaki (✉)
Institute of Vegetable and Floriculture Science, NARO, 360, Kusawa, Ano, Tsu, Mie 514-2392, Japan
e-mail: tkaki@affrc.go.jp

M. Ishida
Institute of Vegetable and Floriculture Science, NARO, 3-1-1 Kannondai, Tsukuba 305-8666, Japan

T. Nishio and H. Kitashiba (eds.), *The Radish Genome*, Compendium of Plant Genomes, DOI 10.1007/978-3-319-59253-4_10

the glucosinolates biosynthesis pathway between species.

The degradation products of glucosinolates, primarily including isothiocyanates, are pungent compounds specific to Brassicaceae plants, which not only influence the taste of vegetables but also act as an attractant or repellent to certain insects. Isothiocyanates including sulforaphane are reported to possess bioactivity beneficial to humans, for example, anticarcinogenic activity or induction of detoxification enzymes. Therefore, the consumption of glucosinolate-containing Brassicaceae vegetables is important in human diet.

Radish (*Raphanus sativus* L., $2n = 18$) is an important Brassicaceae root vegetable, which has been cultivated since ancient times. The composition of glucosinolate shows a simple profile despite the root shape and weight showing large variations among germplasms. The most abundant glucosinolate in roots is glucoraphasatin (4-methylthio-3-butenyl glucosinolate), derived from methionine, and is essential for flavor and nutritional quality of the taproot.

This chapter presents a summary of topics related to the genes involved in glucosinolate biosynthesis in Brassicaceae, especially in radish, along with a summary of the chemical structures, biosynthesis pathways, and importance in plant breeding.

10.2 Structural Variations in Glucosinolates

Glucosinolates are sulfur- and nitrogen-comprising secondary metabolites having a common basic structure containing a β-D-thioglucose group, a sulfonated aldoxime moiety, and a variable side chain (=*R*) derived from amino acids (Fig. 10.1). Based on the precursor amino acid, glucosinolates can be classified into three groups: (1) aliphatic glucosinolates derived from methionine, isoleucine, leucine, or valine; (2) indolic glucosinolates derived from tryptophan; and (3) aromatic glucosinolates derived from phenylalanine or tryptophan. Thus, the side-chain structure primarily results in a diversity of structure; more than 200 structures have been identified (Clarke 2010; Fahey et al. 2001). Glucosinolates from different species show great diversity, with their contents being affected by the environment, cultivation conditions, and genetic background (Fig. 10.2). Genotype is the most important factor affecting glucosinolate profiles, for example, 39 accessions of *A. thaliana* possessed 34 types of glucosinolates containing aliphatic, indolic, and aromatic glucosinolates in leaves and seed, which are useful for the identification of genes encoding the enzymes involved in biosynthesis pathways (Kliebenstein et al. 2001a). *Brassica* vegetables contain various aliphatic glucosinolates that vary in length and modifications of the side chains (Ishida et al. 2014). These variations are explained based on the genomic structure of *Brassica* species.

The number of chromosomes varies in different species of *Brassica*. The genome relationship between three monogenomic species (comprising A, B, and C genomes) and three digenomic species is known as Triangle of U. In general, *Brassica nigra* (BB, $2n = 16$) contains glucosinolates with three carbon side chains derived from a single elongation reaction; *B. oleracea* (CC, $2n = 18$) contains glucosinolates with three or four carbon side chains; and *B. rapa* (AA, $2n = 20$) contains glucosinolates with either four or five carbon side chains. The glucosinolate composition of three amphidiploid *Brassica* species reflects two elementary species, for example, *B. napus* (AACC, $2n = 38$), an amphidiploid having the *B. rapa* and *B. oleracea* genomes, contains three, four, or five carbon side chains.

Glucoraphasatin (4-methylthio-3-butenyl glucosinolate), also known as dehydroerucin, glucodehydroerucin, or 4MTB-GSL, is an aliphatic glucosinolate predominantly found in radish roots, and comprises >90% of the total glucosinolates in Japanese white radish (Ishida et al. 2012). Although the presence of glucoraphasatin was reported in five genera of the Brassicaceae family, such as *Brassica* (Newkirk and Classen 2002; Thacker and Newkirk 2005), *Bunias* (Bennett et al. 2006), *Matthiola* (Bennett et al. 2004), *Raphanus* (Carlson et al. 1985; Ishii et al.

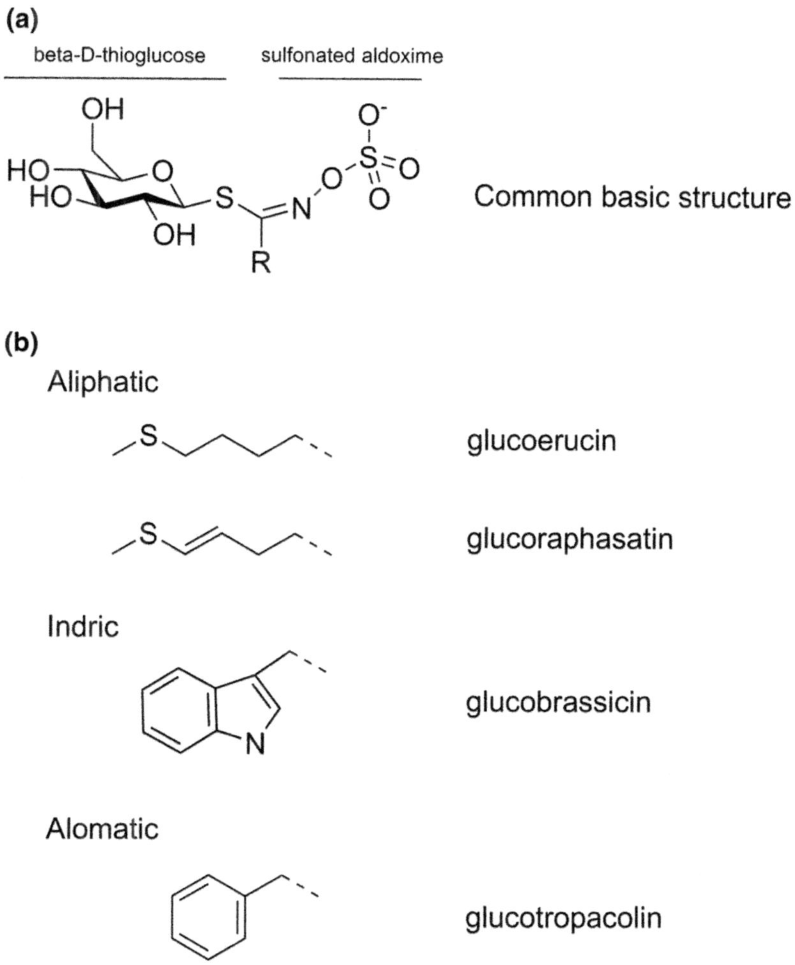

Fig. 10.1 Common basic structure **a** and representative structure of the side chain **b** of glucosinolate. *R*, variable side chain

1989), and *Rapistrum* (Curto et al. 2005), a species with >90% of the total glucosinolate content is restricted to *R. sativus*. In the 632 radish cultivars, glucoraphasatin content ranges from 43.8 to 475.5 μmol g^{-1} dry weight at the cotyledon stage (Ishida et al. 2015). Most cultivars contain two other related glucosinolates, glucoraphenin (4-methylsulfinyl-3-butenyl glucosinolate) and glucoerucin (4-methylthiobutyl glucosinolate), in smaller quantities. Glucoraphasatin is detected in various radish organs such as the seeds, seedlings, mature leaves, stems, and roots (Beevi et al. 2009; Ciska et al. 2008; Griffiths et al. 2001; Yamada et al. 2003).

10.3 Degradation of Glucosinolates

When plant tissue is damaged, glucosinolates are rapidly hydrolyzed by endogenous thioglucosidases called myrosinases (Fig. 10.3) (Rask et al. 2000). Myrosinases are accumulated in myrosin cells, which specifically form along the leaf veins (Andreasson et al. 2001; Rask et al. 2000; Ueda et al. 2006). In contrast, glucosinolates are stored in S-cells that are physically separated from myrosin cells (Koroleva et al. 2000, 2010). Once the tissue is mechanically damaged, glucosinolates are hydrolyzed intensively by myrosinases.

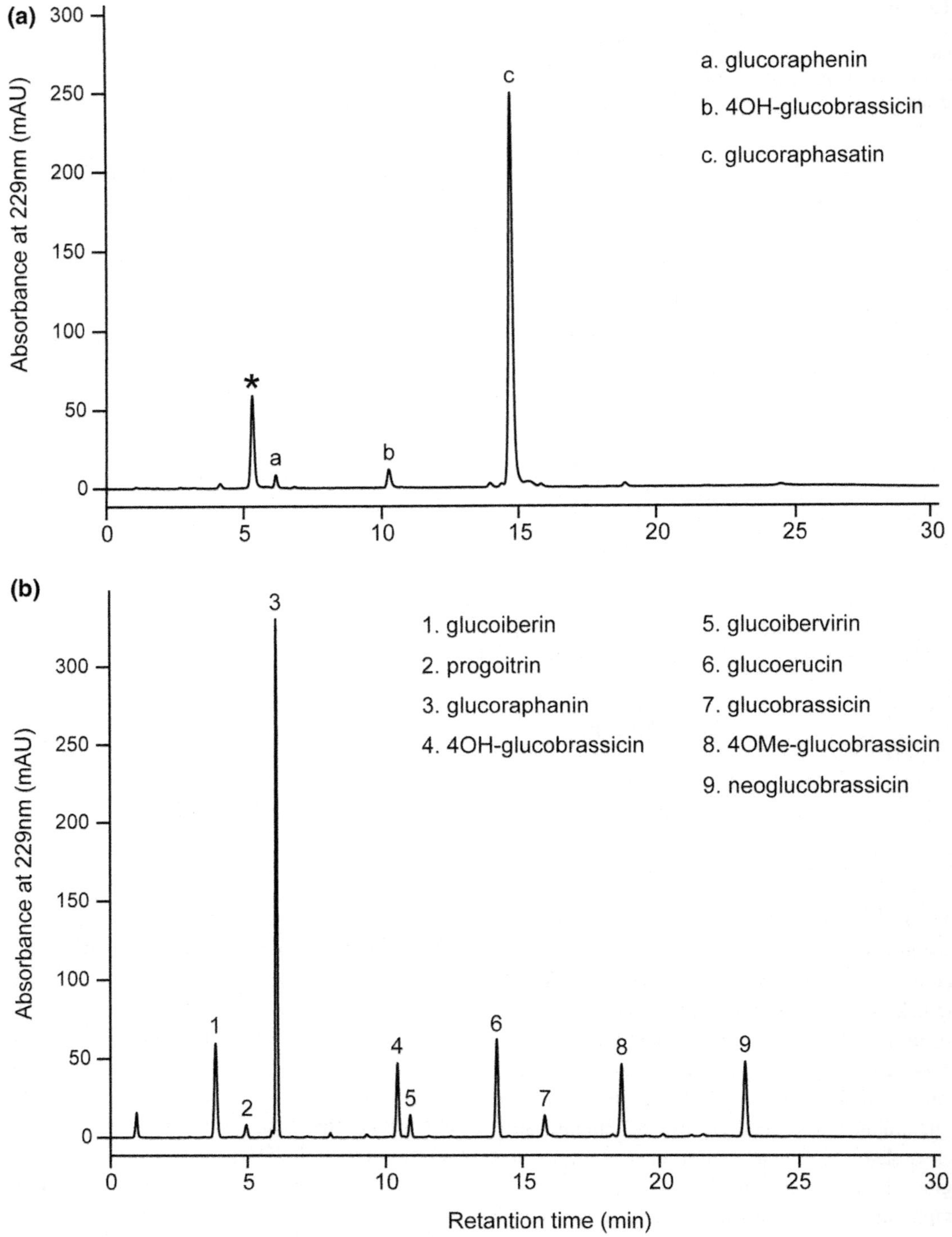

Fig. 10.2 Typical glucosinolate profile of radish (**a**) and broccoli (**b**). Preparation of desulfoglucosinolates with sulfatase digestion was conducted according to the method described by Bjerg and Sørensen (1987) for HPLC analysis. *Asterisk* shows internal control, sinigrin

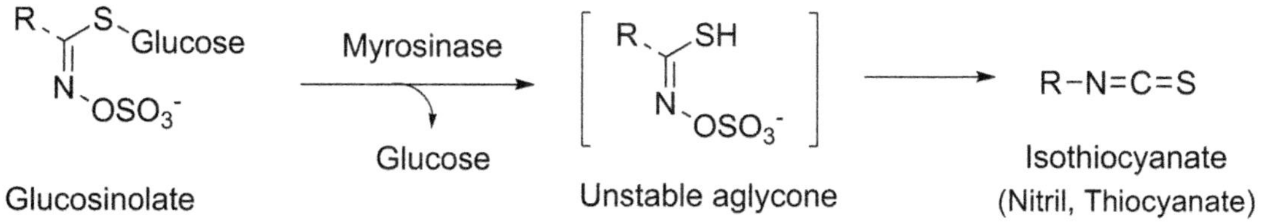

Fig. 10.3 Scheme of glucosinolate hydrolysis and structure of possible degradation products. Glucosinolates are hydrolyzed by myrosinase when tissues are damaged. At neutral pH (pH 5–7), the unstable aglucones change to isothiocyanates. The interaction of epithiospecifier protein (ESP) with myrosinase alters the reaction toward the production of nitriles depending on the glucosinolate structure and pH. Some glucosinolates can be hydrolyzed to thiocyanates. *R*, variable side chain

As a result of myrosinase activity, glucose and sulfate are released and some toxic/bioactive products, such as isothiocyanates, thiocyanates, nitriles, and epithionitriles, are formed depending on the chemical structure of the side chain, reaction conditions such as pH, presence of ferrous ions, and the modifier proteins (Kissen et al. 2009; Wittstock and Halkier 2002). The major degradation products, i.e., isothiocyanates, impart the characteristic taste and flavor to Brassicaceae vegetables. Several in vitro and in vivo studies have shown that sulforaphane (4-methylsulfinylbutane isothiocyanate), a degradation product of glucoraphanin abundant in broccoli sprouts, is associated with potential health-promoting activity (Juge et al. 2007). In radish, isothiocyanates derived from glucoraphasatin are pungent compounds that react with water and produce a yellow pigment and methanethiol, which are involved in the color and smell of pickles made from Asian big radish (Ozawa et al. 1990a, b).

10.4 Biosynthesis Pathway

Genetic and biochemical analyses show that the biosynthesis pathway involves the natural variation in glucosinolate profiles among *A. thaliana* accessions. Here, we focus on the biosynthesis pathway of aliphatic glucosinolate, which is abundant in *A. thaliana*, and outline the function of each enzyme involved in the pathway (Fig. 10.4).

The biosynthesis of aliphatic glucosinolates can be divided into three independent steps. First, the carbon chain of methionine is elongated by a chain elongation cycle in the chloroplast. Second, the core structure, comprising a β-D-thioglucose group and a sulfonated aldoxime moiety, is constructed. Third, the side chain is modified. Polymorphism or the absence/presence of respective genes determines the glucosinolate profile.

10.4.1 Chain Elongation

The side-chain elongation step in *A. thaliana* and *Eruca sativa* was confirmed by a feeding experiment using radio-labeled acetate (Graser et al. 2001, 2000). The side-chain elongation is initiated by transamination of methionine to form the corresponding 2-oxo acid. An *A. thaliana bcat4* mutant shows a decrease in aliphatic glucosinolate levels and an increase in free methionine levels, suggesting that BCAT4 catalyzes this step (Schuster et al. 2006). The newly formed 2-oxo acid is transported to the chloroplast via plastid-localized bile acid transporter 5 (BAT5) and BAT5/bile acid: sodium symporter family protein 5 (BASS5) (Gigolashvili et al. 2009; Sawada et al. 2009b). The transported 2-oxo acid is then involved in a cycle comprising three transformations: condensation with acetyl-CoA catalyzed by a methylthioalkylmalate (MAM) synthase, isomerization by an isopropylmalate isomerase (IPMI), and oxidative

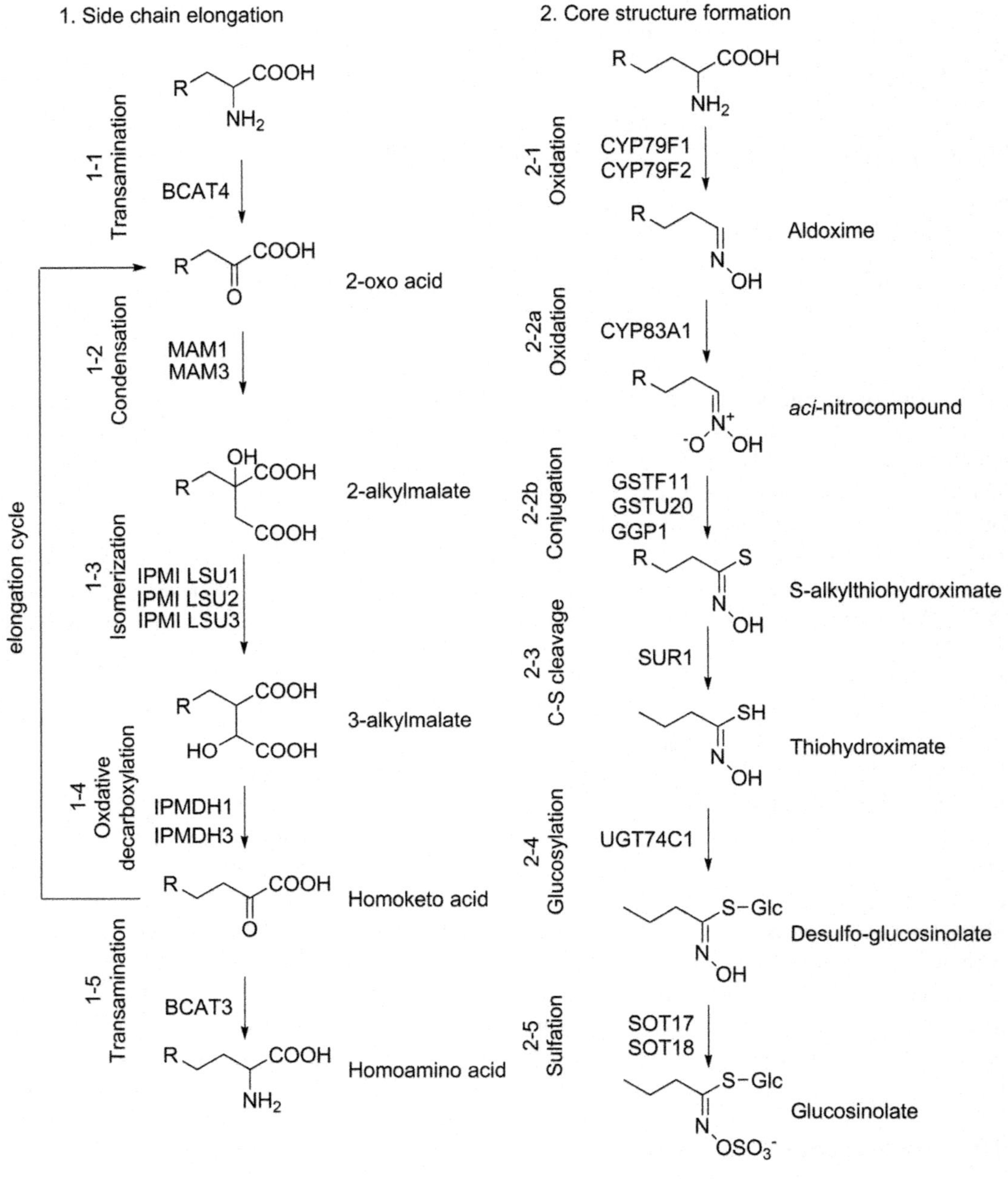

Fig. 10.4 Biosynthesis pathway of aliphatic glucosinolate. **a** Side-chain elongation steps. **b** Biosynthesis of core glucosinolate structure. *A. thaliana* gene identifiers are as follows: *BCAT4* (At3g19710); *MAM1* (At5g23010), *MAM3* (At5g23020); *IPMI LSU1* (At4g13430), *IPMI LSU2* (At2g43100), *IPMI LSU3* (At3g58990); *IPMDH1* (At5g14200), *IPMDH3* (At1g31180); *BCAT3* (At3g49680); *CYP79F1* (At1g16410), *CYP79F2* (At1g16400); *CYP83A1* (At4g13770); *GSTF11* (At3g03190), *GSTU20* (At1g78370), *GGP1* (At4g30530); *SUR1* (At2g20610); *UGT74C1* (At2g31790); *SOT17* (At1g18590), *SOT18* (At1g74090)

decarboxylation by an isopropylmalate dehydrogenase (IPMDH) (Field et al. 2004; Knill et al. 2009; Kroymann et al. 2001; Sawada et al. 2009a; Textor et al. 2007). The final step of the elongation cycle is the transamination by BCAT3 (Knill et al. 2008). The product of these reactions is a 2-oxo acid that is elongated by a single methylene (–CH_2–) moiety at each cycle.

10.4.2 Core Structure Formation

The aliphatic glucosinolate core structure is formed in five biochemical steps catalyzed by 11 enzymes (Grubb and Abel 2006). The first step is oxidation; precursor amino acids, which have elongated side chains, are converted to aldoximes by cytochrome P450 of the CYP79 family. CYP79F1 oxidizes all elongated methionines (1–6 carbons), while CYP79F2 only oxidizes long-chained methionines (5 and 6 carbons) (Chen et al. 2003; Hansen et al. 2001). The aldoximes are oxidized by CYP83A1 of the CYP79 family to an *aci*-nitro compound (Bak and Feyereisen 2001; Hemm et al. 2003). The activated forms are transformed to thiohydroximates via glutathione conjugation and a C–S lyase (SUR1) reaction (Mikkelsen et al. 2004). Thiohydroximates are, in turn, S-glucosylated by glucosyltransferases of the UGT74 family to form desulfoglucosinolates. UGT74C1 was shown to metabolize the methionine-derived thiohydroximates (Grubb et al. 2014). Finally, the thiohydroximates are converted to the glucosinolate structure by S-glucosyltransferases of the sulfotransferases (SOTs) (Piotrowski et al. 2004). After the glucosinolate structure is formed, the side chains are modified by oxygenation, hydroxylation, alkenylation, benzoylation, and methoxylation.

10.4.3 Side-Chain Modification

S-oxygenation is the first modification of aliphatic glucosinolates (Fig. 10.5). *S*-oxygenation of aliphatic glucosinolates is a common modification undergone by flavin-containing monooxygenase ($FMO_{GS\text{-}OXs}$). Phylogenetic analysis of plant FMOs revealed the presence of the

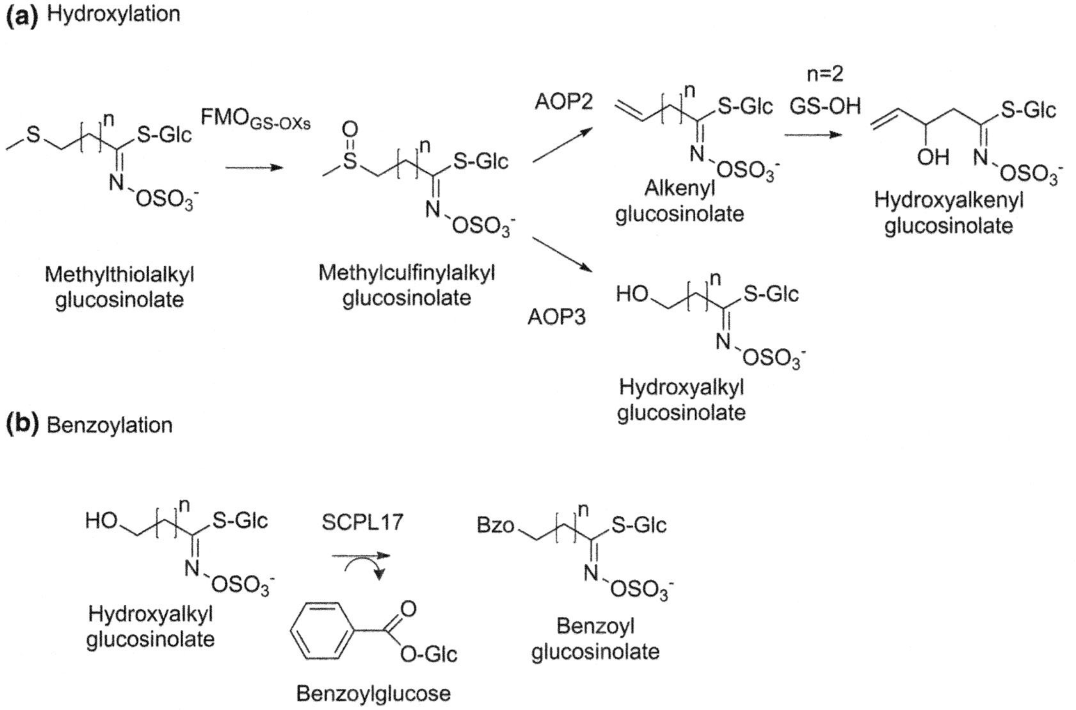

Fig. 10.5 Side-chain modification of aliphatic glucosinolate. *A. thaliana* gene identifiers are as follows: $FMO_{GS\text{-}OX1}$ (At1g65860), $FMO_{GS\text{-}OX2}$ (At1g62540), $FMO_{GS\text{-}OX3}$ (At1g62560), $FMO_{GS\text{-}OX4}$ (At1g62570), $FMO_{GS\text{-}OX5}$ (At1g12140); *AOP2* (At4g03060), *AOP3* (At4g03050), *GS-OH* (At2g25450); *SCPL17* (At3g12203)

Brassicaceae-specific clade, including $FMO_{GS\text{-}OXs}$, which are involved in the *S*-oxygenation of glucosinolates (Hansen et al. 2007; Li et al. 2008). Five $FMO_{GS\text{-}OXs}$ are encoded by the *A. thaliana* genome and are closely linked at chromosome 1. Using a recombinant enzyme assay, it was revealed that $FMO_{GS\text{-}OX1}$–$FMO_{GS\text{-}OX4}$ are able to catalyze the *S*-oxygenation independent of the side-chain length (Hansen et al. 2007; Li et al. 2008). In contrast, $FMO_{GS\text{-}OX5}$ is specific to long-chain (8C) glucosinolate (Li et al. 2008).

The next step of the side-chain modification is performed by three 2-oxoglutarate-dependent dioxygenases, AOP2 and AOP3, and GS-OH. Kliebenstein et al. (2001b) demonstrated the absence of methylsulfinylalkyl side-chain modification in null *aop2* accessions among 39 *A. thaliana* accessions. The over-expression line, which is a transgenic Columbia (Col-0) line having a functional *AOP2* transgene with the non-functional *AOP2* and *AOP3*, accumulated a methylsulfinylalkyl form (Neal et al. 2010). Interestingly, the introduction of a functional *AOP2* into Col-0, which has a null *AOP2*, showed that the total content of aliphatic glucosinolate was increased (Wentzell et al. 2007). AOP3 catalyzes the formation of hydroxyalkyl glucosinolates from methylsulfinylalkyl glucosinolates. None of the 21 *A. thaliana* accessions express both *AOPs* (Kliebenstein et al. 2001b). The GS-OH locus is responsible for the production of 2-hydroxybut-3-butenyl glucosinolate (Mithen et al. 1995; Parkin et al. 1994). The 2-oxoglutarate-dependent dioxygenase, required for 2-hydroxybut-3-butenyl glucosinolate formation, was identified using positional cloning of GS-OH (Hansen et al. 2008).

Another modification is the benzoylation. The high accumulation of benzoylated glucosinolates in *A. thaliana* seeds is known (Kliebenstein et al. 2007). In silico analysis, it was revealed that the three serine carboxypeptidase-like (SCPL) genes are co-expressed with AOP3 and BZO1, which synthesize the benzoate precursor cinnamoyl CoA in the seeds. A benzoate feeding experiment suggested that SCPL17 is involved in the final step of benzoylated glucosinolate biosynthesis (Lee et al. 2012).

10.5 Transcription Factors

The family of avian myeloblastosis virus (MYB) domain transcription factors includes the conserved MYB DNA-binding domain, which can control the transcriptional state of target genes in various plant secondary metabolism pathways (Jin and Martin 1999). In *A. thaliana*, MYB28, 29, and 76 orchestrate the transcription of aliphatic glucosinolates biosynthesis genes, while MYB34, 51, and 122 regulate the indolic glucosinolate pathway (Frerigmann and Gigolashvili 2014; Gigolashvili et al. 2008; Hirai et al. 2007; Malitsky et al. 2008; Sonderby et al. 2010). The functional role of MYB28 and MYB29 was characterized using the omics approach. Co-expression analysis demonstrated the relationship between metabolic pathway genes and transcription factor genes (Hirai et al. 2007). Double mutants in *MYB28* and *MYB29* showed complete absence of short- and long-chain aliphatic glucosinolates, while some indolic glucosinolates increased to a small extent. Interestingly, the biosynthesis of long-chain aliphatic glucosinolates was blocked by the absence of MYB28; however, short-chain aliphatic glucosinolates were reduced by approximately 50% in both the *myb28* and *myb29* single mutants (Beekwilder et al. 2008; Sonderby et al. 2007). In the case of indolic glucosinolate regulation, MYB34, MYB51, and MYB122 reciprocally regulate their expression and perform a central role in the gene expression of indolic glucosinolates pathway under the influence of various signals (Frerigmann and Gigolashvili 2014).

Because glucosinolate metabolism has evolved as a result of plant–environment interaction, its biosynthesis is regulated through various signals, including touch, injury, plant hormones, and insect infestation (Textor and Gershenzon 2008). Although glucosinolates perform the role of a constitutive defense compound during plant life cycle, insect herbivore feeding increases their levels (Wittstock and Halkier 2002); for example, mechanical wounding induces the expression of *MYB29* and *MYB76* within 1 min. Aliphatic glucosinolates are accumulated by the activation of MYB28,

MYB29, and MYB76, with each of these genes upregulating the expression of the others. By contrast, these MYBs downregulate the expression of genes involved in the control of indolic glucosinolate biosynthesis (Gigolashvili et al. 2008). Therefore, external stimuli are believed to be amplified by the expression control of transcription factors.

10.6 Transcriptome Analysis in Radish

Next-generation sequencing-based RNA sequencing of the transcriptome enables comprehensive gene expression analysis for non-model plants such as radish. Two independent studies revealed the expression profile of genes involved in the glucosinolate biosynthesis of radish (Mitsui et al. 2015; Wang et al. 2013). Mitsui et al. (2015) performed a comparison between the transcriptional profiles of glucosinolate biosynthesis in four tissues (root, root tip, cortex, and xylem) at six developmental stages. Interestingly, although the radish transcripts pool contained the majority of genes required for glucosinolate biosynthesis, nine genes (i.e., *RsMAM3*, *RsIPMI-SSU3*, *RsIPMDH3*, *RsCYP79F2*, *RsCYP81F1*, $RsFMO_{GS\text{-}OX3}$, $RsFMO_{GS\text{-}OX4}$, *RsAOP2*, and *RsAOP3*) were not identified. They suggested that the absence of these genes in the genome might explain the specific glucosinolate profile of radish. Wang et al. (2013) constructed a de novo assembly data set at three developmental stages (seedling, taproot thickening, and mature stages). Assembled unigenes included almost ortholog genes for *A. thaliana* glucosinolate biosynthesis genes and their transcription factors, without APOs and GS-OH (Wang et al. 2013). Accumulation of glucoraphasatin is observed in the root tips and outer zones, including the peels (Ishii 1991). Gene expression profiles reflect the distribution of glucoraphasatin. The results indicate that the numerous biosynthesis-related genes were strongly expressed in the root tip and cortex (Mitsui et al. 2015). Moreover, the intensity of gene expression appears to be more likely at a younger stage. These observations indicate that the spacial and developmental controls of biosynthesis genes are involved in glucoraphasatin accumulation. Genetic analysis revealed that three loci containing *RsMAM3*, *RsIPMDH1*, and *RsBCAT4* responsible for controlling the glucosinolate content in taproots were identified by the quantitative trait loci analysis using biparental F_2 populations. Furthermore, a clear difference in gene expression levels of these candidate genes was observed between parental lines at the taproot stage (Zou et al. 2013). From the transcriptome data, it is evident that there are no ortholog genes for 2-oxoglutarate-dependent dioxygenases, *AOP2* and *AOP3*, which catalyze the hydroxylation of side chains, in the radish genome (Mitsui et al. 2015). Thus, the end product of glucosinolate biosynthesis in radish might be glucoraphenin alone. These observations, such as the absence of *AOP2* and *AOP3*, and the presence of desaturase (see below section), could explain the dominant accumulation of 4-carbon side-chain glucosinolates in radish.

10.7 A Radish Mutant with an Altered Glucosinolate Profile

Glucoraphasatin is the predominant glucosinolate found in radish taproots (Carlson et al. 1985) and accounts for >90% of the total GSLs in Japanese cultivars (Ishida et al. 2012). Using a comprehensive analysis of glucosinolate profiles of global radish accessions and cultivars, a spontaneous mutant having significantly low glucoraphasatin was identified, and a completely glucoraphasatin-free line was generated by the self-pollination of the genetic resource (Fig. 10.6). Genetic analysis revealed that the glucoraphasatin-free trait was controlled by a single recessive locus, which is located at the edge of the linkage group 1 (Ishida et al. 2015).

The glucoraphasatin-free line contains glucoerucin, which accounts for >85% of the total glucosinolates, and 2% of glucoraphanin instead of glucoraphasatin and glucoraphenin (Ishida et al. 2015). Glucoraphasatin differs from

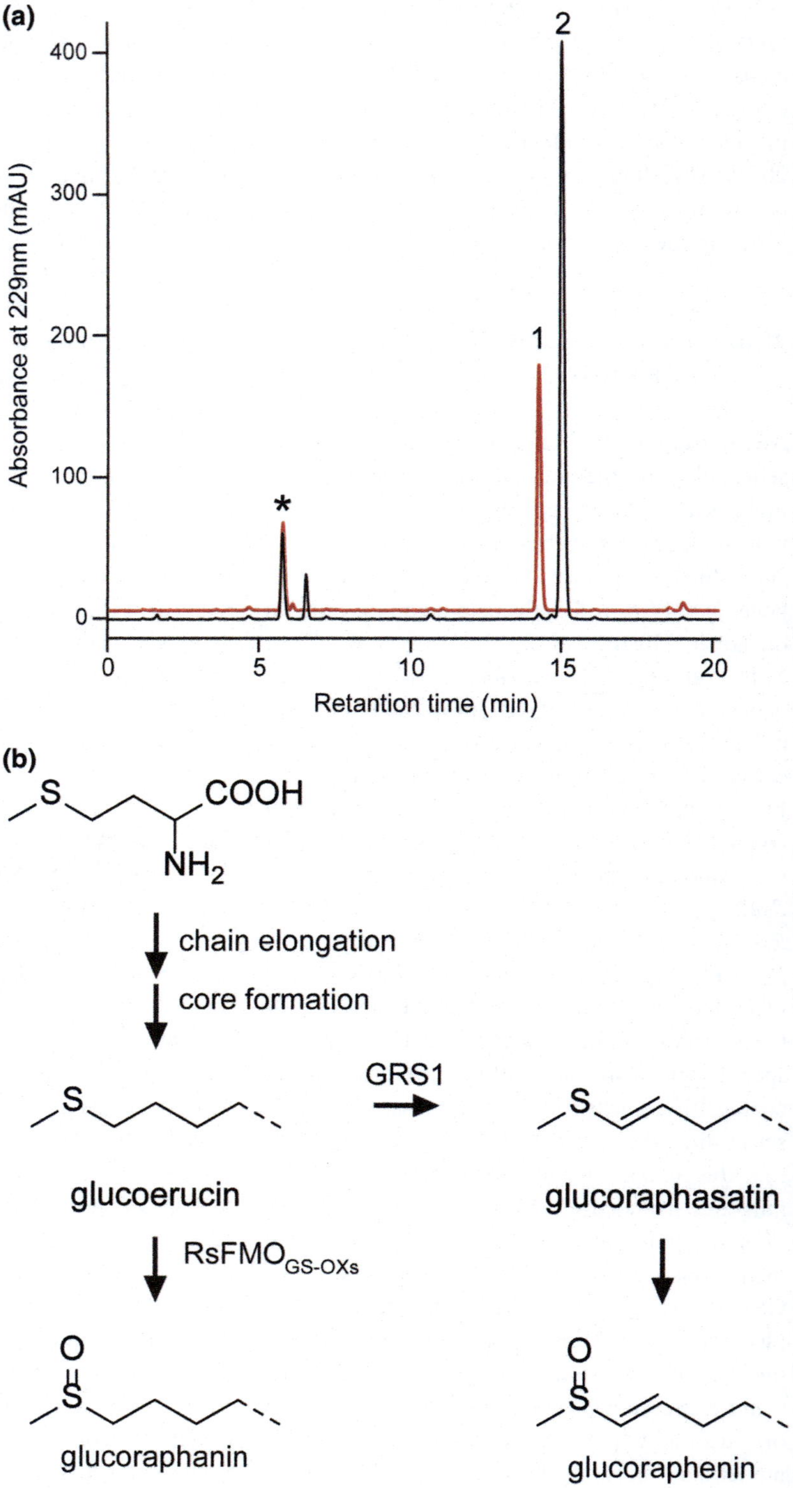

Fig. 10.6 **a** Typical HPLC profile obtained for the wild-type radish (*black line*) and glucoraphasatin-free mutant (*red line*). Peak 1, glucoerucin; peak 2, glucoraphasatin. *Asterisk* shows internal control, sinigrin. **b** A putative pathway of glucosinolate biosynthesis in radish. *GRS1* encoding an enzyme catalyzing the desaturation of glucoerucin side chain is deficient in glucoraphasatin-free mutant

glucoerucin only by the presence of a double bond in the side chain (Fig. 10.1b). These genetic data and glucosinolate profile show that the single gene, which encodes the enzyme involved in the desaturation of the side chain of glucoerucin, has lesions in the mutant line. In other words, radish might have obtained a gene that encodes desaturase, which is not present in other Brassicaceae plants. Similarly, a mutant which predominantly generates 4-methythiobutyl isothiocyanate derived from glucoerucin was identified in Japanese landrace 'Shibori-daikon' (Hori et al. 1999). Recently, we have identified a gene encoding a desaturase of side chain, GLUCORAPHASATIN SYNTHASE 1 (GRS1), by genetic mapping using a mutant that genetically lacks glucoraphasatin (Fig. 10.6b) (Kakizaki et al. 2017). GRS1 is a member of the 2-oxoglutarate and Fe(II)-dependent dioxygenase superfamily and transgenic *A. thaliana* which overexpressed GRS1 cDNA, accumulated glucoraphasatin in the leaves. These data present interesting observation that partly explains the specific GSL profile for radish.

10.8 Conclusion

Radish is an important vegetable that is consumed in a variety of ways: cooked, raw, pickled, brined, or dried. Glucosinolates in the taproots significantly affect the flavor and quality of radish. The glucosinolate profile of radish is found to be amazingly simple when compared with the other Brassicaceae, because glucoraphasatin accounts for more than 90% of the total glucosinolates in Japanese cultivars (Ishida et al. 2012). The breakdown product of glucoraphasatin by myrosinase generates yellow pigment and a methanethiol component that is responsible for the color and flavor of pickles (Takahashi et al. 2015). In our preliminary data, pickles made from the mutant having lesion in *GRS1* was it did not generate appreciable methanethiol. Utilization of non-functional *GRS1* gene should be useful in metabolite engineering for breeding of high-value vegetables.

References

Andreasson E, Jorgensen LB, Hoglund AS, Rask L, Meijer J (2001) Different myrosinase and idioblast distribution in *Arabidopsis* and *Brassica napus*. Plant Physiol 127:1750–1763

Bak S, Feyereisen R (2001) The involvement of two P450 enzymes, CYP83B1 and CYP83A1, in auxin homeostasis and glucosinolate biosynthesis. Plant Physiol 127:108–118

Beekwilder J, van Leeuwen W, van Dam NM, Bertossi M, Grandi V, Mizzi L, Soloviev M, Szabados L, Molthoff JW, Schipper B, Verbocht H, de Vos RCH, Morandini P, Aarts MGM, Bovy A (2008) The impact of the absence of aliphatic glucosinolates on insect herbivory in *Arabidopsis*. PLoS ONE 3:e2068

Beevi SS, Mangamoori LN, Dhand V, Ramakrishna DS (2009) Isothiocyanate profile and selective antibacterial activity of root, stem, and leaf extracts derived from *Raphanus sativus* L. Foodborne Pathog Dis 6:129–136

Bennett RN, Mellon FA, Kroon PA (2004) Screening crucifer seeds as sources of specific intact glucosinolates using ion-pair high-performance liquid chromatography negative ion electrospray mass spectrometry. J Agric Food Chem 52:428–438

Bennett RN, Rosa EA, Mellon FA, Kroon PA (2006) Ontogenic profiling of glucosinolates, flavonoids, and other secondary metabolites in *Eruca sativa* (salad rocket), *Diplotaxis erucoides* (wall rocket), *Diplotaxis tenuifolia* (wild rocket), and *Bunias orientalis* (Turkish rocket). J Agric Food Chem 54:4005–4015

Bjerg B, Sørensen H (1987) Quantitative analysis of glucosinolates and HPLC of intact glucosinolates. In: Wathelet J-P (ed.) Glucosinolates in Rapeseeds: analytical Aspects, Martinus Nijhoff Publishers, Dordrecht, Netherlands, pp. 125–150

Carlson DG, Daxenbichler ME, Vanetten CH, Hill CB, Williams PH (1985) Glucosinolates in radish cultivars. J Am Soc Hortic Sci 110:634–638

Chen SX, Glawischnig E, Jorgensen K, Naur P, Jorgensen B, Olsen CE, Hansen CH, Rasmussen H, Pickett JA, Halkier BA (2003) CYP79F1 and CYP79F2 have distinct functions in the biosynthesis of aliphatic glucosinolates in *Arabidopsis*. Plant J 33:923–937

Ciska E, Honke J, Kozlowska H (2008) Effect of light conditions on the contents of glucosinolates in germinating seeds of white mustard, red radish, white radish, and rapeseed. J Agric Food Chem 56:9087–9093

Clarke DB (2010) Glucosinolates, structures and analysis in food. Anal Methods 2:310–325

Curto G, Dallavalle E, Lazzeri L (2005) Life cycle duration of *Meloidogyne incognita* and host status of Brassicaceae and Capparaceae selected for glucosinate content. Nematology 7:203–212

Fahey JW, Zalcmann AT, Talalay P (2001) The chemical diversity and distribution of glucosinolates and isothiocyanates among plants. Phytochemistry 56:5–51

Field B, Cardon G, Traka M, Botterman J, Vancanneyt G, Mithen R (2004) Glucosinolate and amino acid biosynthesis in *Arabidopsis*. Plant Physiol 135: 828–839

Frerigmann H, Gigolashvili T (2014) MYB34, MYB51, and MYB122 distinctly regulate indolic glucosinolate biosynthesis in *Arabidopsis thaliana*. Mol Plant 7:814–828

Gigolashvili T, Engqvist M, Yatusevich R, Muller C, Flugge UI (2008) HAG2/MYB76 and HAG3/MYB29 exert a specific and coordinated control on the regulation of aliphatic glucosinolate biosynthesis in *Arabidopsis thaliana*. New Phytol 177:627–642

Gigolashvili T, Yatusevich R, Rollwitz I, Humphry M, Gershenzon J, Flugge UI (2009) The plastidic bile acid transporter 5 is required for the biosynthesis of methionine-derived glucosinolates in *Arabidopsis thaliana*. Plant Cell 21:1813–1829

Graser G, Schneider B, Oldham NJ, Gershenzon J (2000) The methionine chain elongation pathway in the biosynthesis of glucosinolates in *Eruca sativa* (Brassicaceae). Arch Biochem Biophys 378:411–419

Graser G, Oldham NJ, Brown PD, Temp U, Gershenzon J (2001) The biosynthesis of benzoic acid glucosinolate esters in *Arabidopsis thaliana*. Phytochemistry 57: 23–32

Griffiths DW, Deighton N, Birch AN, Patrian B, Baur R, Stadler E (2001) Identification of glucosinolates on the leaf surface of plants from the Cruciferae and other closely related species. Phytochemistry 57:693–700

Grubb CD, Abel S (2006) Glucosinolate metabolism and its control. Trends Plant Sci 11:89–100

Grubb CD, Zipp BJ, Kopycki J, Schubert M, Quint M, Lim EK, Bowles DJ, Pedras MS, Abel S (2014) Comparative analysis of *Arabidopsis* UGT74 glucosyltransferases reveals a special role of UGT74C1 in glucosinolate biosynthesis. Plant J 79:92–105

Hansen CH, Wittstock U, Olsen CE, Hick AJ, Pickett JA, Halkier BA (2001) Cytochrome P450 CYP79F1 from *Arabidopsis* catalyzes the conversion of dihomomethionine and trihomomethionine to the corresponding aldoximes in the biosynthesis of aliphatic glucosinolates. J Biol Chem 276:11078–11085

Hansen BG, Kliebenstein DJ, Halkier BA (2007) Identification of a flavin-monooxygenase as the S-oxygenating enzyme in aliphatic glucosinolate biosynthesis in *Arabidopsis*. Plant J 50:902–910

Hansen BG, Kerwin RE, Ober JA, Lambrix VM, Mitchell-Olds T, Gershenzon J, Halkier BA, Kliebenstein DJ (2008) A novel 2-oxoacid-dependent dioxygenase involved in the formation of the goiterogenic 2-hydroxybut-3-enyl glucosinolate and generalist insect resistance in *Arabidopsis*. Plant Physiol 148:2096–2108

Hemm MR, Ruegger MO, Chapple C (2003) The *Arabidopsis ref2* mutant is defective in the gene encoding CYP83A1 and shows both phenylpropanoid and glucosinolate phenotypes. Plant Cell 15:179–194

Hirai MY, Sugiyama K, Sawada Y, Tohge T, Obayashi T, Suzuki A, Araki R, Sakurai N, Suzuki H, Aoki K, Goda H, Nishizawa OI, Shibata D, Saito K (2007) Omics-based identification of *Arabidopsis* Myb transcription factors regulating aliphatic glucosinolate biosynthesis. Proc Natl Acad Sci USA 104:6478–6483

Hori K, Ohisa N, Suzuki M, Sato T, Kikawa A (1999) Chemical structure and quantitative determination of mustard oil in Shibori Daikon (Studies on special food resources in Akita part I). J Jpn Soc Food Sci 46:528–534 (In Japanese)

Ishida M, Nagata M, Ohara T, Kakizaki T, Hatakeyama K, Nishio T (2012) Small variation of glucosinolate composition in Japanese cultivars of radish (*Raphanus sativus* L.) requires simple quantitative analysis for breeding of glucosinolate component. Breed Sci 62:63–70

Ishida M, Hara M, Fukino N, Kakizaki T, Morimitsu Y (2014) Glucosinolate metabolism, functionality and breeding for the improvement of Brassicaceae vegetables. Breed Sci 64:48–59

Ishida M, Kakizaki T, Morimitsu Y, Ohara T, Hatakeyama K, Yoshiaki H, Kohori J, Nishio T (2015) Novel glucosinolate composition lacking 4-methylthio-3-butenyl glucosinolate in Japanese white radish (*Raphanus sativus* L.). Theor Appl Genet 128:2037–2046

Ishii G (1991) Glucosinolate in Japanese Radish, Raphanus-Sativus L. JARQ 24:273–279

Ishii G, Saijo R, Nagata M (1989) The difference of glucosinolate content in different cultivar of daikon roots (*Raphanus sativus* L.). J Jpn Soc Food Sci 36:739–742 (In Japanese)

Jin H, Martin C (1999) Multifunctionality and diversity within the plant MYB-gene family. Plant Mol Biol 41:577–585

Juge N, Mithen RF, Traka M (2007) Molecular basis for chemoprevention by sulforaphane: a comprehensive review. Cell Mol Life Sci 64:1105–1127

Kakizaki T, Kitashiba H, Zou Z, Li F, Fukino N, Ohara T, Nishio T, Ishida M (2017) A 2-oxoglutarate-dependent dioxygenase mediates the biosynthesis of glucoraphasatin in radish. Plant Physiol 173:1583–1593

Kissen R, Pope TW, Grant M, Pickett JA, Rossiter JT, Powell G (2009) Modifying the alkylglucosinolate profile in *Arabidopsis thaliana* alters the tritrophic interaction with the herbivore *Brevicoryne brassicae* and parasitoid *Diaeretiella rapae*. J Chem Ecol 35:958–969

Kliebenstein DJ, Kroymann J, Brown P, Figuth A, Pedersen D, Gershenzon J, Mitchell-Olds T (2001a) Genetic control of natural variation in *Arabidopsis* glucosinolate accumulation. Plant Physiol 126:811–825

Kliebenstein DJ, Lambrix VM, Reichelt M, Gershenzon J, Mitchell-Olds T (2001b) Gene duplication in the diversification of secondary metabolism: tandem 2-oxoglutarate-dependent dioxygenases control glucosinolate biosynthesis in *Arabidopsis*. Plant Cell 13:681–693

Kliebenstein DJ, D'Auria JC, Behere AS, Kim JH, Gunderson KL, Breen JN, Lee G, Gershenzon J, Last RL, Jander G (2007) Characterization of seed-specific benzoyloxyglucosinolate mutations in *Arabidopsis thaliana*. Plant J 51:1062–1076

Knill T, Schuster J, Reichelt M, Gershenzon J, Binder S (2008) *Arabidopsis* branched-chain aminotransferase 3 functions in both amino acid and glucosinolate biosynthesis. Plant Physiol 146:1028–1039

Knill T, Reichelt M, Paetz C, Gershenzon J, Binder S (2009) *Arabidopsis thaliana* encodes a bacterial-type heterodimeric isopropylmalate isomerase involved in both Leu biosynthesis and the Met chain elongation pathway of glucosinolate formation. Plant Mol Biol 71:227–239

Koroleva OA, Davies A, Deeken R, Thorpe MR, Tomos AD, Hedrich R (2000) Identification of a new glucosinolate-rich cell type in *Arabidopsis* flower stalk. Plant Physiol 124:599–608

Koroleva OA, Gibson TM, Cramer R, Stain C (2010) Glucosinolate-accumulating S-cells in *Arabidopsis* leaves and flower stalks undergo programmed cell death at early stages of differentiation. Plant J 64:456–469

Kroymann J, Textor S, Tokuhisa JG, Falk KL, Bartram S, Gershenzon J, Mitchell-Olds T (2001) A gene controlling variation in *Arabidopsis* glucosinolate composition is part of the methionine chain elongation pathway. Plant Physiol 127:1077–1088

Lee S, Kaminaga Y, Cooper B, Pichersky E, Dudareva N, Chapple C (2012) Benzoylation and sinapoylation of glucosinolate R-groups in *Arabidopsis*. Plant J 72:411–422

Li J, Hansen BG, Ober JA, Kliebenstein DJ, Halkier BA (2008) Subclade of flavin-monooxygenases involved in aliphatic glucosinolate biosynthesis. Plant Physiol 148:1721–1733

Malitsky S, Blum E, Less H, Venger I, Elbaz M, Morin S, Eshed Y, Aharoni A (2008) The transcript and metabolite networks affected by the two clades of *Arabidopsis* glucosinolate biosynthesis regulators. Plant Physiol 148:2021–2049

Mikkelsen MD, Naur P, Halkier BA (2004) *Arabidopsis* mutants in the *C–S* lyase of glucosinolate biosynthesis establish a critical role for indole-3-acetaldoxime in auxin homeostasis. Plant J 37:770–777

Mithen R, Clarke J, Lister C, Dean C (1995) Genetics of aliphatic glucosinolates. III. Side-chain structure of aliphatic glucosinolates in *Arabidopsis thaliana*. Heredity 74:210–215

Mitsui Y, Shimomura M, Komatsu K, Namiki N, Shibata-Hatta M, Imai M, Katayose Y, Mukai Y, Kanamori H, Kurita K, Kagami T, Wakatsuki A, Ohyanagi H, Ikawa H, Minaka N, Nakagawa K, Shiwa Y, Sasaki T (2015) The radish genome and comprehensive gene expression profile of tuberous root formation and development. Sci Rep 5:10835

Neal CS, Fredericks DP, Griffiths CA, Neale AD (2010) The characterisation of *AOP2*: a gene associated with the biosynthesis of aliphatic alkenyl glucosinolates in *Arabidopsis thaliana*. BMC Plant Biol 10:170

Newkirk RW, Classen HL (2002) The effects of toasting canola meal on body weight, feed conversion efficiency, and mortality in broiler chickens. Poult Sci 81:815–825

Ozawa Y, Kawakishi S, Uda Y, Maeda Y (1990a) Isolation and identification of a novel beta-carboline derivative in salted radish roots, *Raphanus sativus* L. Agric Biol Chem 54:1241–1245

Ozawa Y, Uda Y, Kawakishi S (1990b) Generation of a beta-carboline derivative, the yellowish precursor of processed radish roots, from 4-methylthio-3-butenyl isothiocyanate and L-tryptophan. Agric Biol Chem 54:1849–1851

Parkin I, Magrath R, Keith D, Sharpe A, Mithen R, Lydiate D (1994) Genetics of aliphatic glucosinolates. II. Hydroxylation of alkenyl glucosinolates in *Brassica napus*. Heredity 72:594–598

Piotrowski M, Schemenewitz A, Lopukhina A, Muller A, Janowitz T, Weiler EW, Oecking C (2004) Desulfoglucosinolate sulfotransferases from *Arabidopsis thaliana* catalyze the final step in the biosynthesis of the glucosinolate core structure. J Biol Chem 279:50717–50725

Rask L, Andreasson E, Ekbom B, Eriksson S, Pontoppidan B, Meijer J (2000) Myrosinase: gene family evolution and herbivore defense in Brassicaceae. Plant Mol Biol 42:93–113

Sawada Y, Kuwahara A, Nagano M, Narisawa T, Sakata A, Saito K, Hirai MY (2009a) Omics-based approaches to methionine side chain elongation in *Arabidopsis:* characterization of the genes encoding methylthioalkylmalate isomerase and methylthioalkylmalate dehydrogenase. Plant Cell Physiol 50:1181–1190

Sawada Y, Toyooka K, Kuwahara A, Sakata A, Nagano M, Saito K, Hirai MY (2009b) *Arabidopsis* bile acid:sodium symporter family protein 5 is involved in methionine-derived glucosinolate biosynthesis. Plant Cell Physiol 50:1579–1586

Schuster J, Knill T, Reichelt M, Gershenzon J, Binder S (2006) branched-chain aminotransferase4 is part of the chain elongation pathway in the biosynthesis of methionine-derived glucosinolates in *Arabidopsis*. Plant Cell 18:2664–2679

Sonderby IE, Hansen BG, Bjarnholt N, Ticconi C, Halkier BA, Kliebenstein DJ (2007) A systems biology approach identifies a R2R3 MYB gene subfamily with distinct and overlapping functions in regulation of aliphatic glucosinolates. PLoS ONE 2: e1322

Sonderby IE, Burow M, Rowe HC, Kliebenstein DJ, Halkier BA (2010) A complex interplay of three R2R3 MYB transcription factors determines the profile of aliphatic glucosinolates in *Arabidopsis*. Plant Physiol 153:348–363

Takahashi A, Yamada T, Uchiyama Y, Hayashi S, Kumakura K, Takahashi H, Kimura N, Matsuoka H (2015) Generation of the antioxidant yellow pigment derived from 4-methylthio-3-butenyl isothiocyanate in salted radish roots (takuan-zuke). Biosci Biotechnol Biochem 79:1512–1517

Textor S, Gershenzon J (2008) Herbivore induction of the glucosinolate-myrosinase defense system: major trends, biochemical bases and ecological significance. Phytochem Rev 8:149–170

Textor S, de Kraker JW, Hause B, Gershenzon J, Tokuhisa JG (2007) MAM3 catalyzes the formation of all aliphatic glucosinolate chain lengths in *Arabidopsis*. Plant Physiol 144:60–71

Thacker PA, Newkirk RW (2005) Performance of growing-finishing pigs fed barley-based diets containing toasted or non-toasted canola meal. Can J Anim Sci 85:53–59

Ueda H, Nishiyama C, Shimada T, Koumoto Y, Hayashi Y, Kondo M, Takahashi T, Ohtomo I, Nishimura M, Hara-Nishimura I (2006) AtVAM3 is required for normal specification of idioblasts, myrosin cells. Plant Cell Physiol 47:164–175

Wang Y, Pan Y, Liu Z, Zhu X, Zhai L, Xu L, Yu R, Gong Y, Liu L (2013) *De novo* transcriptome sequencing of radish (*Raphanus sativus* L.) and analysis of major genes involved in glucosinolate metabolism. BMC Genom 14:836

Wentzell AM, Rowe HC, Hansen BG, Ticconi C, Halkier BA, Kliebenstein DJ (2007) Linking metabolic QTLs with network and *cis*-eQTLs controlling biosynthetic pathways. PLoS Genet 3:1687–1701

Wittstock U, Halkier BA (2002) Glucosinolate research in the *Arabidopsis* era. Trends Plant Sci 7:263–270

Yamada K, Hasegawa T, Minami E, Shibuya N, Kosemura S, Yamamura S, Hasegawa KI (2003) Induction of myrosinase gene expression and myrosinase activity in radish hypocotyls by phototropic stimulation. J Plant Physiol 160:255–259

Zou Z, Ishida M, Li F, Kakizaki T, Suzuki S, Kitashiba H, Nishio T (2013) QTL analysis using SNP markers developed by next-generation sequencing for identification of candidate genes controlling 4-methylthio-3-butenyl glucosinolate contents in roots of radish *Raphanus sativus* L. PLoS One 8:e53541

11 Genes for Bolting and Flowering

Hiroyasu Kitashiba and Shuji Yokoi

Abstract

Floral transition which initiates bolting and flowering is induced in flowering plants by environmental factors such as low temperature and photoperiod and by endogenous factors such as growing stage and Phytohormone. The molecular mechanism in the floral transition has attracted considerable interest, and, in particular, agronomic application by genetic and physiological control of bolting and flowering times has been the subject of many researchers and breeders seeking to develop cultivars efficiently. Study of the floral transition has been remarkably advanced in *Arabidopsis thaliana* as a model plant using a variety of genomic resources such as various ecotypes and information on genome sequences and mutant lines. Four major flowering pathways in *Arabidopsis* have been found, and gene networks in these pathways have been revealed in the past about 20 years. Based on studies in *Arabidopsis*, genetic studies on bolting and flowering times have been conducted in *Brassica* species such as *Brassica rapa* and *Brassica oleracea*. Following these studies, high-density genetic maps and genome sequences have been recently published in radish (*Raphanus sativus* L.), and several studies to identify genetic factors or loci associated with bolting and flowering times have also been attempted. Here, we review genetic studies on bolting and flowering in *Arabidopsis* and *Brassica* species and then introduce recent studies in radish together with results of our current attempts.

11.1 Introduction

Bolting and flowering in flowering plants are observed just after the growing stage of plants transmits from the vegetative phase to the reproductive phase. The floral transition is initiated by several environmental cues, such as day

H. Kitashiba
Graduate School of Agricultural Science, Tohoku University, Sendai, Japan

S. Yokoi (✉)
Graduate School of Life and Environmental Sciences, Osaka Prefecture University,
1-1, Gakuen-cho, Naka-ku, Sakai 599-8531, Japan
e-mail: shyokoi@plant.osakafu-u.ac.jp

T. Nishio and H. Kitashiba (eds.), *The Radish Genome*,
Compendium of Plant Genomes, DOI 10.1007/978-3-319-59253-4_11

length and temperature, and endogenous factors, such as developmental age and Phytohormones. Because *Arabidopsis thaliana* has the smallest size of genome, *A. thaliana* is considered to be a 'model plant' and is much more ahead in genome resources, e.g., information on genome sequences, mutant lines, and so on, than any other plant species (Provart et al. 2016; Redei 1975). These resources have been used to clarify the molecular mechanisms of variation of physiological traits. Studies on the mechanisms of bolting and flowering have also been remarkably advanced by the use of genomic resources. Bolting and flowering in *A. thaliana* are initiated by long-day conditions, and the initiation is promoted by cold exposure (Jack 2004). The former is regulated by a photoperiod pathway and the latter by a vernalization pathway. The number of rosette leaves under short-day conditions increases much more than those under long-day conditions in *Arabidopsis*, and the bolting and flowering times are considerably delayed. This type of flowering initiation is controlled by an autonomous pathway (Soppe et al. 1999). Furthermore, gibberellin, which is a kind of Phytohormone, also induces bolting and flowering through an independent network (gibberellin pathway) (Mouradov et al. 2002). Although a variety of genes participate in the respective pathways, signals from the pathways are integrated into *FT* (*Flowering Locus T*) or *SOC1* (*Suppressor of overexpression of CONSANS 1*) and then transmission of the signals is resumed downstream, prompting floral transition through *AP1* (*Apetala 1*) and *LFY* (*Leafy*) genes (Yanovsky and Kay 2002). In this chapter, first we review the four pathways regulating the bolting and flowering times in *A. thaliana*, including integrators, i.e., *FT* and *SOC1*.

Brassica rapa, *Brassica oleracea*, and *Brassica napus* are also members of Brassicaceae and include important vegetables and oil crops such as turnip, Chinese cabbage, cabbage, and rapeseed. Chinese cabbage, cabbage, and turnip are harvested before flowering. Bolting spoils the quality of these vegetables. In rapeseeds, on the other hand, flowering is required for production, and flowering time is the trait determining suitable production areas. Therefore, the bolting and flowering times are important traits for breeding. In F_1 hybrid breeding of these species, flowering time of parental lines is required to be identical for a crossing. For this purpose, understanding the genetic control of flowering time is necessary for efficient F_1 hybrid breeding. Recently, molecular genetic studies on bolting and flowering have also been conducted in *B. rapa* and *B. oleracea* based on studies in *Arabidopsis*. However, since the multi-copies (paralogous genes) for each gene involved in bolting and flowering times are present in the genomes because of ancestral duplication or triplication of genomes in *B. rapa* and *B. oleracea*, the mechanism in these *Brassica* species is inferred to be more complicated than that in *Arabidopsis*. In the second section, therefore, we review recent studies on the flowering trait in *Brassica* species, especially focusing on the vernalization pathway.

Raphanus sativus, radish, also belongs to the same family as *Brassica* species. Radish is an important vegetable crop in which the F_1 hybrid breeding method is utilized. Although genomic studies in *R. sativus* have lagged behind those in *Arabidopsis* and *Brassica*, several candidates thought to be involved in bolting and flowering times have been identified. More recently, a huge amount of SSR and SNP markers have been developed, and high-density genetic maps have been constructed together with publication of draft genome sequences (Kitashiba et al. 2014). Using basic genome information, we attempt to identify loci and genes responsible for bolting and flowering times. In the third section, therefore, we review recent reports and introduce our latest genetic studies in *Raphanus*.

11.2 Genetic Mechanisms of Flowering in *Arabidopsis*

Followings are brief descriptions of models explaining the different ways in which flowering responses can be induced in *Arabidopsis*. Schematic diagrams of flowering pathways in *Arabidopsis* are shown in Fig. 11.1.

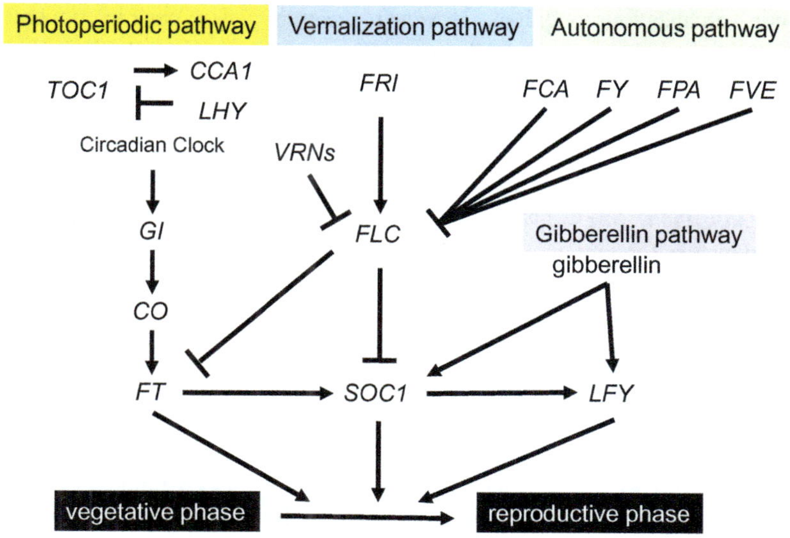

Fig. 11.1 Schematic diagram of flowering pathways in *Arabidopsis*

11.2.1 Photoperiod Pathway

Under long-day conditions, leaves perceive a light signal (step 1), which is transmitted to a circadian clock (step 2), and a series of signal through circadian clock step promote expression of flowering genes, including integrator genes (step 3). Phytochrome A perceives red light, whereas Phytochrome B perceives both red and far-red light (Quail 2002). Blue light is perceived by Cryptochrome (Lin 2000). Based on physiological analyses and genetic analyses using mutants, *Phytochrome A* (*PhyA*), *Phytochrome B* (*PhyB*), and *Cryptochrome 2* (*Cry2*) have been found to be essential for an initiation of flowering (Mockler et al. 2003). In step 2, three genes, i.e., *Timing of CAB expression 1* (*TOC1*), *Circadian Clock Associated 1* (*CCA1*), and *Late Elongated Hypocotyl* (*LHY*), play central roles in the control of circadian rhythm (Huang and Nusinow 2016). In the morning, as the transcripts of both *CCA1* and *LHY* increase, the transcript of *TOC1* is suppressed. To the contrary, as the transcripts of both *CCA1* and *LHY* decrease in the afternoon, the transcripts of *TOC1* are up-regulated. The up- and down-regulations produce a circadian rhythm in *Arabidopsis*, which is transmitted to the bolting and flowering steps. In step 3, firstly *GIGANTEA* (*GI*) (Huq et al. 2000) perceives the signals from step 1 and step 2, and then up-regulation and down-regulation of *GI* gene expression also are repeated, responding to the circadian rhythm. The rhythmical regulation of the *GI* gene controls the expression pattern of the *CONSTANS* (*CO*) gene (Hayama and Coupland 2003; Putterill et al. 1995; Simpson and Dean 2002; Yanovsky and Kay 2002). Finally, CO activates *Flowering Locus T* (*FT*) and *Suppressor of overexpression of CO 1* (*SOC1*) genes, resulting in the promotion of the floral transition (Song et al. 2016).

11.2.2 Vernalization Pathway

Among *Arabidopsis* accessions, there are natural variations in bolting and flowering times, i.e., winter-annual type (late flowering) and rapid-cycling type (early flowering). The former requires long-term cold exposure for floral transition and this response is called vernalization. The floral transition of the latter type accessions is induced without vernalization. Using these genetic resources together with several mutants, *Flowering Locus C* (*FLC*), which is a MADS-box gene, has been found to be a key gene for regulation of floral transition (Michaels and Amasino 1999). Transcripts of the *FLC* gene are highly accumulated in shoot apical meristem cells of the winter-annual accession under non-vernalization conditions. However, the transcripts are decreased gradually, depending on a term of cold exposure, resulting in no transcript at all (Sheldon et al. 2000). Molecular genetic and physiological studies have revealed that the suppression of *FLC* is regulated epigenetically by *Vernalization Insensitive 3* (*VIN3*), which is responsible for initiating the modification of the *FLC* chromatin structure, and that *Vernalization 1* (*VRN1*) and *VRN2*, which are the genes for a DNA-binding protein and a polycomb-group protein, respectively, stably maintain the *FLC* suppression (Bond et al. 2009). Under normal conditions, the activity of CArG-box present in the promoter of *SOC1*, which is an integrator of the bolting and flowering pathway, is repressed by FLC protein (Hepworth et al. 2002). However, as the transcripts of *FLC* are decreased by cold exposure, the expression of *SOC1* becomes up-regulated and accelerates the floral transition under long-day conditions.

Another key gene in the vernalization pathway is *FRIGIDA* (*FRI*), which has been found to enhance *FLC* function as an activator (Johanson et al. 2000). A dominant functional allele and a recessive non-functional allele of *FRI* are reported to be present in *A. thaliana* accessions (Gazzani et al. 2003). Since *flc* mutant having the functional allele of *FRI* did not show late flowering, the functional *FRI* is suggested to suppress the floral transition depending on the presence of functional *FLC* allele as an enhancer (Michaels and Amasino 2001).

11.2.3 Autonomous Pathway

Many rapid-cycling type accessions have a functional *FLC* together with a recessive non-functional *FRI* (Johanson et al. 2000). However, they show early flowering because of suppression of the *FLC* gene through an autonomous pathway (Moon et al. 2005). Genetic analysis has revealed that this autonomous pathway consists of four major genes, i.e., *FCA* (Macknight et al. 1997), *FY* (Simpson et al. 2003), *FPA* (Schomburg et al. 2001), and *FVE* (Ausin et al. 2004), and that these genes suppress the expression of the *FLC* gene in cooperation with three minor genes, i.e., *LUMINIDEPENDENS* (*LD*) (Lee et al. 1994), *Flowering Locus K* (*FLK*) (Lim et al. 2004), and *Flowering Locus D* (*FLD*) (Sanda and Amasino 1996). Other biochemical studies have examined molecular mechanisms of suppression of *FLC*. *FCA* encodes a protein containing RNA-binding domain and a protein-binding WW domain (Macknight et al. 1997). This gene is remarkably expressed in stem and root meristem cells, and the FCA protein negatively regulates *FLC* through direct interaction with *FLC* mRNAs (Macknight et al. 1997). FY protein has a WD40 domain and suppresses *FLC* gene expression through DNA-binding in cooperation with *FCA* (Simpson et al. 2003). In addition, *FPA*, which encodes a protein containing the RNA-binding domain (Schomburg et al. 2001), and *FVE,* which encodes a protein containing the WD40 domain, have been found to modify the chromatin structure of the *FLC* gene in cooperation with *FCA*, causing a suppression of transcription (Lim et al. 2004). Finally, the autonomous pathway epigenetically suppresses the *FLC* gene together with the vernalization pathway in control of bolting and flowering times, resulting in initiation of floral transition.

11.2.4 Gibberellin Pathway

Under the short-day conditions, bolting and flowering times in *A. thaliana* are delayed. In a molecular event of this phenomenon, REPRESSOR of *ga1-3* (RGA) protein, which is a member of the DELLA-family, is accumulated in leaf cells and shoot apical meristem cells and suppresses expressions of the integrator genes, i.e., *LFY* and *SOC1*. However, synthesis of active GA, i.e., GA_4, in the shoot apical meristem or exogenous GA_4 treatment induces a degradation of RGA protein through the 26S proteasome pathway (Cheng et al. 2004), resulting in the up-regulation of integrator genes (Achard et al. 2007). In *LFY*, induction of transcription is initiated by AtMYB33 (Gocal et al. 2001). After degradation of RGA protein, the expression of the *SOC1* gene is elevated, and a heterodimer of

SOC1 and AGL24 promotes transcription of *LFY* (Lee et al. 2008; Liu et al. 2008). This gibberellin pathway is independent of the pathways via *FLC*.

11.3 Genetic Studies in *Brassica* Species

Based on studies in *Arabidopsis*, a variety of genetic studies have been advanced in *B. rapa*, *B. oleracea*, and *B. napus*, which are major crops in *Brassica* species. To identify loci associated with bolting and flowering times in *B. rapa*, Ajisaka et al. (2001) and Nishioka et al. (2005) conducted QTL analyses and then detected one locus under a vernalized condition and ten loci in total under five conditions with or without vernalization, respectively. However, these loci cannot be assigned to the reference sequences of *B. rapa* (Wang et al. 2011), because the markers used in the studies are anonymous markers. Li et al. (2009) conducted QTL analyses for the bolting and flowering times with vernalization and non-vernalization using F_2 populations from a cross between a doubled haploid line of cv. 'Osome,' which requires vernalization, and an inbred line of cv. 'Yellow Sarson,' which is an early-flowering Indian oilseed rape, and detected two significant QTLs located on A02 and A10. In *B. rapa*, four paralogous *FLC* genes (*BrFLC1*, *BrFLC2*, *BrFLC3*, and *BrFLC5*) are known to be distributed on three chromosomes, i.e., A02, A03, and A10, according to the published pseudomolecule sequences (Wang et al. 2011). One QTL on A02 contained *BrFLC2* and another on A10 contained *BrFLC1*. The *BrFLC2* locus, especially, has been reproducibly detected as a QTL for bolting and flowering times in several reports (Kitamoto et al. 2014; Lou et al. 2007; Schranz et al. 2002; Zhao et al. 2010). Transcripts of *BrFLC2* in 'Yellow sarson' were much lower than those in 'Osome' (Li et al. 2009), similar to the report by Zhao et al. (2010), possibly due to several polymorphisms in the 5' region (Li et al. 2009). A large insertion (ca. 5 kb) was also observed in the first intron of *BrFLC2* of the late bolting cultivar 'Tsukena No. 2' (Kitamoto et al. 2014). Expression of the gene during vernalization was not so much decreased in 'Tsukena No. 2' (Kitamoto et al. 2014). From these reports, *BrFLC2* is suggested to be an important gene for bolting and flowering times in *B. rapa*.

In *B. oleracea*, Schranz et al. (2002) identified three *FLC* paralogous genes (*BoFLC1*, *BoFLC3*, and *BoFLC5*) and a fourth gene (*BoFLC2*) was cloned by Okazaki et al. (2007). A genetic study by QTL analysis using broccoli (non-vernalization type) and cabbage (vernalization type) showed that only the *BoFLC2* locus is associated with the bolting and flowering times. Further genomic analyses revealed that a frameshift occurred in exon 4 of the gene derived from broccoli because of 1 bp deletion, and the gene was deleted in the genome of rapid-cycling lines (Okazaki et al. 2007). Regardless of the *Brassica* species, a common derivative gene, *FLC2*, is suggested to be important for bolting and flowering times.

In *B. napus*, Zou et al. (2012) identified nine *FLC* homologues distributed on six chromosomes, BnaA02, BnaA03, BnaA10, BnaC02, BnaC03, and BnaC09. Among them, the expression pattern of the *BnFLC.A3b* gene which exhibited high similarity with *BrFLC5* and *BoFLC5* was different between winter type and semi-winter type cultivars, and a greater number of incorrectly spliced transcripts were detected in the semi-winter type than in the winter type. These results suggest a significant association of *BnFLC.A3b* with bolting and flowering times in *B. napus* (Zou et al. 2012).

11.4 Current Genetic Studies in Radish

Although the genetic mechanism for the bolting and flowering times in radish has been little revealed, several approaches have been attempted to find loci or genes associated with the traits.

11.4.1 Genes Involved in Bolting and Flowering Network in Radish

To know networks of genes responsible for bolting and flowering regulation, transcripts in leaves at proper growing stages in late-bolting radish have been comprehensively surveyed by a high-throughput RNA sequencing technology (Nie et al. 2016). Among 53,642 unigenes of radish, 24 candidate genes associated with the plant circadian clock mechanism, such as *CO* and *GI*, were identified. In addition, 142 potential genes, including flowering pathway integrators such as *FT*, *SOC1*, and *LFY*, and their regulators such as *FLC* were also identified as genes associated with bolting and flowering. A similar experiment conducted by Jung et al. (2016), using two inbred lines with different flowering times, demonstrated that the expression patterns of genes related to the vernalization pathway showed expected correlations with the phenotypes of the two lines, suggesting that the vernalization pathway is conserved between radish and *Arabidopsis*. In addition to analysis of expressed unigenes, microRNAs (miRNAs) related to bolting and flowering times have been collected (Nie et al. 2015). Forty-two known miRNAs and seventeen novel miRNAs were identified as bolting-related miRNAs.

Among the candidate genes described above, several genes have been investigated to elucidate their function in radish. In the *GI* gene of radish, transgenic radish plants with low expression of *GI* caused by the introduction of an antisense *GI* sequence driven by *CaMV35S* promoter were produced (Curtis et al. 2002), and a significant delay in both bolting time and flowering time (about 18 days) was observed in the transgenic plants compared with wild type radish plants. This result suggests that *GI* is obviously involved in both traits and that the bolting time and flowering time could be controlled by regulation of the *GI* expression level.

As described in the above paragraphs, *FLC* is the most significant gene for bolting and flowering times in *Arabidopsis* and *Brassica* species. In radish, three *FLC* paralogous genes, i.e., *RsFLC1*, *RsFLC2*, and *RsFLC3*, which are orthologous to *BrFLC1*, *BrFLC2*, and *BrFLC3*, respectively, have been identified in radish by de novo assembled transcriptome analysis (Yi et al. 2014; Nie et al. 2015). According to our search based on the information of radish genome sequences published by Kitashiba et al. (2014) and Jeong et al. (2016), *RsFLC1*, *RsFLC2*, and *RsFLC3* are distributed on three chromosomes, i.e., Rs7, Rs2, and Rs3 (chromosome number is defined in Chap. 3), respectively. Each gene has a genome structure similar to that in *A. thaliana* and *Brassica* species, i.e., seven exons and six introns (Yi et al. 2014) with high similarity. Although all the *RsFLC* genes were expressed in radish under non-vernalized conditions, the respective expression levels were decreased in proportion to the duration of cold exposure (Yi et al. 2014). In particular, the expression patterns of *RsFLC1* and *RsFLC3* exhibited strong correlations with bolting and flowering times. Overexpression of all the *RsFLC* genes caused late-flowering phenotypes in *Arabidopsis*, indicating that every single *RsFLC* gene functions as a regulator of bolting and flowering times.

11.4.2 QTL Analysis for Bolting and Flowering

In *B. rapa* and *B. oleracea*, several QTL analyses have been carried out to find the genome regions responsible for bolting time and flowering time, whereas little QTL analysis for the traits in radish has been reported so far. Recently, we attempted QTL analyses to find genome regions for bolting time and flowering time, using two populations.

First, 'Sayatori' (rat's tail radish) inbred line (Fig. 11.2a), which is an early-bolting type not requiring vernalization for bolting, and 'Aokubi' inbred line (Fig. 11.2a), which is a late-bolting type requiring vernalization for bolting, were crossed with each other to produce an F_2 population. This parental combination is the same as those reported by Li et al. (2011) and Kitashiba et al. (2014). The F_2 individual plants were grown in a field of the Field Center of Tohoku University, Kawatabi, Japan, under the same conditions as reported by Li et al. (2011). Bolting states at 76 days after sowing in November 2007 were scored from 0 to 4 (Fig. 11.2b): 0, no floral bud; 1, floral buds at the base; 2, floral stem shorter than 1/3 of the average height of floral stem having open flowers; 3, floral stems shorter than 2/3 of the average height; and 4, floral stems higher than 2/3 of the average height. The bolting score of 'Aokubi' was 0 (n = 3), while that of 'Sayatori' was 3.8 (n = 5). The score of F_1 plants was 1.0 (n = 5), suggesting that non-bolting is dominant. Scores in the 159 F_2 plants exhibited a continuous distribution (Fig. 11.2c). On the other hand, F_2 plants were genotyped using 171 DNA markers including 10 SCAR (sequence characterized amplified region), 9 CAPS, and 152 SNP markers, resulting in a linkage map consisting of 9 linkage groups which corresponded to the respective linkage groups reported by Li et al. (2011). QTL analysis was carried out using both phenotype and genotype data, and then two QTLs were detected (Table 11.1). One, designated SA-BS1 ('Sayatori-Aokubi'-Bolting Score 1), was detected in a region between RS2CL5281s and RS2CL3345s (Kitashiba et al. 2014) in Rs7 with the highest genetic effect, i.e., 33.7% explained phenotypic variance (Table 11.1). Another QTL, designated as SA-BS2, was detected in the region between RS2CL1710s and RS2CL3017s (Kitashiba et al. 2014) in Rs2 with 19.7% explained phenotypic variance (Table 11.1).

Second, we attempted another QTL analysis using another population derived from a cross between a 'Yumehomare' inbred line, which is a late flowering type, and a 'Taibyosobutori' inbred line. 'Taibyosobutori' is much earlier than 'Yumehomare' in bolting and flowering times (Fig. 11.3a, b). One hundred eighty-five F_2 plants were cultivated in a field of Iwate University, Morioka, Japan. Seeds of F_2 were sown in November 2010, and then days from sowing to appearance of floral buds at the base (bolting time) and days from sowing to opening of the fifth flower (flowering time) were recorded. As well as the above population, a continuous distribution was observed in both traits (Fig. 11.3a, b). Additionally, quite a high correlation (R^2 = 0.92) was observed between both phenotype data. A linkage map was constructed by genotyping of F_2 individual plants with 114 SNP markers produced by Li et al. (2011). Using both phenotype and genotype data, QTL analyses were carried out. Two significant QTLs for bolting time, YT-BT1 ('Yumehomare-Taibyosobutori'-Bolting Time 1) and YT-BT2, were detected in regions between RS2CL9061s and RS2CL5452s (Kitashiba et al. 2014) in Rs2 and between RS2CL4386s and RS2CL2344s in Rs7, respectively (Table 11.1). Regarding flowering time, YT-FT1 ('Yumehomare-Taibyosobutori'-Flowering Time 1) was detected in the same region as YT-BT1 in Rs2 (Table 11.1). Interestingly, the YT-BT1 and YT-FT1 QTL regions overlapped with the above-mentioned SA-BS2, although the parental lines used in the two studies were different from each other. These results suggest that the overlapped region including SA-BS2 is significantly associated with bolting and flowering in radish (Fig. 11.4).

Based on the comparison of syntenic genome regions between *R. sativus* and *A. thaliana*, the SA-BS2 region was found to show a high synteny with a part of chromosome 5 of *A. thaliana*, from At5g01530 locus to At5g13760 locus. The syntenic region contained the *FLC* gene (At5g10140). The *RsFLC2* gene is confirmed to be present in the SA-BS2 region. Additionally, the SA-BS1 region also showed a high synteny with a part of chromosome 5 containing *FLC*. Actually, the *FLC* marker which was produced based on *BrFLC1* sequences (Li et al. 2011) was mapped in the SA-BS1 region, and the location of *RsFLC1* was confirmed in the SA-BS1 region by the published information of genome sequences (Jeong et al. 2016; Kitashiba et al. 2014). These results exhibited that the *FLC* gene is also a key regulatory gene for initiation of bolting and flowering in radish as well as in *A. thaliana* and *Brassica* species. A part of the third QTL between RS2CL4386s and RS2CL2344s included a high syntenic region with a region from At1g66070 to At1g78320 (Kitashiba et al. 2014; Li et al. 2011) where photoperiod genes such as *FLM* (*Flowering Locus M*) (At1g77080) and *TEM2* (*Tempranillo 2*) (At1g68840) are located. This suggests that *FLM* and *TEM2* might

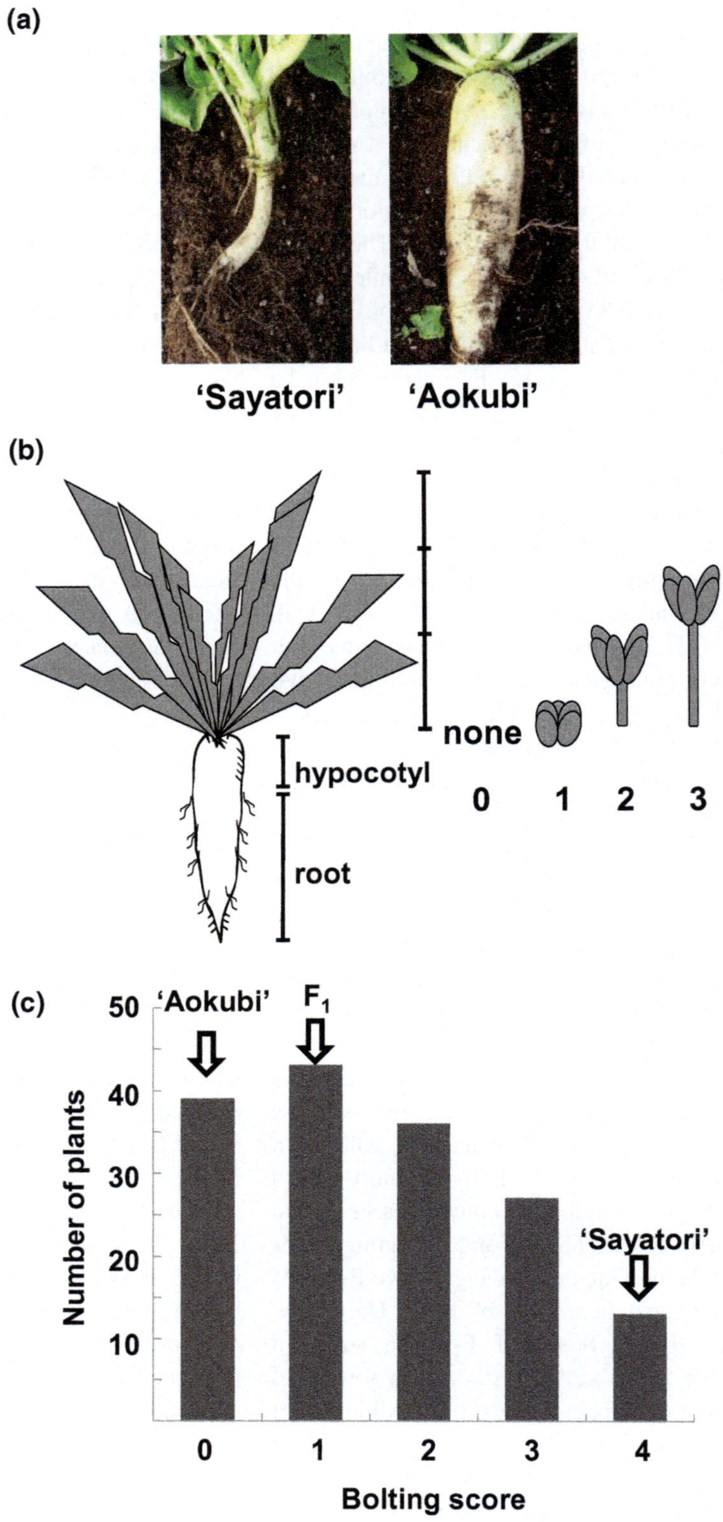

Fig. 11.2 Plant materials and measurement of phenotypic scores. **a** Root shape of parental inbred lines. **b** Scores of bolting conditions defined in this study. **c** Distribution of bolting scores in F_2 plants

Table 11.1 QTLs for bolting and flowering times in F_2 populations

Populations	Traits	QTL names	Linkage groups (Li et al. 2011)	Linkage groups (Chap. 3)	Regions	LOD scores	Additive effects	Explained phenotypic variance
'Sayatori' × 'Aokubi'	Bolting score	SA-BS1	LG9	Rs7	RS2CL5281s–RS2CL3345s	13.4	1.0[a]	0.34
		SA-BS2	LG4	Rs2	RS2CL1710s–RS2CL3017s	8.5	0.8[a]	0.19
'Yumehomare' × 'Taibyosobutori'	Bolting time	YT-BT1	LG4	Rs2	RS2CL9061s–RS2CL5452s	32.2	20.1[b]	0.56
		YT-BT2	LG9	Rs7	RS2CL4386s–RS2CL2344s	5.8	5.0[b]	0.08
	Flowering time	YT-FT1	LG4	Rs2	RS2CL9061s–RS2CL5452s	42.4	22.2[b]	0.69

[a]Additive effect of the 'Sayatori' allele
[b]Additive effect of the 'Yumehomare' allele

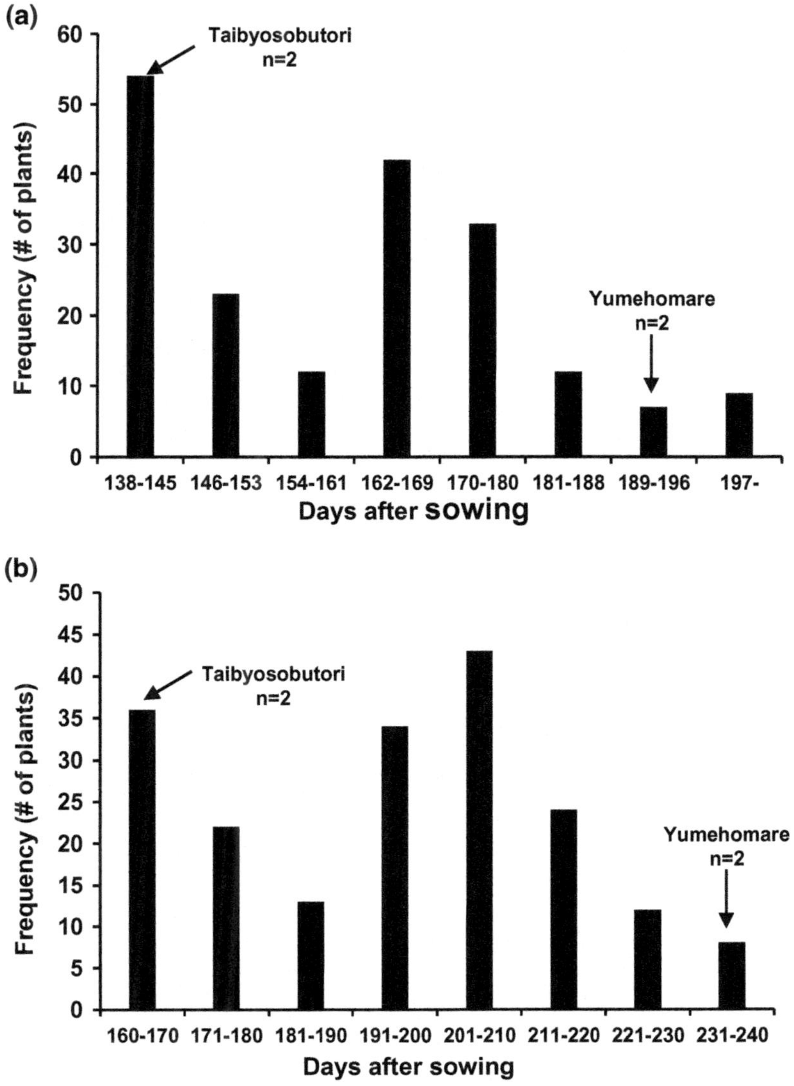

Fig. 11.3 Frequency distribution of bolting time (**a**) and flowering time (**b**) in F_2 plants (185 individuals) and parental lines ('Yumehomare' and 'Taibyosobutori'). Positions and # of plants of the parental lines in the distributions are indicated by *arrows*

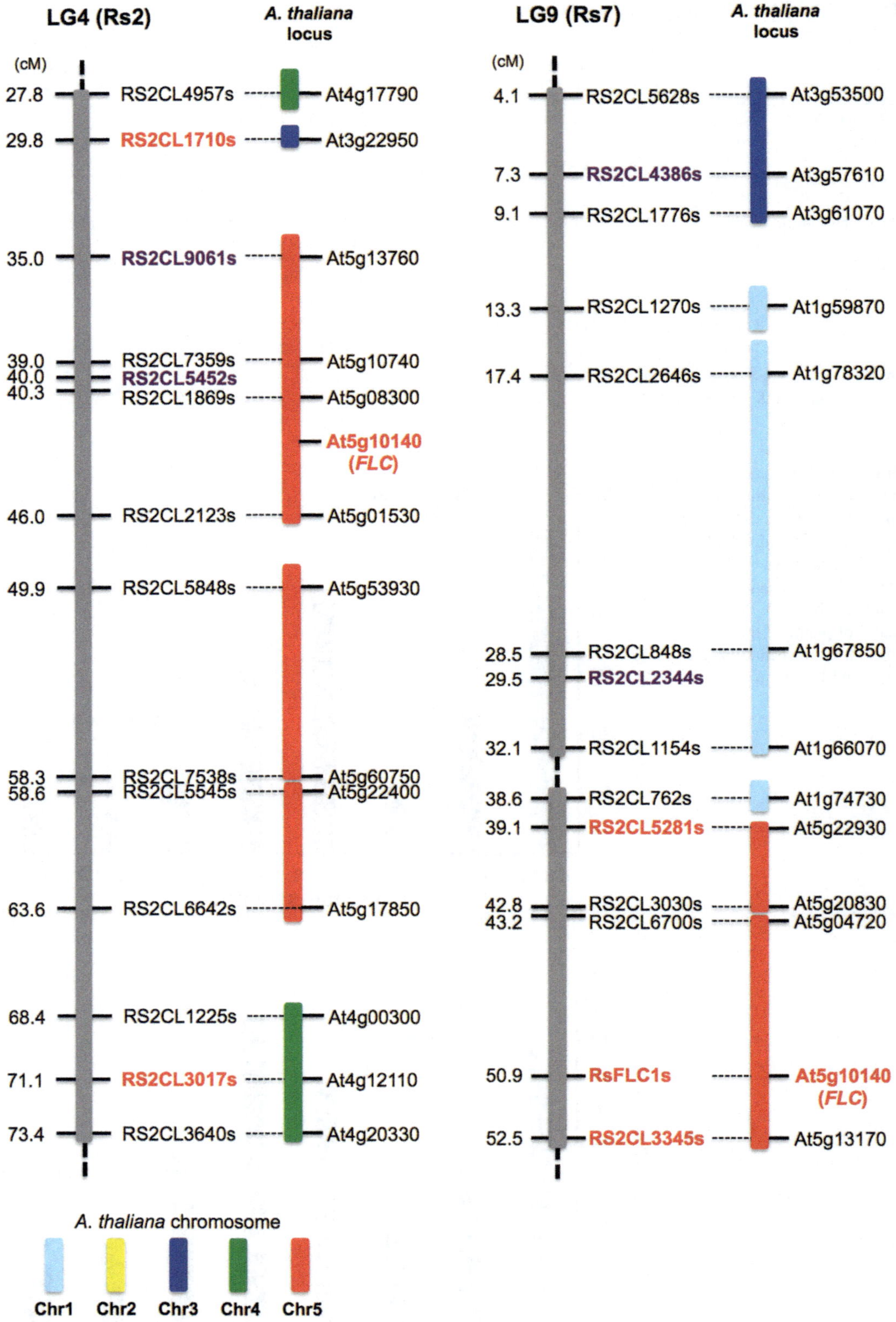

Fig. 11.4 Comparison of syntenic regions in detected QTLs between radish and *Arabidopsis*

also be significant genes for bolting and flowering times in radish.

11.5 Future Prospects for the Development of Breeding Techniques for Bolting and Flowering Times

Based on findings of recent genetic studies in radish, the *FLC* gene was inferred to play an important role in the floral transition. A variety of polymorphisms, such as SNPs in the 6th intron (Li et al. 2011) and the 7th exon of *RsFLC1* causing no-substitution of amino acid residue (our unpublished data) between 'Sayatori' and 'Aokubi,' and a number of variations of *RsFLC2* and *RsFLC3* among nine accessions from RDA-GENEBANK collection (Yi et al. 2014) were found. Although Yi et al. (2014) reported no association of any DNA polymorphisms with flowering time, further studies using a greater number of cultivars and considering a combination of *RsFLC* alleles would yield useful information. Additionally, associations of genes other than *FLC,* such as *FT*, *SOC1*, and *LFY*, should also be considered for radish breeding because radish can be grown all year round and the bolting and flowering times are inferred to be influenced by photoperiod conditions and other internal and external factors. Considering studies in *Arabidopsis*, *FLC* is also inferred to be little involved in the photoperiodic pathway in radish. Investigation of polymorphisms in genes responsible for the circadian clock mechanism and the integrator together with *GI* and *CO* among cultivars is also necessary. These further studies will enable clarification of the mechanisms for bolting and flowering and development of MAS (marker assisted selection) techniques for radish breeding in the near future.

References

Achard P, Baghour M, Chapple A, Hedden P, Straeten DVD, Genschik P, Moritz T, Harberd NP (2007) The plant stress hormone ethylene controls floral transition via DELLA-dependent regulation of floral meristem-identity genes. Proc Natl Acad Sci USA 104:6484–6489

Ajisaka H, Kuginuki Y, Yui S, Enomoto S, Hirai M (2001) Identification and mapping of a quantitative trait locus controlling extreme late bolting in Chinese cabbage (*Brassica rapa* L. ssp. *pekinensis* syn. *campestris* L.) using bulked segregant analysis. Euphytica 118:75–81

Ausin I, Alonso-Blanco C, Jarillo JA, Ruiz-Garcia L, Martinez-Zapater JM (2004) Regulation of flowering time by *FVE*, a retinoblastoma-associated protein. Nat Genet 36:162–166

Bond DM, Dennis ES, Pogson BJ, Finnegan EJ (2009) Histone acetylation, VERNALIZATION INSENSITIVE 3, FLOWERING LOCUS C, and the vernalization response. Mol Plant 2:724–737

Cheng H, Qin L, Lee S, Fu X, Richards DE, Cao D, Luo D, Harberd NP, Peng J (2004) Gibberellin regulates *Arabidopsis* floral development via suppression of DELLA protein function. Development 131:1055–1064

Curtis IS, Nam HG, Yun JY, Seo KH (2002) Expression of an antisense *GIGANTEA* (*GI*) gene fragment in transgenic radish causes delayed bolting and flowering. Transgenic Res 11:249–256

Gazzani S, Gendall AR, Lister C, Dean C (2003) Analysis of the molecular basis of flowering time variation in *Arabidopsis* accessions. Plant Physiol 132:1107–1114

Gocal GFW, Sheldon CC, Gubler F, Moritz T, Bagnall DJ, MacMillan CP, Li SF, Parish RW, Dennis ES, Weigel D, King RW (2001) GAMYB-like genes, flowering, and gibberellin signaling in *Arabidopsis*. Plant Physiol 127:1682–1693

Hayama R, Coupland G (2003) Shedding light on the circadian clock and the photoperiodic control of flowering. Curr Opin Plant Biol 6:13–19

Hepworth S, Valverde F, Ravenscroft D, Mouradov A, Coupland G (2002) Antagonistic regulation of flowering-time gene *SOC1* by *CONSTANS* and *FLC* via separate promoter motifs. EMBO J 21:4327–4337

Huq E, Tepperman JM, Quail PH (2000) GIGANTEA is a nuclear protein involved in phytochrome signaling in *Arabidopsis*. Proc Natl Acad Sci USA 97:9789–9794

Huang H, Nusinow DA (2016) Into the evening: Complex interactions in the *Arabidopsis* circadian clock. Trends Genet 32:674–686

Jack T (2004) Molecular and genetic mechanisms of floral control. Plant Cell 16(Suppl):S1–17

Jeong YM, Kim N, Ahn BO, Oh M, Chung WH, Chung H, Jeong S, Lim KB, Hwang YJ, Kim GB, Baek S, Choi SB, Hyung DJ, Lee SW, Sohn SH, Kwon SJ, Jin M, Seol YJ, Chae WB, Choi KJ, Park BS, Yu HJ, Mun JH (2016) Elucidating the triplicated ancestral genome structure of radish based on chromosome-level comparison with the *Brassica* genomes. Theor Appl Genet 129:1357–1372

Johanson U, West J, Lister C, Michaels S, Amasino R, Dean C (2000) Molecular analysis of *FRIGIDA*, a major determinant of natural variation in *Arabidopsis* flowering time. Science 290:344–347

Jung WY, Park HJ, Lee A, Lee SS, Kim YS, Cho HS (2016) Identification of flowering-related genes responsible for differences in bolting time between two radish inbred lines. Front Plant Sci 7:1844

Kitamoto N, Yui S, Nishikawa K, Takahata Y, Yokoi S (2014) A naturally occurring long insertion in the first intron in the *Brassica rapa FLC2* gene causes delayed bolting. Euphytica 196:213–223

Kitashiba H, Li F, Hirakawa H, Kawanabe T, Zou Z, Hasegawa Y, Tonosaki K, Shirasawa S, Fukushima A, Yokoi S, Takahata Y, Kakizaki T, Ishida M, Okamoto S, Sakamoto K, Shirasawa K, Tabata S, Nishio T (2014) Draft sequences of the radish (*Raphanus sativus* L.) genome. DNA Res 21:481–490

Lee I, Aukerman MJ, Gore SL, Lohman KN, Michaels SD, Weaver LM, John MC, Feldmann KA, Amasino RM (1994) Isolation of *LUMINIDEPENDENS-A* gene involved in the control of flowering time in *Arabidopsis*. Plant Cell 6:75–83

Lee J, Oh M, Park H, Lee I (2008) SOC1 translocated to the nucleus by interaction with *AGL24* directly regulates *leafy*. Plant J 55:832–843

Li F, Hasegawa Y, Saito M, Shirasawa S, Fukushima A, Ito T, Fujii H, Kishitani S, Kitashiba H, Nishio T (2011) Extensive chromosome homoeology among *Brassiceae* species were revealed by comparative genetic mapping with high-density EST-based SNP markers in radish (*Raphanus sativus* L.). DNA Res 18:401–411

Li F, Kitashiba H, Inaba K, Nishio T (2009) A *Brassica rapa* linkage map of EST-based SNP markers for identification of candidate genes controlling flowering time and leaf morphological traits. DNA Res 16:311–323

Lim MH, Kim J, Kim YS, Chung KS, Seo YH, Lee I, Kim J, Hong CB, Kim HJ, Park CM (2004) A new *Arabidopsis* gene, *FLK*, encodes an RNA binding protein with K homology motifs and regulates flowering time via *FLOWERING LOCUS C*. Plant Cell 16:731–740. doi:10.1105/tpc.019331

Lin C (2000) Plant blue-light receptors. Trends Plant Sci 5:337–342

Liu C, Chen H, Er HL, Soo HM, Kumar PP, Han JH, Liou YC, Yu H (2008) Direct interaction of *AGL24* and *SOC1* integrates flowering signals in *Arabidopsis*. Development 135:1481–1491

Lou P, Zhao J, Kim JS, Shen S, Del Carpio DP, Song X, Jin M, Vreugdenhil D, Wang X, Koornneef M, Bonnema G (2007) Quantitative trait loci for flowering time and morphological traits in multiple populations of *Brassica rapa*. J Exp Bot 58:4005–4016

Macknight R, Bancroft I, Page T, Lister C, Schmidt R, Love L, Westphal L, Murphy G, Sherson S, Cobbett C, Dean C (1997) *FCA*, a gene controlling flowering time in *Arabidopsis*, encodes a protein containing RNA-binding domains. Cell 89:737–745

Michaels SD, Amasino RM (1999) *FLOWERING LOCUS C* encodes a novel *MADS* domain protein that acts as a repressor of flowering. Plant Cell 11:949–956

Michaels SD, Amasino RM (2001) Loss of *FLOWERING LOCUS C* activity eliminates the late-flowering phenotype of *FRIGIDA* and autonomous pathway mutations but not responsiveness to vernalization. Plant Cell 13:935–941

Mockler T, Yang H, Yu X, Parikh D, Cheng Y-C, Dolan S, Lin C (2003) Regulation of photoperiodic flowering by *Arabidopsis* photoreceptors. Proc Natl Acad Sci USA 100:2140–2145

Moon J, Lee H, Kim M, Lee I (2005) Analysis of flowering pathway integrators in *Arabidopsis*. Plant Cell Physiol 46:292–299

Mouradov A, Cremer FDR, Coupland G (2002) Control of flowering time: interacting pathways as a basis for diversity. Plant Cell 14:S111–S130

Nie S, Li C, Xu L, Wang Y, Huang D, Muleke EM, Sun X, Xie Y, Liu L (2016) *De novo* transcriptome analysis in radish (*Raphanus sativus* L.) and identification of critical genes involved in bolting and flowering. BMC Gen 17: 389

Nie S, Xu L, Wang Y, Huang D, Muleke EM, Sun X, Wang R, Xie Y, Gong Y, Liu L (2015) Identification of bolting-related microRNAs and their targets reveals complex miRNA-mediated flowering-time regulatory networks in radish (*Raphanus sativus* L.). Sci Rep 5: 14034

Nishioka M, Tamura J, Hayashi M, Fujimori Y, Ohkawa Y, Kuginuki Y, Harada K (2005) Mapping of QTLs for bolting time in *Brassica rapa* (syn. *campestris*) under different environmental conditions. Breed Sci 55:127–133

Okazaki K, Sakamoto K, Kikuchi R, Saito A, Togashi E, Kuginuki Y, Matsumoto S, Hirai M (2007) Mapping and characterization of *FLC* homologs and QTL analysis of flowering time in *Brassica oleracea*. Theor Appl Genet 114:595–608

Provart NJ, Alonso J, Assmann SM, Bergmann D, Brady SM, Brkljacic J, Browse J, Chapple C, Colot V, Cutler S, Dangl J, Ehrhardt D, Friesner JD, Frommer WB, Grotewold E, Meyerowitz E, Nemhauser J, Nordborg M, Pikaard C, Shanklin J, Somerville C, Stitt M, Torii KU, Waese J, Wagner D, McCourt P (2016) 50 years of Arabidopsis research: highlights and future directions. New Phytol 209:921–944

Putterill J, Robson F, Lee K, Simon R, Coupland G (1995) The *CONSTANS* gene of *Arabidopsis* promotes flowering and encodes a protein showing similarities to zinc finger transcription factors. Cell 80:847–857

Quail PH (2002) Phytochrome photosensory signalling networks. Nat Rev Mol Cell Biol 3:85–93
Redei GP (1975) *Arabidopsis* as a genetic tool. Ann Rev Gen 9:111–127
Sanda SL, Amasino RM (1996) Ecotype-specific expression of a flowering mutant phenotype in *Arabidopsis thaliana*. Plant Physiol 111:641–644
Schomburg FM, Patton DA, Meinke DW, Amasino RM (2001) *FPA*, a gene involved in floral induction in *Arabidopsis*, encodes a protein containing RNA-recognition motifs. Plant Cell 13:1427–1436
Schranz ME, Quijada P, Sung S-B, Lukens L, Amasino R, Osborn TC (2002) Characterization and effects of the replicated flowering time gene *FLC* in *Brassica rapa*. Genetics 162:1457–1468
Sheldon CC, Rouse DT, Finnegan EJ, Peacock WJ, Dennis ES (2000) The molecular basis of vernalization: the central role of *FLOWERING LOCUS C* (*FLC*). Proc Natl Acad Sci USA 97:3753–3758
Simpson GG, Dean C (2002) *Arabidopsis*, the Rosetta stone of flowering time? Science 296:285–289
Simpson GG, Dijkwel PP, Quesada V, Henderson I, Dean C (2003) *FY* is an RNA 3′ end-processing factor that interacts with *FCA* to control the *Arabidopsis* floral transition. Cell 113:777–787
Song YH, Shim JS, Kinmonth-Schultz HA, Imaizumi T (2016) Photoperiodic flowering: time measurement mechanisms in leaves. Ann Rev Plant Biol 66:441–464
Soppe W, Bentsink L, Koornneef M (1999) The early-flowering mutant *efs* is involved in the autonomous promotion pathway of *Arabidopsis thaliana*. Development 126:4763–4770
Wang J, Lydiate DJ, Parkin IA, Falentin C, Delourme R, Carion PW, King GJ (2011) Integration of linkage maps for the Amphidiploid *Brassica napus* and comparative mapping with *Arabidopsis* and *Brassica rapa*. BMC Gen 12:101
Xu L, Hu K, Zhang Z, Guan C, Chen S, Hua W, Li J, Wen J, Yi B, Shen J, Ma C, Tu J, Fu T (2016) Genome-wide association study reveals the genetic architecture of flowering time in rapeseed (*Brassica napus* L.). DNA Res 23:43–52
Yanovsky MJ, Kay SA (2002) Molecular basis of seasonal time measurement in *Arabidopsis*. Nature 419:308–312
Yi G, Park H, Kim JS, Chae WB, Park S, Huh JH (2014) Identification of three *FLOWERING LOCUS C* genes responsible for vernalization response in radish (*Raphanus sativus* L.). Hort Environ Biotechnol 55:548–556
Zhao J, Kulkarni V, Liu N, Del Carpio DP, Bucher J, Bonnema G (2010) *BrFLC2* (*FLOWERING LOCUS C*) as a candidate gene for a vernalization response QTL in *Brassica rapa*. J Exp Bot 61:1817–1825
Zou X, Suppanz I, Raman H, Hou J, Wang J, Long Y, Jung C, Meng J (2012) Comparative analysis of *FLC* homologues in *Brassicaceae* provides insight into their role in the evolution of oilseed rape. PLoS One 7:e45751

Molecular Mapping of Disease Resistance Genes

12

Xiaona Yu, Su Ryun Choi and Yong Pyo Lim

Abstract

Radish is susceptible to all kinds of diseases, and to breed cultivars with high levels of diseases resistance is important for reducing chemical sprays in order to avoid effect on environment. In this chapter, a summary is given on genetic studies of resistance genes in radish. First, we provide an overview of fungi, virus, bacteria, and nematode disease-related QTL studies which were published in recent years and also discuss about the candidate genes/orthologous that were found by comparative QTL/gene mapping on disease resistance loci of Brassicaceae family. Finally, we suggest a few strategies to identify and utilize the resistance genes in radish breeding.

12.1 Introduction

In radish (*Raphanus sativus* L.; $2n = 18$), diseases caused by fungi, viruses, bacteria, and nematodes, which are typically hard to control, can be severe and significantly reduce yield and quality. A reliable and environmentally friendly approach to disease control is the use of resistant cultivars. Thus, breeding crops with enhanced disease resistance is an important goal to improve yield and quality to increase economic efficiency. Utilizing resistant resources for breeding requires a comprehensive understanding of the genomics, genetics, and evolution of the genes controlling disease resistance. Many resistance traits have been identified from different *Brassica* species, and the genes controlling disease resistance have been mapped and cloned (Voorrips et al. 1997; Saito et al. 2006; Kato et al. 2013; Qian et al. 2013; Shimizu et al. 2015). Most of the disease resistance genes (*R* genes) contain conserved sequence motifs, such as nucleotide-binding sites (NBSs) and leucine-rich repeats (LRRs), and this class of R genes mediates gene-for-gene resistance against fungal, viral, bacterial, or nematode diseases. The conserved regions between different NBS-LRR resistance genes provide

X. Yu · S.R. Choi · Y.P. Lim (✉)
Molecular Genetics and Genomics Laboratory, Department of Horticulture, Chungnam National University, Daejeon 305-764, Korea
e-mail: yplim@cnu.ac.kr

T. Nishio and H. Kitashiba (eds.), *The Radish Genome*, Compendium of Plant Genomes, DOI 10.1007/978-3-319-59253-4_12

possibilities for marker-based strategies to isolate and map the R gene family members or analogs. New molecular technologies and developed molecular tools are being used to accelerate disease-resistant plant breeding programs. Recently, the mapping of resistance genes, as well as quantitative trait loci (QTL) linked to resistance reactions, has been studied extensively in radish (Table 12.1).

Radish is perhaps the least characterized in terms of the genetics that governs disease resistance traits among the Brassicaceae crops. The comparative mapping of genetic regions in different *Brassica* species provides a strategy for gene identification in radish. The radish genome size is 573 Mb (Johnston et al. 2005). Through comparing with markers, genetic maps, and genomes with *Arabidopsis thaliana*, *Brassica rapa*, and *Brassica oleracea* revealed that *Raphanus* genomes have undergone relatively few genome rearrangements and duplications and that they share similar gene content and order. The mapping of candidate genes on radish chromosomes could be performed owing to the homology of the chromosomes among these *Brassica* species. This comparative work provided useful information for the molecular mapping of resistance genes and the development of related markers in radish.

In this chapter, we summarize the genetic progress and molecular mapping studies associated with disease resistance in radish, as well as the comparative mapping within *Brassica* species and features applicable for breeding application.

Table 12.1 Detail of disease resistance QTL/gene mapping studies in radish

Parental lines (source)	No. of locus/QTLs	Locus symbol/candidate gene	Mapping population	Marker type	Disease	Reference
Chinese radish × Japanese radish	Single locus	*Crs1*	F_2	AFLP, SSR	Clubroot	Kamei et al. (2010)
HA2 × GSK3-1	One QTL		BC_1S_1	AFLP, RAPD	Fusarium wilt	Kaneko et al. (2007)
835 × B2	One major and seven minor QTLs		F_2	SSR, EST-SSR	Fusarium wilt	Yu et al. (2013)
NAU-dhp08 × NAU-qtbjq-06	Single dominant locus	*RsDmR*	F_2	SRAP, RAPD, RGA, EST-SSR, ISSR	Downy mildew	Xu et al. (2014)
Q07-12 × KB07-10	2 QTLs		F_2	SRAP, SSR	black rot	Li (2009)
KB10Q-22 × PP10Q-36	14QTLs		F_2	SNP, SSR	black rot	Duan (2014)
KB10Q-22 × KB10Q-33	Fine mapping	*RAR1*	F_2	SNP, SSR	black rot	Duan (2014)
Q07-12 × KB07-10	2 QTLs		F_2	SRAP, SSR	TuMV	Li (2009)
'Pegletta × Silettanova'	A major QTL		F_2	RAPD, AFLP	Beet cyst nematode	Mousa et al. (2004)
'Pegletta' × 'Siletta Nova'	A dominant major QTL	$Hs1^{Rph}$	F_2	RAPD, dpRAPD[a], AFLP, SSR	Beet cyst nematode	Budahn et al. (2009)

[a]Budahn et al. (2008)

12.2 Genetic Analysis and Mapping of Disease Resistance in Radish

12.2.1 Mapping of Clubroot Resistance (CR)

Clubroot disease is caused by the fungus *Plasmodiophora brassicae* and is a harmful disease in Brassicaceae crops, especially in China, Korea, and Japan. The pathogen infects the roots of plants, causing root swelling and abnormal growth; as a result, club-shaped roots form, instead of the normal roots (Ingram and Tommerup 1972), and the aerial parts of the host plants become weak and withered. *P. brassicae* was first reported in Russia in 1878 by Woronin (see Karling 1968). Genetic resources with CR were found in European turnips (Yoshikawa 1993), and the genetic basis of CR has been well studied in *B. rapa*. Seven major genes and one minor gene associated with CR have been reported in *B. rapa*: *Crr1* (Kuginuki et al. 1997; Suwabe et al. 2003), *Crr2* (Suwabe et al. 2003), *CRc* (Sakamoto et al. 2008), and *Crr4* (Suwabe et al. 2006) on the A08, A01, A02, and A06 linkage groups of Chinese cabbage, respectively. *Crr3* (Hirai et al. 2004; Saito et al. 2006), *CRa* (Ueno et al. 2012), *CRb* (Piao et al. 2004), and *CRk* (Sakamoto et al. 2008) were mapped to A03. The radish is a close relative of *Brassica,* and most research focuses on the identification and comparison of CR to *Brassica* species due to the existence of conserved genomic information. Akaba et al. (2009) used *Brassica napus*–radish chromosome additional lines (MALs) to analyze CR and found that one chromosome additional line, the c-type, showed a major CR gene against the Ano-01 isolate, which may be located on the radish chromosome. Afterward, Kamei et al. (2010) performed a QTL analysis for clubroot resistance from an F_2 radish population derived from a crossing of a CR Japanese radish and a clubroot-susceptible Chinese radish. They identified one CR locus named *CR locus of R. sativus 1* (*Crs1*). The markers linked to *Crs1* had sequence homology to the genomic region of chromosome 3 in *Arabidopsis*. In addition, the markers BN142, BrSTS-33, BrSTS-61, and BrSTS-72, which were mapped to the *Crs1* locus, were linked to *Crr3*, a CR locus in *B. rapa* (Kamei et al. 2010) (Fig. 12.1a). This synteny analysis suggests that this region in radish and the *Crr3* region in *B. rapa* (Saito et al. 2006) originated from the same region of their ancestral genome (Poczai et al. 2013).

12.2.2 Mapping of Fusarium Wilt Resistance

Fusarium wilt (FW) is caused by the soil-borne fungal pathogen *Fusarium oxysporum* and is one of major diseases being threats to the productivity of cruciferous crops. In radish, the FW disease is affected by *F. oxysporum* f. sp. *raphani* (Armstrong and Armstrong 1952; Bosland and Williams 1987; Garibaldi et al. 2006). This pathogen can survive enduringly without any plant host. Symptoms include yellowing and dropping that firstly appear in leaves, followed by dark browning, vascular discoloration, severe wilting, and eventually plant death (Kendrick and Snyder 1942). Many genomic resources in *Brassica* and *A. thaliana* have been utilized for genetic diversity analyses, FW-resistant traits identification, and genetic maps construction (Diener and Ausubel 2005; Pu et al. 2011). In *A. thaliana*, *RFO1* is the dominant resistance gene against *F. oxysporum* that encodes a wall-associated non-LRR receptor-like kinase (Diener and Ausubel 2005), while *RFO2* encodes an LRR-RLP (Shen and Diener 2013), and *RFO3* encodes an *S*-receptor-like kinase protein (Cole and Diener 2013). The diverse group of genes conferring resistance to different isolates of the same pathogen indicates the complexity of the *F. oxysporum* infection process. In *B. rapa*, the differential expression at the whole genome level between resistant and susceptible inbred lines using RNA sequencing has been performed. Bra012688 and Bra012689 were selected as candidates for Fusarium yellows resistance in Chinese cabbage (Shimizu et al. 2014). In *B. oleracea*, QTL mapping identified a dominant resistance QTL/gene, *FocBol*, in

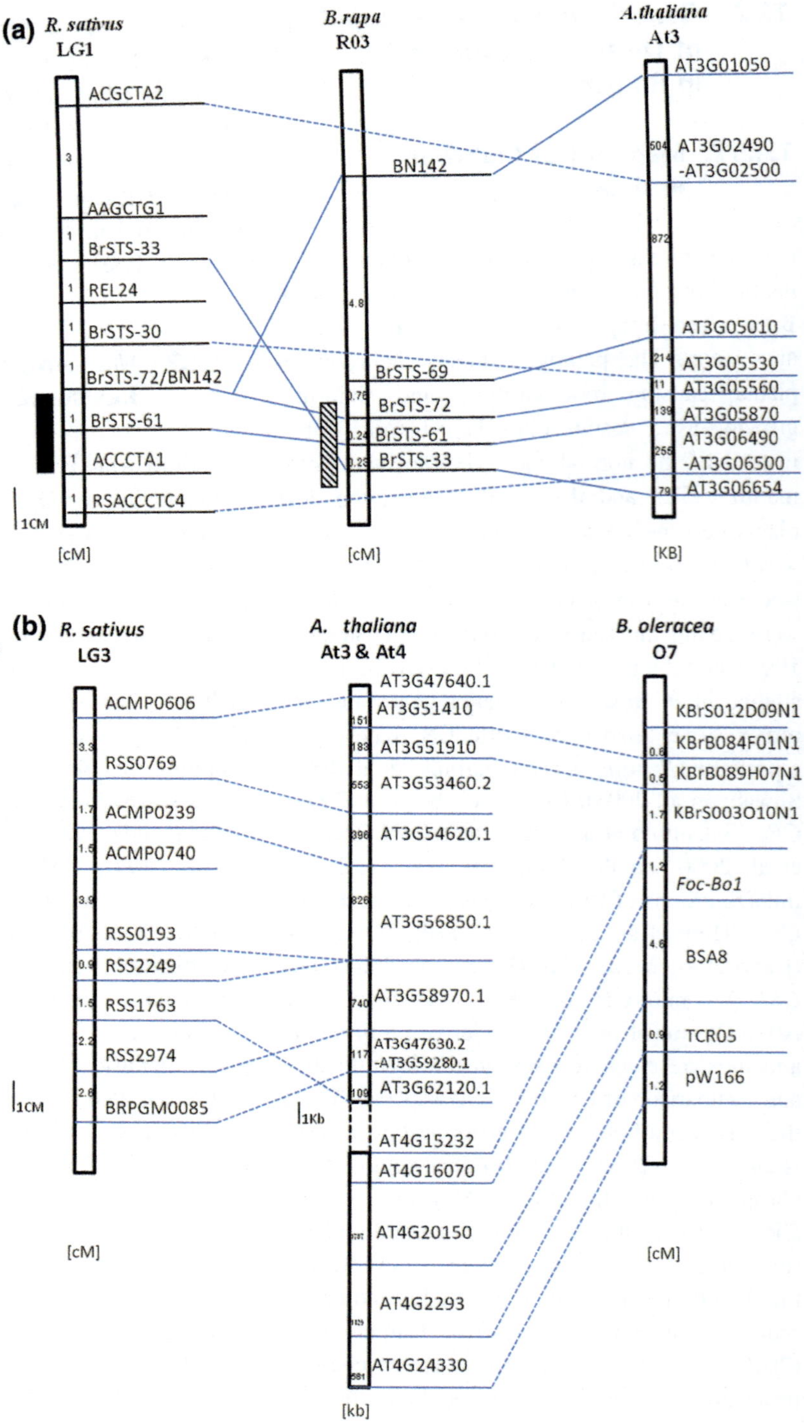

Fig. 12.1 a Synteny of the radish genomic region around the clubroot resistance locus, *Crs1*. The top of chromosome 3 of *Arabidopsis* (*left*), the *Crr3* region of linkage group (LG) R03 of *Brassica rapa* (*middle*), and the *Crs1* region of LG 1 of *Raphanus sativus* (*right*) are shown. The *bar* to the *left* of LG R03 indicates the expected genomic region of *Crr3* (Saito et al. 2006) (From Kamei et al. 2010). **b** *R. sativus* genomic region around the major fusarium wilt resistant QTL Locus *Fwr1* shows synteny with the *A. thaliana* chromosome 3, 4 and *B. oleracea* O7. The *markers* of radish that are homologous with *A. thaliana* genes or *B. oleracea Foc-Bol* gene are connected by *broken lines* (From Yu et al. 2013)

linkage group 7 (O7), and after fine mapping using 139 recombinant F_2 plants, the location of *FocBo1* was delimited within 1.00 cM. An orthologous resistant gene of Bra012688 was identified, and it is also known as candidate gene for conferring yellows resistance in Chinese cabbage. In radish, Kaneko et al. (2007) used cultivated Japanese radish inbred lines, HA2 and

GSK3-1; the single F_1 plant was crossed to HA2 to develop a BC_1 population and identified a QTL-associated resistance to FW that was linked to the RAPD markers, OPJ14 and CC18 locus on linkage group 1 corresponding to Rs7 defined in Chap. 3. Furthermore, Shirasawa et al. (2011) predicted that this locus corresponded to the long arm of *A. thaliana* Chr1 using a comparative map between *R. sativus* and *A. thaliana*. Yu et al. (2013) developed a genetic map of *R. sativus* using an F_2 population, and QTL mapping detected a total of eight loci conferring FW resistance that were distributed on LG 2 (Rs6), LG 3 (Rs5), LG 4 (Rs4) (2 QTLs each), LG 6 (Rs9), and LG 7 (Rs8) (1 QTL each). Of the detected QTLs, the major QTL *qFW4*, flanked by ACMP0609 and cnu_mBRPGM0085, on LG 3 (Rs5) was constitutively detected throughout the 2-year-long experiment and had a comparatively higher logarithm of the odds value, as well as percentage of phenotypic variation. A synteny analysis using the linked markers to this QTL showed homology to *A. thaliana* chromosome 3, which contained disease resistance gene clusters that provide significant opportunities to identify genes with resistance functions in this QTL region. The fine mapping of the major QTL *qFW4* is in progress. Moreover, the FW QTL region of *B. oleracea* in linkage group 7 (Pu et al. 2011) was aligned with the corresponding region of chromosomes 4 and 3 from *A. thaliana* and Chinese cabbage, respectively (Fig. 12.1b). The comparative study suggested that new resistance gene(s) exist in radish or that the resistant genes for FW are distributed in different genomic regions after the rearrangement occurred in the chromosome (Wubben et al. 2001). The evolution of resistance mechanisms may have diverged in *Brassica* species. A cluster of NBS-LRR genes was identified in the corresponding chromosomal region of *A. thaliana* through the synteny analysis with yellows resistance in the radish map, which suggests that a member of the gene family may be responsible for *F. oxysporum* resistance in *R. sativus*.

12.2.3 Mapping of Downy Mildew Resistance

Downy mildew (DM) caused by the fungus *Peronospora parasitica* is a destructive fungal disease of Brassicaceae crops worldwide (Constantinescu and Fatehi 2002). It is a seed-borne, air-borne, and soil-borne disease. The pathogen can not only infect young seedlings, but also damage crops at all stages through maturity (Molinero-Ruiz et al. 2003; Koike and Ochoa 2007). DM also causes damage to the leaves and roots of radish, especially in cool, damp weather. Symptoms include yellow spots on the upper surfaces, a grayish white growth on the undersides of leaves, and dark discolored areas extending down the root from the crown (Xu et al. 2014). DM is a serious disease, and resistance sources and QTLs/genes have been identified in *Brassica* vegetables (Farinho et al. 2004). Farnham et al. (2002) showed DM resistance at the cotyledon stage and detected a major locus of a dominant resistance allele in a cross of rapid cycling kale and broccoli. In *B. rapa*, Kim et al. (2011) reported that resistance to DM at the adult plant stage was controlled by a single dominant gene, *BrRHP1*, based on evaluations of F_2 and BC_1 populations. In radish, Xu et al. (2014) used F_2 populations derived from a resistant line, NAU-dhp08, and a susceptible line, NAU-qtbjq-06, to perform QTL mapping, and they determined that resistance to DM at the seedling stage in radish was controlled by a single dominant allele designated as *RsDmR*. Three markers (Em9/ga24370, NAUISSR826700, and Me7/em10400) linked to the *RsDmR* locus within a 10.0-cM distance were identified by a bulked segregant analysis (BSA). Especially, the SRAP marker Em9/ga24370, derived from the resistance parent NAU-dhp08, was most tightly linked to RsDmR locus, suggesting that the marker has a great potential to be used for marker-assisted integration of this resistance gene into novel elite cultivars in radish breeding programs.

12.2.4 Mapping of Black Rot Resistance

Black rot is a bacterial disease caused by *Xanthomonas campestris* pv. *campestris* (*Xcc*). The earliest description of the disease on *B. oleracea* crops was by Garman in 1894 in Kentucky, USA. The disease can attack all cruciferous crops, and it has been identified worldwide (Williams 1972; Vicente and Holub 2013). Radish is susceptible to the pathogen *X. campestris*. Symptoms of black rot include V-shaped chlorotic, marginal leaf chlorosis, necrosis, and the darkening of leaf veins and vascular tissues within the stem. Finally, dwarfing and plant death occur in the advanced stage of disease development (Taylor et al. 2002; Soengas et al. 2007).

Nine races of *X. campestris* have been identified, with races 1 and 4 being predominant globally and more destructive on *B. oleracea* and *B. rapa,* respectively (Vicente et al. 2001; Tonu et al. 2013). Thus, most of the studies related to *Xcc* resistance have focused on *B. oleracea* (C genome) , *B. rapa* (A genome), and *B. napus* (AC). However, limited sources of resistance were identified (Soengas et al. 2007; Singh et al. 2011; Kifuji et al. 2013; Lee et al. 2015). Bain (1952) found that the Japanese cabbage cultivar 'Early Fuji' had a high level of resistance to *Xcc*, and he later reported that this resistance was controlled by one or two dominant genes (Williams et al. 1972). Camargo et al. (1995) identified four *Xcc* resistance-associated QTLs in a mapping population derived from the cross of a resistant cabbage and a susceptible broccoli. Soengas et al. (2007) studied the genetics of broad-spectrum resistance in *B. rapa* using an F_2 population derived from a cross involving the broad-spectrum resistance line B162 and a susceptible line R-o-18 that was subjected to *Xcc* races 1 and 4. The QTL analysis detected four QTLs, of which two clustered on A06, forming a major QTL.

In radish, Li (2009) first detected the two QTLs (named *RB3* and *RB5*) in an F_2 segregation population of two radish inbred lines (Q07-12P1 and KB07-10P2), which differed in the resistance to both black rot and turnip mosaic virus (TuMV). The QTL *RB3* enhanced QTL located in LG3 and *RB5* inhibited QTL located in LG5, which cannot be corresponded to chromosomes in Chap. 3. The contribution rates were 26.6 and 45.3%, respectively. Meanwhile, Li (2009) found that 12 lines of radish materials appeared to be resistant to different avirulence gene strains. This suggested the existence of corresponding resistance genes. Then, the BSA method with a molecular tagging marker system was used to find the molecular marker (CuMe6F/CoEm11R-260) linkage with a resistance gene, which corresponded to the virulence gene Xcc_3176 of the black rot pathogen in a segregating F_2 population, constructed from inbred lines KB07-3 and KB07-10. Duan (2014) analyzed QTLs and the fine mapping of the resistance to black rot using two radish F_2 populations of an inbred line resistant to black rot (KB10Q-22) and two susceptible inbred lines (KB10Q-33 and PP10Q-36). There are 14 QTLs associated with resistance to the black rot pathogen Xcc8004 that were mapped using an F_2 population of 120 plants from the cross of KB10Q-22 and PP10Q-36. Four QTLs (*qBRR-9-1*, *qBRR-6-1*, *qBRR-6-2*, and *qBRR-6-3*) had higher LOD values and contributions relative to the other QTLs. *qBRR-9-1* was located in LG9, and *qBRR-6-1*, *qBRR-6-2*, and *qBRR-6-3* were located in LG6. Another F_2 population of 900 plants was constructed using KB10Q-22 and the inbred line KB10Q-33. In total, 11 pairs of primers linked to a black rot resistance gene were screened using the BSA method. The resistance gene was mapped between QTL1-21 and M26. Then, QTLs of the resistance to black rot were also mapped between QTL1-21 and M26 based on the map and resistance data. The genetic distance between the two markers was 0.5 cM, and the physical distance was 59 kb. There was a resistance signal gene *RAR1* in this interval.

12.2.5 Mapping of TuMV Resistance

Turnip mosaic virus TuMV is an RNA virus belonging to the *Potyviridae* family that causes serious economic damages. It has been ranked among the five most damaging vegetable viruses worldwide and has long been the focus of disease resistance studies in *Brassica* crops (Tomlinson 1987; Shattuck 1992). The genetic mapping of resistance genes against strain-specific TuMV has been performed in *B. napus* and *B. rapa*. Five resistance genes have been reported in *B. napus*. The dominant resistance genes *TuRB01* and *TuRB03* were observed on linkage group N6, and the *TuRB02* locus was found on linkage group N14 of *B. napus* (Walsh et al. 1999; Hughes et al. 2003). Dominant resistance genes *TuRB04* and *TuRB05* interacted with the viral necrotic response (Jenner et al. 2002). In *B. rapa*, many genes have been identified and mapped on different chromosomes. Recessive resistance gene *retr01* and dominant gene *ConTR01* were located on chromosome A04 and A08, respectively (Rusholme et al. 2007), and both dominant resistance gene *TuRB01b* and recessive resistance gene *rnt1* were located in A06 (Rusholme et al. 2007; Fujiwara et al. 2011). Recessive resistance gene *retr02* was detected on chromosome A04 (Qian et al. 2013).

Through molecular genetics analyses, many R genes against viruses have been identified. Most dominant R genes control the resistance against a narrow spectrum of TuMV isolates. In addition, the combined actions of a recessive (retr01) and a dominant (ConTR01) gene can control broad-spectrum resistance to TuMV (Walsh and Jenner 2002; Rusholme et al. 2007). Among the R genes, the largest class is that of the NBS-LRR genes. In some recessive genes, the eukaryotic initiation factor 4E plays an important role during the viral infection of the *Potyviridae* family. In radish, Kaneko et al. (1996) found that the TuMV resistance gene(s) was located on the *f* chromosome (from radish) of kale using a *R. sativus–B. oleracea* monosomic addition line series. Li (2009) established an F_2 segregation population using two radish inbred lines (Q07-12P1 and KB07-10P2) that differed in the resistance to both TuMV and black rot. Two QTLs were related to TuMV resistance. One had a positive effect with an explained variance of 7.3%, while the other had a negative effect of 11.7%. In addition, the resistance gene to TuMV was tagged with the help of a BSA. The molecular marker CoMe7F/BEm12R-120 was linked to a resistance gene with a genetic linkage distance of 7.9 cM. Cheng (2013) cloned two eukaryotic translation initiation factor genes [*RseIF4E* and *RseIF(iso)4E*] in radish and suggested that they may be involved in interactions between host and virus. Proteins encoded by the two genes were predicted to be hydrophilic and present in the cytoplasm, according to a bioinformatics analysis, and qRT-PCR results indicated that the genes were highly expressed in TuMV-infected leaves.

12.2.6 Mapping of Nematode Resistance

Nematodes cause extensive damage to plant crops every year. The sedentary beet cyst nematode (BCN) *Heterodera schachtii* SCHM is a plant pathogenic nematode, which is a widespread parasite with a host range of more than 200 different plants, including *Brassica* vegetable crops (radishes, cabbages, Chinese cabbages, and turnips) (Taha and Raski 1969; Stirling et al. 1992). Once fields are infested with BCN, they may have one or more localized area of infestation, resulting in elliptical areas where plant growth is poor. BCN can parasitize the roots of plants of all ages. For seedlings, this may result in severe injury or even induce plant death. When older plants are attacked, it reduces the plant's ability to take up nutrients and water, and they remain stunted until harvested (Roberts et al. 1981; Paul et al. 1987).

Because of the BCN's ability to broadly infect host plants and the long persistence of the cysts, control of BCN is a problem. Thus, the most favored strategy is the breeding of resistant cultivars. Resistance to BCN has been reported in radish, and the oil radish (*Raphanus sativus* var. *oleiferus*) is resistant to various pests and abiotic stress factors (Thierfelder et al. 1991; Lelivelt and Krens 1992), which indicates that radish is good gene resource for BCN resistance. Voss et al. (1999) introduced resistance to nematodes from radish to rape by creating a radish chromosome addition line of rape using interspecific hybrids. They identified the rapeseed, and it showed resistance expression (Voss et al. 1999). Then, Peterka et al. (2004) determined the specific radish chromosome that carried nematode resistance gene(s) using chromosome-specific randomly amplified polymorphic DNA (RAPD) markers. Mousa et al. (2004) first reported that QTL(s) for BCN resistance using a segregating F_2 population of the oil radish derived from 'Pegletta' (nematode resistant) × 'Siletta Nova' (susceptible). A BSA revealed 16 RAPD and amplified fragment length polymorphic (AFLP) markers that differentiated the resistance and the susceptible lines based on band intensity. One major QTL and two minor QTLs were detected. Moreover, the RAPD marker OPD-04-568 and the AFLP marker E39M48-268(D) were candidates flanking the QTL of BCN resistance. In another experiment, genes for BCN resistance in oilseed radish were mapped using an intraspecific F_2 mapping population derived from 'Pegletta' 'Siletta Nova'. A dominant major QTL, $Hs1^{Rph}$, explaining 46.4% of the phenotypic variability, was detected between markers E41M59-297 and OPI19-471. The location of $Hs1^{Rph}$ on chromosome *d* was in agreement with the results obtained from testing added radish chromosomes in rapeseed (Budahn et al. 2009). Magdi (2012) scored the phenotypes of four oil radish populations, and *B. napus* cv. 'Madora' as the susceptible control. They were analyzed to reveal the genetic behavior of BCN resistance in radish. All of the F_1 plants appeared in the resistant parents' class, indicating the dominant behavior of the resistant gene(s). The F_2 individuals were distributed in two phenotypic classes, following the assumption of the monogenic dominant inheritance of BCN resistance in oil radish (Magdi 2012).

12.3 Conclusions and Perspectives

The molecular mapping and locating of resistance genes are crucial first steps in genetic mechanism research and the use of disease resistance in plant breeding (Riaz et al. 2011; Xu et al. 2014). Fine mapping allows the precise identification of QTL sites within a chromosomal region and is necessary for positional cloning. Disease resistance is a very important trait in plant breeding, and compared with other *Brassica* species, relatively little work has been done in the mapping of disease resistance genes in radish. Only a few QTL/gene (s) linked to molecular markers are being used in marker-assisted breeding programs because the map was built using an unstable F_2 population having low genetic continuity, and the map is low resolution. With the release of the radish genome sequence (Kitashiba et al. 2014), it is important to accelerate disease resistance-associated gene discovery and radish breeding. Depending on the genomic information, significant associations between single nucleotide polymorphisms and particular phenotypes should be identified in genome-wide association studies, which have the advantage in mapping quantitative traits and can be widely used to locate genes associated with important traits in radish.

Basically, three different strategies have been used for disease resistance gene mapping in radish: (1) co-segregation of markers with disease resistance genes in populations, (2) QTL mapping and map-based cloning, and (3) genome-wide association analyses. These approaches have proven to be highly successful in different plants. Recent data on genes identified in radish and their mapping to chromosomes are reviewed annually. Molecular geneticists may take advantage of molecular markers and genetic maps to explore the most useful gene functional studies. This will allow more pragmatic breeding programs. The identification of molecular markers linked to QTLs for disease resistance is a laborious and long-term strategy. However, once a QTL has been clarified, then the breeding efficiency will be improved.

Acknowledgements This work was supported by Korea Institute of Planning and Evaluation for Technology in Food, Agriculture, Forestry and Fisheries (IPET) through Golden Seed Project, funded by Ministry of Agriculture, Food and Rural Affairs (MAFRA) (No:213006-05-1-SBO20).

References

Akaba M, Kaneko Y, Hatakeyama K, Ishida M, Bang SW, Matsuzawa Y (2009) Identification and evaluation of clubroot resistance of radish chromosome using a *Brassica napus-Raphanus sativus* monosomic addition line. Breed Sci 59:203–206

Armstrong GM, Armstrong JK (1952) Physiologic races of Fusaria causing wilts of the Cruciferae. Phytopathology 42:255–257

Bain DC (1952) Reaction of *Brassica* seedlings to black rot. Phytopathology 42:497–500

Bosland PW, Williams PH (1987) An evaluation of *Fusarium oxysporum* from crucifers based on pathogenicity, isozyme polymorphism, vegetative compatibility, and geographic origin. Can J Bot 65:2067–2073

Budahn H, Schrader O, Peterka H (2008) Development of a complete set of disomic rape-radish chromosome-addition lines. Euphytica 162:117–128

Budahn H, Peterka H, Mousa MA, Ding Y, Zhang S, Li J (2009) Molecular mapping in oil radish (*Raphanus sativus* L.) and QTL analysis of resistance against beet cyst nematode (*Heterodera schachtii*). Theor Appl Genet 118:775–782

Camargo LEA, Williams PH, Osborn TC (1995) Mapping of quantitative trait loci controlling resistance of *Brassica oleracea* to *Xanthomonas campestris* pv. *campestris* in the field and greenhouse. Phytopathology 85:1296–1300

Cheng D (2013) Cloning of turnip mosaic virus resistance related genes and development of trap marker in radish (*Raphanus sativus L.*). Dissertation, Nanjing Agricultural University

Cole SJ, Diener AC (2013) Diversity in receptor-like kinase genes is a major determinant of quantitative resistance to *Fusarium oxysporum* f.sp. *matthioli*. New Phytol 200:172–184

Constantinescu O, Fatehi J (2002) *Peronospora*-like fungi (Chromista, Peronosporales) parasitic on Brassicaceae and related hosts. Nova Hedwigia 74:291–338

Diener AC, Ausubel FM (2005) *RESISTANCE TO FUSARIUM OXYSPORUM 1*, a dominant Arabidopsis disease-resistance gene, is not race specific. Genetics 171:305–321

Duan Y (2014) Genetic analysis and QTL fine mapping of the resistance to black rot in radish (*Raphanus sativus L.*) germplasm. Dissertation, Chinese Academy of Agricultural Sciences

Farinho M, Coelho P, Carlier J, Svetleva D, Monteiro A, Leitao J (2004) Mapping of a locus for adult plant resistance to downy mildew in broccoli (*Brassica oleracea* convar. *italica*). Theor Appl Genet 109:1392–1398

Farnham MW, Wang M, Thomas CE (2002) A single dominant gene for downy mildew resistance in broccoli. Euphytica 128:405–407

Fujiwara A, Inukai T, Kim BM, Masuta C (2011) Combinations of a host resistance gene and the CI gene of turnip mosaic virus differentially regulate symptom expression in *Brassica rapa* cultivars. Arch Virol 156:1575–1581

Garibaldi A, Gilardi G, Gullino ML (2006) Evidence for an expanded host range of *Fusarium oxysporum* f. sp. *raphani*. Phytoparasitica 34:115–121

Hirai M, Harada T, Kubo N, Tsukada M, Suwabe K, Matsumoto S (2004) A novel locus for clubroot resistance in *Brassica rapa* and its linkage markers. Theor Appl Genet 108:639–643

Hughes S, Hunter P, Sharpe A, Kearsey M, Lydiate D, Walsh J (2003) Genetic mapping of the novel *Turnip mosaic virus* resistance gene *TuRB03* in *Brassica napus*. Theor Appl Genet 107:1169–1173

Ingram DS, Tommerup IC (1972) The life history of *Plasmodiophora brassicae* Woron. Proc R Soc Lond Ser B 180:103–112

Jenner CE, Tomimura K, Ohshima K, Hughes SL, Walsh JA (2002) Mutations in *Turnip mosaic virus* P3 and cylindrical inclusion proteins are separately required to overcome two *Brassica napus* resistance genes. Virology 300:50–59

Johnston JS, Pepper AE, Hall AE, Chen ZJ, Hodnett G, Drabek J, Lopez R, Price HJ (2005) Evolution of genome size in Brassicaceae. Ann Bot 95:229–235

Kamei A, Tsuro M, Kubo N, Hayashi T, Wang N, Fujimura T, Hirai M (2010) QTL mapping of clubroot resistance in radish (*Raphanus sativus* L.). Theor Appl Genet 120:1021–1027

Kaneko Y, Natsuaki T, Bang SW, Matsuzawa Y (1996) Identification and evaluation of Turnip mosaic virus (TuMV) resistance gene in kale monosomic addition lines of radish. Breed Sci 46:117–124

Kaneko Y, Kimizuka-Takagi C, Bang SW, Matsuzawa Y (2007) Radish. In: Kole C (ed) Genome mapping and molecular breeding in plants, vol 5. Springer, Heidelberg, pp 141–160

Karling JS (1968) The Plasmodiophorales: including a complete host index, bibliography, and a description of diseases caused by species of this order, 2nd edn. Hafner, New York

Kato T, Hatakeyama K, Fukino N, Matsumoto S (2013) Fine mapping of the clubroot resistance gene *CRb* and development of a useful selectable marker in *Brassica rapa*. Breed Sci 63:116–124

Kendrick JB, Snyder WC (1942) Fusarium wilt of radish. Phytopathology 32:1031–1033

Kifuji Y, Hanzawa H, Terasawa Y, Ashutosh Nishio T (2013) QTL analysis of black rot resistance in cabbage using newly developed EST-SNP markers. Euphytica 190:289–295

Kim SG, Song YH, Lee JY, Choi SR, Dhandapani V, Jang CS, Lim YP, Han TH (2011) Identification of the BrRHP1 locus that confers resistance to downy mildew in Chinese cabbage (*Brassica rapa* ssp *pekinensis*) and development of linked molecular markers. Theor Appl Genet 123:1183–1192

Kitashiba H, Li F, Hirakawa H, Kawanabe T, Zou Z, Hasegawa Y, Tonosaki K, Shirasawa S, Fukushima A, Yokoi S, Takahata Y, Kakizaki T, Ishida M,

Okamoto S, Sakamoto K, Shirasawa K, Tabata S, Nishio T (2014) Draft sequences of the radish (*Raphanus sativus* L.) genome. DNA Res 21:481–490

Koike ST, Ochoa OE (2007) Downy mildew caused by *Bremia lactucae* on strawflower (*Helichrysum bracteatum*) in California. Plant Dis 91:326

Kuginuki Y, Ajisaka H, Yui M, Yoshikawa H, Hida K, Hirai M (1997) RAPD markers linked to a clubroot-resistance locus in *Brassica rapa* L. Euphytica 98:149–154

Lee J, Izzah NK, Jayakodi M, Perumal S, Joh HJ, Lee HJ, Lee SC, Park JY, Yang KW, Nou IS, Seo J, Yoo J, Suh Y, Ahn K, Lee JH, Choi GJ, Yu Y, Kim H, Yang TJ (2015) Genome-wide SNP identification and QTL mapping for black rot resistance in cabbage. BMC Plant Biol 15:32

Lelivelt CL, Krens FA (1992) Transfer of resistance to the beet cyst nematode (*Heterodera schachtii* Schm.) into the *Brassica napus* L. gene pool through intergeneric somatic hybridization with *Raphanus sativus* L. Theor Appl Genet 83:887–894

Li HS (2009) Genetic dissection of resistance to turnip mosaic virus and black rot in radish (*Raphanus Sativus L.*). Dissertation, Chinese Academy of Agricultural Sciences

Magdi AAM (2012) Genetic behavior of resistance to the beet cyst nematode (*Heterodera schachtii* Schm) in radish (*Raphanus sativus* L.). African J Microbiol Res 6:3755–3760

Molinero-Ruiz ML, Melero-Vara JM, Dominguez J (2003) Inheritance of resistance to two races of sunflower downy mildew (*Plasmopara halstedii*) in two *Helianthus annuus* L. lines. Euphytica 131:47–51

Mousa M, Kandeel N, Mohamed M, Nasr MA, Budahn H, Peterka H (2004) Mapping QTL (s) for resistance to the beet cyst nematode (*Heterodera schachtii* Schm.) in radish (*Raphanus sativus* L.). Ass Univ Bull Environ Res 7:1–18

Paul H, Zijlstra C, Leeuwangh JE, Krens FA, Huizing HJ (1987) Reproduction of the beet cyst nematode *Heterodera schachtii* Schm. on transformed root cultures of *Beta vulgaris* L. Plant Cell Rep 6:379–381

Peterka H, Budahn H, Schrader O, Ahne R, Schutze W (2004) Transfer of resistance against the beet cyst nematode from radish (*Raphanus sativus*) to rape (*Brassica napus*) by monosomic chromosome addition. Theor Appl Genet 109:30–41

Piao ZY, Deng YQ, Choi SR, Park YJ, Lim YP (2004) SCAR and CAPS mapping of *CRb*, a gene conferring resistance to *Plasmodiophora brassicae* in Chinese cabbage (*Brassica rapa* ssp. *pekinensis*). Theor Appl Genet 108:1458–1465

Poczai P, Varga I, Laos M, Cseh A, Bell N, Valkonen JP, Hyvonen J (2013) Advances in plant gene-targeted and functional markers: a review. Plant Methods 9:6

Pu ZJ, Shimizu M, Zhang YJ, Nagaoka T, Hayashi T, Hori H, Matsumoto S, Fujimoto R, Okazaki K (2011) Genetic mapping of a fusarium wilt resistance gene in *Brassica oleracea*. Mol Breed 30:809–818

Qian W, Zhang S, Zhang S, Li F, Zhang H, Wu J, Wang X, Walsh JA, Sun R (2013) Mapping and candidate-gene screening of the novel Turnip mosaic virus resistance gene retr02 in Chinese cabbage (*Brassica rapa* L.). Theor Appl Genet 126:179–188

Riaz S, Tenscher AC, Ramming DW, Walker MA (2011) Using a limited mapping strategy to identify major QTLs for resistance to grapevine powdery mildew (*Erysiphe necator*) and their use in marker-assisted breeding. Theor Appl Genet 122:1059–1073

Roberts PA, Thomason IJ, McKinney HE (1981) Influence of nonhosts, Crucifers, and fungal parasites on field populations of *Heterodera schachtii*. J Nematol 13:164–171

Rusholme RL, Higgins EE, Walsh JA, Lydiate DJ (2007) Genetic control of broad-spectrum resistance to turnip mosaic virus in *Brassica rapa* (Chinese cabbage). J Gen Virol 88:3177–3186

Saito M, Kubo N, Matsumoto S, Suwabe K, Tsukada M, Hirai M (2006) Fine mapping of the clubroot resistance gene, *Crr3*, in *Brassica rapa*. Theor Appl Genet 114:81–91

Sakamoto K, Saito A, Hayashida N, Taguchi G, Matsumoto E (2008) Mapping of isolate-specific QTLs for clubroot resistance in Chinese cabbage (*Brassica rapa* L. ssp. *pekinensis*). Theor Appl Genet 117:759–767

Shattuck V (1992) The biology, epidemiology, and control of turnip mosaic virus. Hortic Rev 14:199–238

Shen Y, Diener AC (2013) *Arabidopsis thaliana RESISTANCE TO FUSARIUM OXYSPORUM 2* implicates tyrosine-sulfated peptide signaling in susceptibility and resistance to root infection. PLoS Genet 9:e1003525

Shirasawa K, Oyama M, Hirakawa H, Sato S, Tabata S, Fujioka T, Kimizuka-Takagi C, Sasamoto S, Watanabe A, Kato M, Kishida Y, Kohara M, Takahashi C, Tsuruoka H, Wada T, Sakai T, Isobe S (2011) An EST-SSR linkage map of *Raphanus sativus* and comparative genomics of the Brassicaceae. DNA Res 18:221–232

Shimizu M, Fujimoto R, Ying H, Pu ZJ, Ebe Y, Kawanabe T, Saeki N, Taylor JM, Kaji M, Dennis ES, Okazaki K (2014) Identification of candidate genes for fusarium yellows resistance in Chinese cabbage by differential expression analysis. Plant Mol Biol 85:247–257

Shimizu M, Pu ZJ, Kawanabe T, Kitashiba H, Matsumoto S, Ebe Y, Sano M, Funaki T, Fukai E, Fujimoto R, Okazaki K (2015) Map-based cloning of a candidate gene conferring Fusarium yellows resistance in *Brassica oleracea*. Theor Appl Genet 128:119–130

Singh D, Dhar S, Yadava DK (2011) Genetic and pathogenic variability of Indian strains of *Xanthomonas campestris* pv. *campestris* causing black rot disease in crucifers. Curr Microbiol 63:551–560

Soengas P, Hand P, Vicente JG, Pole JM, Pink DA (2007) Identification of quantitative trait loci for resistance to *Xanthomonas campestris* pv. *campestris* in *Brassica rapa*. Theor Appl Genet 114:637–645

Stirling GR, Stanton JM, Marshall JW (1992) The importance of plant-parasitic nematodes to Australian and New Zealand agriculture. Australas Plant Pathol 21:104

Suwabe K, Tsukazaki H, Iketani H, Hatakeyama K, Fujimura M, Nunome T, Fukuoka H, Matsumoto S, Hirai M (2003) Identification of two loci for resistance to clubroot (*Plasmodiophora brassicae* Woronin) in *Brassica rapa* L. Theor Appl Genet 107:997–1002

Suwabe K, Tsukazaki H, Iketani H, Hatakeyama K, Kondo M, Fujimura M, Nunome T, Fukuoka H, Hirai M, Matsumoto S (2006) Simple sequence repeat-based comparative genomics between *Brassica rapa* and *Arabidopsis thaliana*: the genetic origin of clubroot resistance. Genetics 173:309–319

Taha AH, Raski DJ (1969) Interrelationships between root-nodule bacteria, plant-parasitic nematodes and their leguminous host. J Nematol 1:201–211

Taylor JD, Conway J, Roberts SJ, Astley D, Vicente JG (2002) Sources and origin of resistance to *Xanthomonas campestris* pv. *campestris* in *Brassica* Genomes. Phytopathology 92:105–111

Thierfelder A, Hackenberg E, Nichterlein K et al (1991) Development of nematode-resistant rapeseed genotypes via interspecific hybridization. In: Proceedings of the GCIRC 8th international rapeseed congress, pp 269–273

Tomlinson J (1987) Epidemiology and control of virus diseases of vegetables. Ann Appl Biol 110:661–681

Tonu NN, Doullah MA-u, Shimizu M, Karim MM, Kawanabe T, Fujimoto R, Okazaki K (2013) Comparison of positions of QTLs conferring resistance to *Xanthomonas campestris*; pv. *campestris*; in *Brassica oleracea*. Amer J Plant Sci 04:11–20

Ueno H, Matsumoto E, Aruga D, Kitagawa S, Matsumura H, Hayashida N (2012) Molecular characterization of the *CRa* gene conferring clubroot resistance in *Brassica rapa*. Plant Mol Biol 80:621–629

Vicente JG, Conway J, Roberts SJ, Taylor JD (2001) Identification and origin of *Xanthomonas campestris* pv. *campestris* races and related pathovars. Phytopathology 91:492–499

Vicente JG, Holub EB (2013) *Xanthomonas campestris* pv. *campestris* (cause of black rot of crucifers) in the genomic era is still a worldwide threat to brassica crops. Mol Plant Pathol 14:2–18

Voorrips RE, Jongerius MC, Kanne HJ (1997) Mapping of two genes for resistance to clubroot (*Plasmodiophora brassicae*) in a population of doubled haploid lines of *Brassica oleracea* by means of RFLP and AFLP markers. Theor Appl Genet 94:75–82

Voss A, Lühs WW, Snowdon RJ, Friedt W (1999) Development and molecular characterisation of nematode-resistant rapeseed (*Brassica napus* L.). In: Mugnozza GTS, Porceddu E, Pagnotta MA (eds) Genetics and breeding for crop quality and resistance: proceedings of the XV EUCARPIA congress, Viterbo, Italy, September 20–25, 1998. Springer, Dordrecht, pp 195–202

Walsh J, Sharpe A, Jenner C, Lydiate D (1999) Characterisation of resistance to turnip mosaic virus in oilseed rape (*Brassica napus*) and genetic mapping of *TuRB01*. Theor Appl Genet 99:1149–1154

Walsh JA, Jenner CE (2002) Turnip mosaic virus and the quest for durable resistance. Mol Plant Pathol 3: 289–300

Williams PH, Staub T, Sutton JC (1972) Inheritance of resistance in cabbage to black rot. Phytopathology 62:247–252

Wubben MJ 2nd, Su H, Rodermel SR, Baum TJ (2001) Susceptibility to the sugar beet cyst nematode is modulated by ethylene signal transduction in *Arabidopsis thaliana*. Mol Plant-Microbe Interact 14:1206–1212

Xu L, Jiang QW, Wu J, Wang Y, Gong YQ, Wang XL, Limera C, Liu LW (2014) Identification and molecular mapping of the *RsDmR* locus conferring resistance to downy mildew at seedling stage in radish (*Raphanus sativus* L.). J Integr Agr 13:2362–2369

Yoshikawa H (1993) Studies on breeding of clubroot resistance in cole crops. Bull Natl Res Inst Veg Ornam Plants Tea Jpn Ser A 7:1–165

Yu X, Choi SR, Ramchiary N, Miao X, Lee SH, Sun HJ, Kim S, Ahn CH, Lim YP (2013) Comparative mapping of *Raphanus sativus* genome using *Brassica* markers and quantitative trait loci analysis for the Fusarium wilt resistance trait. Theor Appl Genet 126:2553–2562

Polymorphism of Self-Incompatibility Genes

13

Takeshi Nishio and Koji Sakamoto

Abstract

Radish (*Raphanus sativus* L.) has a self-incompatibility system similar to those of *Brassica rapa* and *Brassica oleracea*. Self-recognition specificities of pollen and stigmas are determined by multiple alleles of *SP11* and *SRK*, respectively, both of which are located on the *S* locus. As in *B. rapa* and *B. oleracea*, stigmas in radish also have *S* glycoprotein encoded by *SLG* closely linked to *SP11* and *SRK*. The set of alleles of *SP11*, *SRK*, and *SLG* is called *S* haplotype, and *S* haplotypes are numbered as *S-1*, *S-2*, *S-3*, etc. Although radish belongs to a genus different from *Brassica*, nucleotide sequences of *SP11*, *SRK*, and *SLG* alleles of radish and *Brassica* are intermingled in phylogenetic trees of *SP11*, *SRK*, and *SLG*, respectively, indicating that diversification of these alleles predates speciation of these genera. *SP11*, *SRK*, and *SLG* alleles of some *S* haplotypes in radish are highly similar to those of some *S* haplotypes in *Brassica*, and one *S* haplotype in radish has been revealed to have the same recognition specificity as that of one *S* haplotype in *B. rapa*. Comparison of nucleotide sequences of *SP11* and *SRK* alleles and recognition specificities between similar *S* haplotypes of radish and *Brassica* may provide information useful for understanding the molecular structures of SP11 and SRK proteins. The self-incompatibility system is used for seed production in F_1 hybrid cultivars of radish, in which characterization of each *S* haplotype, especially the strength of the self-incompatibility phenotype, is required. However, the numbering of *S* haplotypes in radish varies by researchers, and nucleotide sequence information on *S* haplotypes is thus confusing. The importance of unification of *S* haplotype nomenclature is herein discussed.

T. Nishio
Graduate School of Agricultural Science, Tohoku University, 468-1, Aza-Aoba, Aramaki, Aoba-Ku, Sendai 980-0845, Japan

K. Sakamoto (✉)
Takii Plant Breeding and Experiment Station, Konan, Shiga 520-3231, Japan
e-mail: k-sakamoto@takii.co.jp

T. Nishio and H. Kitashiba (eds.), *The Radish Genome*, Compendium of Plant Genomes, DOI 10.1007/978-3-319-59253-4_13

13.1 Introduction

Radish (*Raphanus sativus* L.) has a self-incompatibility (SI) system similar to those of *Brassica rapa* and *Brassica oleracea*, which inhibits pollen germination on and pollen tube penetration into stigmas after self-pollination. Since *B. rapa* and *B. oleracea* include important vegetables and can be easily handled in pollination experiments, the SI system in Brassicaceae has long been studied using *B. rapa* or *B. oleracea* and revealed to be controlled by a single locus named *S*, which has multiple alleles (Thompson and Taylor 1966; Hinata and Nishio 1978). Stigmas of different *S* homozygotes of these *Brassica* species have been found to have glycoproteins with different isoelectric points named *S* glycoproteins (Hinata and Nishio 1978). The gene encoding the *S* glycoprotein has been cloned and named *SLG* (*S-Locus Glycoprotein*) (Nasrallah et al. 1988). Stein et al. (1991) have identified a gene encoding a receptor-like protein kinase having an extracellular domain similar to the *S* glycoprotein, which is closely linked to *SLG* on the *S* locus, and named it *SRK* (*S-locus Receptor Kinase*). *SRK* has been demonstrated to be responsible for self-recognition specificity of the stigma by plant transformation (Takasaki et al. 2000). A highly polymorphic gene specifically expressed in anthers has been found near the locus of *SRK* in *B. rapa* (Suzuki et al. 1999; Schopfer et al. 1999) and demonstrated to be the gene determining self-recognition specificity of pollen (Schopfer et al. 1999; Takayama et al. 2000). This gene was named *SP11* (*S-locus Protein 11*) and *SCR* (*S-locus Cysteine Rich protein*) by Suzuki et al. (1999) and Schopfer et al. (1999), respectively.

SLG, *SRK*, and *SP11/SCR* (*SP11* hereafter) have multiple alleles that are inherited by progeny as one set without meiotic recombination between them. A set of *SLG*, *SRK*, and *SP11* alleles is termed *S* haplotype. Distances between *SLG*, *SRK*, and *SP11* and their orientations on the *S* locus differ between different *S* haplotypes (Fukai et al. 2003). There are no homologies in the intergenic regions between *SLG*, *SRK*, and *SP11*, thus suppressing the meiotic recombination between these genes. Some *S* haplotypes lack *SLG* (Okazaki et al. 1999; Sato et al. 2002), and some others contain nonfunctional *SLG* alleles (Suzuki et al. 2000), indicating that *SLG* is dispensable for SI. Based on nucleotide sequence similarities of *SLG* and *SRK* alleles, *S* haplotypes have been classified into two groups, i.e., class I and class II, in *Brassica* (Nasrallah et al. 1991). Class-II *S* haplotypes are generally recessive to class-I *S* haplotypes in pollen. The dominance relationship between *S* haplotypes in pollen does not correspond to that in stigmas (Thompson and Taylor 1966). Expression of *SP11* of class-II *S* haplotypes is suppressed in class-I/class-II *S* heterozygotes (Shiba et al. 2002), and a small RNA sequence near class-I *SLG* has been reported to be responsible for suppression of the expression of class-II *SP11* alleles (Tarutani et al. 2010).

Arabidopsis thaliana is self-compatible, whereas closely related *Arabidopsis lyrata* is self-incompatible. *SRK* and *SCR* alleles in *A. lyrata* have been cloned and sequenced, and *SRK* and *SCR* homologs in *A. thaliana* have been revealed to be nonfunctional (Kusaba et al. 2001). The *S* locus of *A. lyrata* does not have *SLG*. It has been revealed that a set of *SRK* and *SCR* alleles of *A. lyrata* introduced into *A. thaliana* can make *A. thaliana* self-incompatible (Nasrallah et al. 2004). *SRK* alleles in *A. lyrata* are more polymorphic than those in *Brassica* and do not belong to class I or class II (Schierup et al. 2001).

Although radish belongs to the genus *Raphanus*, it is closely related to *B. rapa* and *B. oleracea* and has a triplicated genome like the genus *Brassica* (Kitashiba et al. 2014). Presence of *S* glycoprotein has been reported by Okazaki and Hinata (1984), and nucleotide sequences of *SLG* alleles have been revealed to be similar to those in *Brassica* (Niikura and Matsuura 1997; Sakamoto et al. 1998; Lim et al. 2002), suggesting that radish has an SI system very similar to that of *Brassica*. *S* haplotypes in radish can be classified into class I and class II as in *Brassica* (Sakamoto et al. 1998; Okamoto et al. 2004). Since radish is relatively far from *B. rapa* and *B. oleracea* in phylogenetic relationship (Inaba and Nishio 2002; Warwick and Sauder 2005), study of SI in radish may shed light on the study of *Brassica* SI. However, there is confusion concerning *S* haplotype nomenclature. In this chapter, we compare published nucleotide sequences of the *S*-locus genes and discuss the importance of the unification of *S* haplotype names for sharing information on characteristics of each *S* haplotype in studies of the molecular structures of SP11 and SRK proteins determining self-recognition specificity and the use of the SI system in F_1 hybrid breeding of radish.

13.2 Polymorphism of *S*-Locus Genes

Nucleotide sequences of *SLG* alleles of one *S* haplotype, i.e., *S-201*, and eight *S* haplotypes from *S-1* to *S-8* have been published by Niikura and Matsuura (1997) and Sakamoto et al. (1998), respectively (Table 13.1). Comparison of deduced amino acid sequences has revealed that they are highly similar to those of *Brassica SLG* alleles, having twelve conserved cysteine residues and three hypervariable regions (Sakamoto et al. 1998). Similarities of deduced amino acid sequences of *S* glycoproteins of *S* haplotypes from *S-1* to *S-8* to that of *B. oleracea S-6* haplotype range from 77.8 to 87.4%, comparable to variations within *B. oleracea*. In a phylogenetic tree constructed with deduced amino acid sequences of *S* glycoprotein of radish, *B. oleracea*, and *B. rapa*, there are no species-specific clusters, and sequences of different species are intermingled, indicating that diversification of *SLG* alleles predates speciation of these species.

Using the same materials as Sakamoto et al. (1998), Okamoto et al. (2004) have determined nucleotide sequences of the second exon (encoding the whole mature protein) of ten *SP11* alleles, i.e., *S-1*, *S-2*, *S-4*, *S-6*, *S-8*, *S-9*, *S-14*, *S-18*, *S-19*, and *S-21*, and the first exon (encoding the *S* domain) of nine *SRK* alleles, i.e., *S-1*, *S-2*, *S-4*, *S-6*, *S-8*, *S-9*, *S-18*, *S-19*, and *S-21* (Table 13.1). *SP11* and *SRK* have also been found to have structures similar to those in *B. rapa* and *B. oleracea*. Interestingly, the *SP11* allele of *S-21* in radish has been found to be highly similar, i.e., 89.5% amino acid sequence identity, to that of *S-9* in *B. rapa*. As the second most similar sequence in the database is *SP11-9* in *B. oleracea* having an amino acid sequence identity of 48.8%, the high similarity between *RsSP11-21* and *BrSP11-9* is remarkable.

Table 13.1 *S* haplotypes with published nucleotide sequences of *SP11*, *SRK*, and *SLG* alleles identified in Japan

S haplotype	Class	*SP11*	*SRK*		*SLG*
			SD	KD	
RsS-1	I	Okamoto	Okamoto		Sakamoto
RsS-2	I	Okamoto	Okamoto		Sakamoto
RsS-3	I				Sakamoto
RsS-4	I	Okamoto	Okamoto		Sakamoto
RsS-5	I				Sakamoto
RsS-6	I	Okamoto	Okamoto		Sakamoto
RsS-7	I				Sakamoto
RsS-8	I	Okamoto	Okamoto		Sakamoto
RsS-9	II	Okamoto	Okamoto		
RsS-18	I	Okamoto	Okamoto		
RsS-19	I	Okamoto	Okamoto		
RsS-21	I	Okamoto	Okamoto		
RsS-201	I				Niikura

Sakamoto: Sakamoto et al. (1998)
Okamoto: Okamoto et al. (2004)
Niikura: Niikura and Matsuura (1997)

Furthermore, the *SRK* allele of *S-21* in radish is also similar, i.e., 91.0% amino acid sequence identity, to that of *S-9* in *B. rapa*. The correspondence of *SP11* and *SRK* alleles between *S-21* in radish and *S-9* in *B. rapa* suggests that these *S* haplotypes originated from the same *S* haplotype in a common ancestral species (Okamoto et al. 2004). Such pairs of *S* haplotypes having similar *SLG*, *SRK*, and *SP11* alleles have also been identified in many combinations of *S* haplotypes between *B. rapa* and *B. oleracea* (Kusaba and Nishio 1999; Sato et al. 2002) and are called interspecific pairs.

The commonality of recognition specificity of *S* haplotypes in several interspecific pairs between *B. rapa* and *B. oleracea* has been demonstrated using interspecific hybrids, transgenic plants, or recombinant proteins of SP11 (Kimura et al. 2002; Sato et al. 2003). Introducing *RsSP11-21* into *B. rapa* by *Agrobacterium*-mediated transformation, Sato et al. (2004) have revealed that *S-21* in radish has the same recognition specificity as that of *S-9* in *B. rapa*. However, some class-II *S* haplotypes in interspecific pairs between *B. rapa* and *B. oleracea* have been reported to have different recognition specificities, suggesting that recognition specificities were altered after speciation (Sato et al. 2006). Radish is considered to have diverged from *B. rapa* or *B. oleracea* earlier than the divergence of *B. oleracea* from *B. rapa*. As can be expected, *S* haplotypes in interspecific pairs between radish and *Brassica* have more amino acid differences in SP11 and SRK proteins than those between *B. rapa* and *B. oleracea*. Common recognition specificities between *S* haplotypes in interspecific pairs must have been lost in more interspecific pairs between radish and *Brassica*, and more new recognition specificities must have been acquired by them than those between *B. rapa* and *B. oleracea* (Fig. 13.1). Such interspecific pairs are interesting materials for studying the molecular

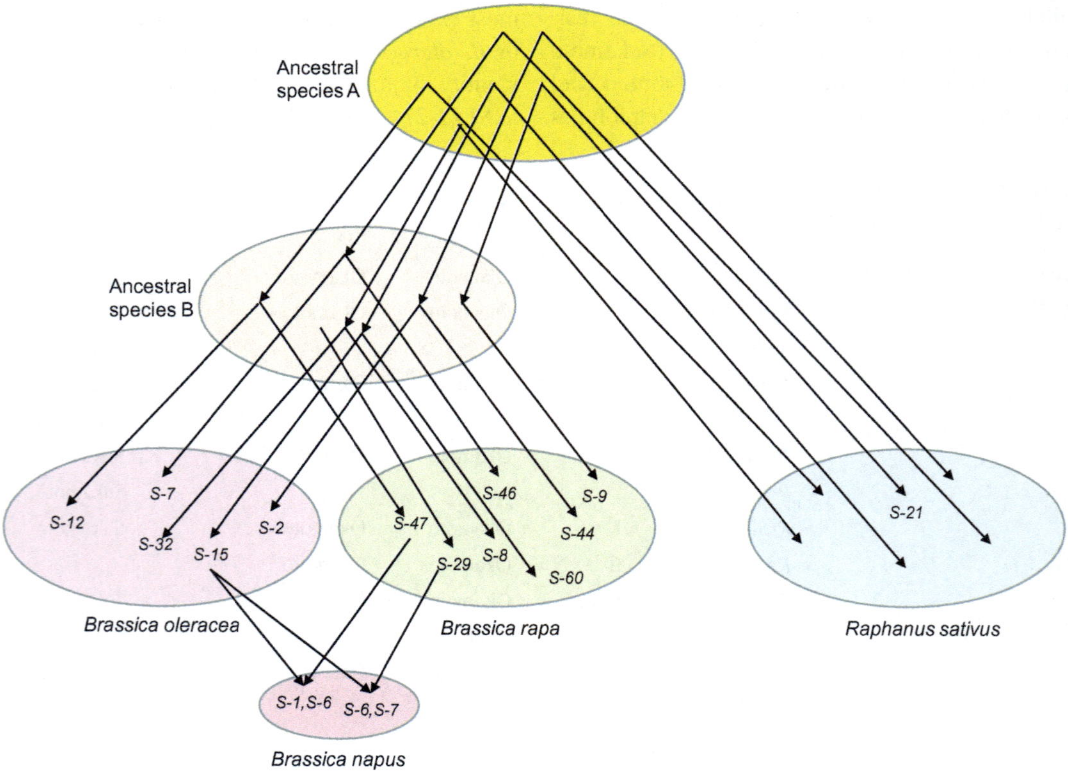

Fig. 13.1 Evolution of *S* haplotypes in *Brassica* species and *Raphanus*

structures of SP11 and SRK proteins that determine self-recognition specificity and the process of generation of new *S* haplotypes.

13.3 Confusion of *S* Haplotype Names

Lim et al. (2002) have reported nucleotide sequences of *SLG* alleles of *S-1*, *S-2*, *S-4*, *S-5*, *S-6*, *S-7*, and *S-10* in radish. Although they cited the paper of Sakamoto et al. (1998), they presumed to deposit different *SLG* allele sequences with the same names as those of Sakamoto et al. (1998). The *SLG* sequences deposited by Sakamoto et al. (1998) were not included in sequence comparison or phylogenetic analysis by Lim et al. (2002). Furthermore, Lim's group have deposited nucleotide sequences of eight *SLG* alleles, i.e., *S-11*, *S-12*, *S-16*, *S-18*, *S-21*, *S-24*, *S-26*, and *S-27* in 2004 (Accession Nos. AY527399.1–AY527402.1 with no paper) (Table 13.2). The *SLG* nucleotide sequence of *S-2* of Lim et al. (2002) corresponds to that of *S-2* of Sakamoto et al. (1998), but those of *S-7* of Lim et al. (2002) and *S-18* of Lim's group in 2004 are the same as *S-4* and *S-6* of Sakamoto et al. (1998), respectively (Table 13.3). Furthermore, the *SLG* nucleotide sequence of *S-12* of Lim's group in 2004 is 99.9% identical to that of

Table 13.2 *S* haplotypes with published nucleotide sequences of *SRK* and *SLG* alleles identified by Lim et al. (2002) and Lim's group in 2004

S haplotype	Class	*SP11*	*SRK*		*SLG*
			SD	KD	
RsS-1	I			2002	2002
RsS-2	I			2002	2002
RsS-4	II				2002
RsS-5	II				2002
RsS-6	I			2002	2002
RsS-7	I			2002	2002
RsS-8	I			2002	
RsS-9	I			2002	
RsS-10	I			2002	2002
RsS-11	I			2004	2004
RsS-12	I				2004
RsS-15	I			2004	
RsS-16	I			2004	
RsS-17	I				2004
RsS-18	I			2004	2004
RsS-20	I			2004	
RsS-21	II				2004
RsS-22	I			2004	
RsS-23	I			2004	
RsS-24	II				2004
RsS-26	II				2004
RsS-27	I			2004	2004
RsS-29	I			2004	
RsS-30	I			2004	
RsS-31	I			2004	

Table 13.3 Nucleotide and amino acid sequences similar between different *S* haplotypes

Alleles of *S* haplotypes in Japan	Class	Alleles with similar nucleotide sequences	Nucleotide sequence identity (%)	Amino acid sequence identity (%)
		Lim et al. (2002) and Lim's group in 2004		
RsSLG-1	I	*RsSLG-6*	99.8	99.8
RsSLG-2	I	*RsSLG-2*	100	100
RsSLG-3	I	*RsSLG-12*	99.9	100
RsSLG-4	I	*RsSLG-7*	100	100
RsSLG-6	I	*RsSLG-18*	100	100

S-3 of Sakamoto et al. (1998) and encodes the same amino acid sequence as the latter. The nucleotide sequence and deduced amino acid sequence of *SLG-6* of Lim et al. (2002) have 99.8% identities with those of *SLG-1* of Sakamoto et al. (1998) (Fig. 13.2). Since some variations of nucleotide sequences of *SLG* within the same *S* haplotypes have been observed in *S-2* and *S-13* in *B. oleracea* (Kusaba et al. 2000), *S-6* and *S-12* of Lim et al. (2002) are considered to be the same *S* haplotypes as *S-1* and *S-3* of Sakamoto et al. (1998), respectively.

Kim's group have determined and deposited nucleotide sequences of five class-II alleles of *SP11* in radish, i.e., *S-1* to *S-5* in 2003 (Accession Nos. AY422009.1–AY422013.1 with no paper) (Table 13.4), which are different from those of Sakamoto et al. (1998) and Lim et al. (2002). Deduced amino acid sequences of SP11 encoded by these sequences have the common structures of SP11. SP11 of *S-4* has only one different amino acid residue at the C terminal from that of *S-5* (Fig. 13.3). Since this position is not considered to be important for recognition specificity of pollen (Sato et al. 2004), the difference of SP11 between *S-4* and *S-5* might be a variation within the same *SP11* allele. However, they have different *S* haplotype names. If *S-4* has recognition specificity different from that of *S-5*, these would be interesting materials for studying recognition specificities of *SP11*. The *SP11* nucleotide sequence of *S-5* of Kim's group in 2003 is 100% identical to *S-9* of Okamoto et al. (2004) (Table 13.5).

Nucleotide sequences from the fourth exon to the seventh exon of 18 *SRK* alleles encoding the C terminal region of the kinase domain have been determined and deposited by Lim et al.

```
                         Hyper variable region I
RsSLG-1        NSDDPSSGEISYKLDTQRGMPEFYILKDGLRSHRSGPWNGIRFSGIPEDQKSSYMVYSFT
RsSLG-6_Lim    NSDDPSSGEISYKLDTQRGMPEFYILKDGLRSHRSGPWNGIRFSGIPEDQKSSYMVYSFT
               ************************************************************
                                              Hyper variable region II
RsSLG-1        ENSEEVAYTFRMTNSSIYSRLKISSEGFLERWTTTLESIPWNLFWSAPVDLKCDVYKTCG
RsSLG-6_Lim    ENSEEVAYTFRMTNSSIYSRLKISSEGFLERWT-TLESIPWNLFWSAPVDLKCDVYKTCG
               ********************************* **************************
                          Hyper variable region III
RsSLG-1        PYSYCDLNTSPLCNCIQGFMPSNVQQRDLRDPSGGCIRRARLSCSGDGFTRMRNMKLPET
RsSLG-6_Lim    PYSYCDLNTSPLCNCIQGFMPSNVQQRDLRDPSGGCIRRARLSCSGDGFTRMRNMKLPET
               ************************************************************

RsSLG-1        TMAIVDRSIGVKECEKRCLSDCNCTAFANADIRNGGTGCVIWTGELEDIRTYLADGQDLY
RsSLG-6_Lim    TMAIVDRSIGVKECEKRCLSDCNCTAFANADIRNGGTGCVIWTGELEDIRTYLADGQDLY
               ************************************************************
```

Fig. 13.2 Alignment of deduced amino acid sequences of *RsSLG-1* determined by Sakamoto et al. (1998) and *RsSLG-6* determined by Lim et al. (2002)

Table 13.4 *S* haplotypes with published nucleotide sequences of *SRK* and *SLG* alleles identified by Kim's group in 2003

S haplotype	Class	*SP11*	*SRK*		*SLG*
			SD	KD	
RsS-1	II	2003			
RsS-2	II	2003			
RsS-3	II	2003			
RsS-4	II	2003			
RsS-5	II	2003			

```
                 I              II           III      IV        V         VI
RsSP11-9       LTYVQALDVGAWNCPEGVTISNPILGRCFNSRSRQCGENCRCCDYSKHNRRITCYCCKVK
RsSP11-4_Kim   LTYVQALDVGAWNCPEGVTISNPILGRCFNSRSRQCGENCRCCDYSKHNRRITCYCCKVQ
RsSP11-5_Kim   LTYVQALDVGAWNCPEGVTISNPILGRCFNSRSRQCGENCRCCDYSKHNRRITCYCCKVK
               ***********************************************************:
```

Fig. 13.3 Alignment of deduced amino acid sequences of *RsSP11-9* determined by Okamoto et al. (2004), *RsSP11-4*, and *RsSP11-5* determined by Kim's group in 2003

Table 13.5 Nucleotide and amino acid sequences of *SP11* similar between different *S* haplotypes

Alleles of *S* haplotypes in Japan	Class	Alleles with similar nucleotide sequences	Nucleotide sequence identity (%)	Amino acid sequence identity (%)
		Kim's group in 2003		
RsSP11-9	II	*RsSP11-4*	97.7	98.9
RsSP11-9	II	*RsSP11-5*	100	100

(2002) and Lim's group in 2004 (Table 13.2). Despite being different *S* haplotypes, *S-11*, *S-16*, and *S-29* have highly similar sequences, i.e., more than 98% identity in nucleotide sequences. Since their sequence variation level is comparable to that of *SLG* within an *S* haplotype (Kusaba et al. 2000), it is difficult to identify *S* haplotypes using these sequences. A similar result has been reported by Nishio et al. (1997), i.e., 100% identity from the second exon to the fifth exon, encoding the transmembrane domain and the upper half of the kinase domain, between *S-13* and *S-32* in *B. oleracea*. Such sequence similarity of the kinase domain of SRK has been inferred to have resulted from recombination in the *S* locus, but such recombination is considered to be disfavored by the region from *SP11* to the first exon of *SRK* encoding the *S* domain (Takuno et al. 2007).

Using a large number of radish lines, Niikura and Matsuura (1999) have identified 37 *S* haplotypes, which have been named *S-201* to *S-237*, by test cross and PCR-RFLP analysis of *SLG*-like sequences. Some of these *S* haplotypes must correspond to those from *S-1* to *S-18* by Sakamoto et al. (1998), from *S-1* to *S-10* by Lim et al. (2002), from *S-11* to *S-31* by Lim's group in 2004, and from *S-1* to *S-5* by Kim's group in 2003. Such confusion of *S* haplotype nomenclature has been inevitable because *S* homozygous lines in radish are important breeding materials and difficult to share by different research groups. For studying recognition specificity and evolution of *S* haplotypes, unification of *S* haplotype names is important. In *B. oleracea* and *B. rapa*, *S* tester lines, which are *S* homozygous lines used for test cross for *S* haplotype identification as standards, have been developed (Ockendon 2000; Nishio et al. 2006), and *S* haplotype names given by an English group (Thompson and Taylor 1966; Ockendon 2000) and a Japanese group (Nou et al. 1993), respectively, are commonly used worldwide. However, inbreeding depression makes maintenance of these *S* tester lines difficult, and the number of seeds obtained by hand pollination in radish is much smaller than that in *B. oleracea* and *B. rapa*. Instead of producing and sharing *S* tester lines in radish, we here propose sharing *S* heterozygous populations, each having two *S* haplotypes, e.g., an *S-1/S-2* line, an *S-3/S-4* line, and an *S-5/S-6* line, between researchers of radish SI and radish breeders worldwide. Information on *S* genotypes of commercially available F_1 hybrid cultivars may also be useful for sharing the same *S* haplotypes by many researchers. Here, we show our unpublished data on *S* genotypes of radish cultivars marketed in Japan (Table 13.6). *S* homozygous plants can be obtained from such *S* heterozygous lines or F_1 hybrid cultivars. For efficient selection of the *S* homozygous plants, a simple reliable method for *S* haplotype identification is required.

13.4 Identification of *S* Haplotypes of Radish

Although isoelectric focusing analysis of *S* glycoproteins (Nishio and Hinata 1980) and genomic Southern-blot analysis using an *SLG* probe (Nasrallah et al. 1988) have previously been proposed as methods for *S* haplotype identification, the most popular method for *S* haplotype identification is PCR-RFLP analysis, also called CAPS, of *SLG* or *SLG*-like sequences. Brace et al. (1993) first reported that PCR-RFLP analysis of *SLG*-like sequences is useful for *S* haplotype identification. The primer sequences used for PCR by them (Primer A: 5′-AGAACACTTGTATCTCCCGGT-3′ and Primer B: 5′-CAATCTGACATAAAGATCTTG-3′) are possessed by both *SLG* and the first exon of *SRK* and can amplify both *SLG* and *SRK* or one of them. These primers can also amplify *SLR1* and *SLR2*, which are *SLG*-like genes not linked with the *S* locus, and result in complicated band patterns by PCR-RFLP. To avoid amplification of *SLR1*, use of Primer D (5′-AAGGTCAGCAG(G/C)AGCCAATC-3′) instead of Primer B has been suggested (Brace et al. 1993), but Primer D sequence is also homologous to *SRK*. Nishio et al. (1994) used *SLG*-specific primers, i.e., PS1 (5′-AGATGAAAGGAGTAAGAAAA-3′) and PS2 (5′-ATCCGTGCTATTTTAAGA-3′) for amplification of class-I *S* haplotypes, and restriction endonucleases recognizing tetranucleotide sequences, e.g., *Mbo*I, *Afa*I, and *Msp*I, for high-resolution PCR-RFLP analysis. Since amplification of class-I *SLG* alleles by PS1 and PS2 is sometimes unstable, PS5 (5′-ATGAAAGGCGTAAGAAAAACCTA-3′) and PS15 (5′-CCGTGTTTTATTTTAAGAGAAAGAGCT-3′) for class-I *S* haplotypes and PS3 (5′-ATGAAAGGGGTACAGAACAT-3′) and PS21 (5′-CTCAAGTCCCACTGCTGCGG-3′) for class-II *S* haplotypes were selected as *SLG*-specific primer pairs to avoid amplification of *SRK* and other *SLG*-like sequences from various combinations of primer pairs (Nishio et al. 1996). In PCR-RFLP analysis of radish *S* haplotypes, Niikura and Matsuura (1998) used sequences slightly shifted from Primer A and Primer B (5′-ACAGAACACTTGTATCTCCAGG-3′ and 5′-AGCCAATCTGACATAAAGATC-3′) as primers, which are also possessed by *SRK*, whereas Sakamoto et al. (1998) used PS5 and PS15, showing clear polymorphism between different *S* haplotypes. The same sequences as those of PS5 and PS15 and those of PS3 and PS21 have been used by Lim et al. (2002), and the high ability of PCR-RFLP analysis with these primers has been confirmed.

PCR-RFLP of *SLG* is useful for distinguishing different *S* haplotypes, but identification of an *S* haplotype among many *S* haplotypes in a species is not easy by band patterns of electrophoretic analysis. Furthermore, some *S* haplotypes lack an *SLG* gene (Sato et al. 2002), and some others have sequence variation in *SLG* (Kusaba et al. 2000). Since *SP11* is responsible for the recognition specificity of *S* haplotypes (Schopfer et al. 1999; Takayama et al. 2000) and *SP11* alleles are highly polymorphic between different *S* haplotypes (Watanabe et al. 2000; Sato et al. 2002), dot-blot hybridization of genomic DNA using probes of *SP11* alleles has been suggested to be usable for *S* haplotype

Table 13.6 *S* haplotypes of radish F_1 hybrid cultivars in Japan

S haplotype (class)	Commercial cultivars	Number of cultivars	Frequency (%)
RsS-1 (I)	Sinkisei, Kiryu-Sohbuto, Akishino, YR-Shintomioh, YR-Atago, YR-Tengu, Kanpaku, Shinhasshu	8	5.3
RsS-2 (I)	Kenka, Kenka No.37, Fuyumine, Seitoku, Fukutenka, Fukuaji, Daiyakushin, Jinya-Nagafuto, Kaishin-Sohbuto, Kurazukuri-Sohbuto, Natsuotoko, Shinjin-Sohbuto, Shin-uchi-Sohbuto, YR-Atago, Choh-un, Kinoene, Kensei-Sohbutori, Natsunyudoh, Satojiman, Natsusodachi, Haruka-Sohbutori, Taibyou-Sohbutori, Natsuminowase No.3, Ohkura, Sinhasshu	25	16.4
RsS-3 (I)	YR-Itsukamachi, YR-Hatsuratsu, Kaiyoh, Taibyo-Kifune, Hakutenyu, Natsuminowase No.3, Akizumari, Hayabutori-Ohkura, Fuyudori-Ohkura	9	5.9
RsS-4 (I)	Natsutenka, YR-Itsukamachi, Taibyo-Kifune, Fuyushokunin, Hakutenyu, Natori-Risoh, YR-Tengu	7	4.6
RsS-6 (I)	Akimine, Akioh, Seitoku, Fukuaji, Fukuaji No.2, Aogaki, Seikurabe, Hiragiyama, YR-Akishino, Natsutsukasa, Daiyakushin, Jinya-Nagafuto, Kurazukuri-Sohbutso, Natsuotoko, Kenshi-Sohbutori, YR-Hatsuratsu, Katsumi, Choh-un, Yakushayokochoh, Aozumari No.2, Aozumari No.3, Genbu, Natsusodachi, Haruka-Sohbuto	24	15.8
RsS-9 (II)	Fukubijin, Taibyo-Sohbutori, Ohkura, Shigatsuwase	4	2.6
RsS-11 (II)	Taibyo-Kifune, Hakutenyu, Natori-Risoh, Hayabutori-Ohkura, Fuyudori-Ohkura	5	3.3
RsS-14 (I)	Kenka No.37, Akimine, Natsusakari, Akisakari	4	2.6
RsS-15 (II)	Kenka, Kenka No37, Akimine, Fuyumine, Akioh, Seitoku, Fukutenka, Natsutenka, Ajitenka, Fukuaji No.2, Aogaki, Seikurabe, Hiragiyame, Jinya-Nagafuto, Kenshi-Sohbutori, Katsumi, Kaiyoh, Fuyushokunin, Yakushayokochoh, Yuta, Kenfuku, Natsunyudoh, Aozumari No.2, Aozumari No.3, Genbu, Satojiman, Taibyo-Sohbutori	27	17.8
RsS-17 (I)	Kinoene, Kenfuku, Shigatsuwase	3	2.0
RsS-21 (I)	Fuyumini, YR-Shintomioh, YR-Tengu, Kanpaku, Shinhashu, Shigatsuwase	6	3.9
RsS-25 (I)	Sachinomiya, Fukubijin, Yuta	3	2.0
RsS-26 (II)	Kenka, Natsusakari, Akisakari, YR-Shintomioh, Kanpaku, Fuyudori-Ohkura	6	3.9
RsS-30 (I)	Fukuaji No2, Kiryu-Sohbuto, Aogaki, Akishino, Sekurabe, Hiragiyama, YR-Akishino, Daiyakushin, Kaishin-Sohbuto, Shinjin-Sohbuto, Shinuchi-Sohbuto, Kenshi-Sohbuto, YR-Hatsuratsu, Katsumi, Kaiyoh, Taibyo-Kifune, Yakushayokochoh, Kensei-Sohbuto, Genbu, Natsusodachi, Haruka-Sohbuto	21	13.8

identification (Fujimoto and Nishio 2003). Genomic dot-blot hybridization for detection of a single-copy gene requires experimental expertise, and therefore, Takuno et al. (2010) modified this method using *SP11* DNA fragments amplified by multiplex PCR for dot blotting. Oikawa et al. (2011) have designed allele-specific probes and primers of multiplex PCR for identification of 40 *S* haplotypes in *B. rapa* and 33 *S* haplotypes in *B. oleracea*. After determination of the nucleotide sequences of *SP11* alleles of more *S* haplotypes, the same method will be applied to identification of *S* haplotypes of radish.

The modified dot-blot hybridization of *SP11* alleles is simple and sensitive and has been used for *S* haplotype identification of a large number of plants (Takuno et al. 2010), but preparation of primer sets for multiplex PCR is a bit troublesome, and dot-blot analysis is slightly time-consuming. Bead hybridization of *SRK* alleles, in which PCR products of *SRK* captured by streptavidin-coated magnetic beads are hybridized with allele-specific probes, has been proposed as an alternative method for *S* haplotype identification (Tonosaki et al. 2013). In this method, primers are designed for amplification of a region containing hypervariable region I of *SRK* alleles, and nucleotide sequences of hypervariable region I are used as probes. Although *SLG* alleles can be amplified by this primer pair, the probes are highly specific, detecting only *SRK*. For use of this method to become widespread for *S* haplotype identification of radish, determination of nucleotide sequences of *SRK* alleles of many *S* haplotypes in radish is required.

13.5 Use of *S* Haplotype Information in Radish Breeding

In most Brassicaceae vegetables, e.g., cabbage, broccoli, cauliflower, Chinese cabbage, turnip, and radish, F_1 hybrids, which are vigorous, uniform, and resistant to diseases and environmental stresses, are major cultivars in the seed market. Hand pollination is commonly carried out for F_1 seed production in fruit vegetables, e.g., tomato, cucumber, melon, and watermelon, but the seed production by hand pollination is unprofitable in cabbage and Chinese cabbage, which set 15–20 seeds per pollinated flower, and in radish setting 1–6 seeds per pollinated flower. In 1950 and 1951, F_1 hybrid cultivars of Chinese cabbage and cabbage, respectively, were developed using the SI system to avoid selfing and marketed in Japan first in the world. Since the number of seeds per pollinated flower in radish is extremely smaller than those in other Brassicaceae vegetables, the release of the first F_1 hybrid cultivar in radish, i.e., 'Spring Cross,' which is a double-cross hybrid, was delayed ten years. In breeding programs of double-cross hybrids and three-way cross hybrids using the SI system, it is important to know the *S* genotypes, the strength of the SI phenotype, and the dominance–recessiveness relationships between *S* haplotypes of parental inbred lines (Haruta 1962).

In F_1 hybrid seed production using the SI system, a weak or unstable SI phenotype causes contamination of more selfed seeds, resulting in production of low-quality F_1 hybrid seeds. Since the SI phenotype is influenced by physiological conditions of plants, e.g., plant vigor, plant age, and flower age, and environmental factors, e.g., temperature and water potential, the purity of F_1 hybrid seeds depends on environmental factors and is sometimes lowered to an unsalable level. Therefore, the use of cytoplasmic male sterility is increasing in F_1 hybrid breeding of Brassicaceae vegetables. However, the SI system is still preferred because the yield of F_1 seeds is higher in seed production using the SI system than that using the male sterility system. In F_1 hybrid breeding using the SI system, it is important to select inbred lines having a strong and stable SI phenotype.

Strength of SI phenotype in *Brassica* vegetables depends on *S* haplotypes and other genetic factors. Class-II *S* haplotypes have been reported to show weaker SI phenotypes than those of class-I *S* haplotypes (Nasrallah et al. 1991). *S-15* in *B. oleracea*, which is a class-II *S* haplotype, is known to show weak SI. Interestingly, the SI phenotype of *S-60* in *B. rapa*, which is highly similar in nucleotide sequences of *SP11*, *SRK*, and *SLG* to those of *S-15* in *B. oleracea* and has the same recognition specificity as *S-15* in *B. oleracea*, is not weak (Sato et al. 2006). From our experience, we know that the SI phenotype of *S-8* in *B. rapa* is weaker than that of *S-9* in *B. rapa*, both of which are class-I *S* haplotypes. A QTL for the number of seeds obtained with self-pollination by bees has been detected in a region containing the *S* locus in *B. rapa* (Hatakeyama et al. 2010), supporting the finding that *S* haplotypes are responsible for the strength of the SI phenotype. However, the relationship between the strength of the SI phenotype and *S* genotypes has not been detected in radish (Niikura and Matsuura 1999). Although evaluation of the strength of the SI phenotype is not easy due to the influence of environmental factors, characterization of each *S* haplotype, especially the strength of the SI phenotype, is required. For the international exchange of such important information for radish breeding, unification of *S* haplotype names is indispensable.

References

Brace J, Ockendon DJ, King GJ (1993) Development of a method for the identification of *S* alleles in *Brassica oleracea* based on digestion of PCR-amplified DNA with restriction endonucleases. Sex Plant Reprod 6:133–138

Fujimoto R, Nishio T (2003) Identification of *S* haplotypes in *Brassica* by dot blot analysis of *SP11* alleles. Theor Appl Genet 106:1433–1437

Fukai E, Fujimoto R, Nishio T (2003) Genomic organization of the *S* core region and the *S* flanking regions of a class-II *S* haplotype in *Brassica rapa*. Mol Genomics Genet 269:361–369

Haruta T (1962) Studies on the genetics of self- and cross-incompatibility in cruciferous vegetable. Res Bull Takii Plant Breed Exp Stn 2:1–169 (in Japanese)

Hatakeyama K, Horisaki A, Niikura S, Narusaka Y, Abe H, Toshiaki H, Ishida M, Fukuoka H, Matsumoto S (2010) Mapping of quantitative trait loci for high level of self-incompatibility in *Brassica rapa* L. Genome 53:257–265

Hinata K, Nishio T (1978) *S*-allele specificity of stigma proteins in *Brassica oleracea* and *B. campestris*. Heredity 41:93–100

Inaba R, Nishio T (2002) Phylogenetic analysis of *Brassiceae* based on the nucleotide sequences of the *S*-locus related gene, *SLR1*. Theor Appl Genet 105:1159–1165

Kimura R, Sato K, Fujimoto R, Nishio T (2002) Recognition specificity of self-incompatibility maintained after the divergence of *Brassica oleracea* and *Brassica rapa*. Plant J 29:215–223

Kitashiba H, Li F, Hirakawa H, Kawanabe T, Zou Z, Hasegawa Y, Tonosaki K, Shirasawa S, Fukushima A, Yokoi S, Takahata Y, Kakizaki T, Ishida M, Okamoto S, Sakamoto K, Shirasawa K, Tabata S, Nishio T (2014) Draft sequences of the radish (*Raphanus sativus* L.) genome. DNA Res 21:481–490

Kusaba M, Nishio T (1999) Comparative analysis of *S* haplotypes with very similar *SLG* alleles in *Brassica rapa* and *Brassica oleracea*. Plant J 17:83–91

Kusaba M, Matsushita M, Okazaki K, Satta Y, Nishio T (2000) Sequence and structural diversity of the *S* locus genes from different lines with the same self-recognition specificities in *Brassica oleracea*. Genetics 154:413–420

Kusaba M, Dwyer K, Hendershot J, Vrebalov J, Nasrallah JB, Nasrallah ME (2001) Self-incompatibility in the genus *Arabidopsis*: characterization of the *S* locus in the outcrossing *A. lyrata* and its autogamous relative *A. thaliana*. Plant Cell 13:627–643

Lim SH, Cho HJ, Lee SJ, Cho YH, Kim BD (2002) Identification and classification of *S* haplotypes in *Raphanus sativus* by PCR-RFLP of the *S* locus glycoprotein (*SLG*) gene and the *S* locus receptor kinase (*SRK*) gene. Theor Appl Genet 104:1253–1263

Nasrallah JB, Yu S-M, Nasrallah ME (1988) Self-incompatibility genes of *Brassica oleracea*: expression, isolation, and structure. Proc Natl Acad Sci USA 85:5551–5555

Nasrallah JB, Nishio T, Nasrallah ME (1991) The self-incompatibility genes of *Brassica*: expression and use in genetic ablation of floral tissues. Annu Rev Plant Physiol Plant Mol Biol 42:393–422

Nasrallah ME, Liu P, Sherman-Broyles S, Boggs NA, Nasrallah JB (2004) Natural variation in expression of self-incompatibility in *Arabidopsis thaliana*: implications for the evolution of selfing. Proc Natl Acad Sci USA 101:16070–16074

Niikura S, Matsuura S (1997) Genomic sequence and expression of *S*-locus gene in radish (*Raphanus sativus* L.). Sex Plant Reprod 10:250–252

Niikura S, Matsuura S (1998) Identification of self-incompatibility alleles (*S*) by PCR-RFLP in radish (*Raphanus sativus* L.). Euphytica 102:379–384

Niikura S, Matsuura S (1999) Genetic variation of the *S*-alleles and level of self-incompatibility in the Japanese cultivated radish (*Raphanus sativus* L.). Breed Res 1:211–220 (In Japanese)

Nishio T, Hinata K (1980) Rapid detection of *S*-glycoproteins of self-incompatible crucifers using Con-A reaction. Euhytica 29:217–221

Nishio T, Sakamoto K, Yamaguchi J (1994) PCR-RFLP of *S* locus for identification of breeding lines in cruciferous vegetables. Plant Cell Rep 13:546–550

Nishio T, Kusaba M, Watanabe M, Hinata K (1996) Registration of *S* alleles in *Brassica campestris* L by the restriction fragment sizes of *SLG*s. Theor Appl Genet 92:388–394

Nishio T, Kusaba M, Sakamoto K, Ockendon DJ (1997) Polymorphism of kinase domain of the *S*-locus receptor kinase gene (*SRK*) in *Brassica oleracea* L. Theor Appl Genet 95:335–342

Nishio T, Izumida A, Hanzawa H, Sakamoto K (2006) Development of *S* tester lines of *Brassica oleracea*, *Brassica rapa*, and *Raphanus sativus* as genetic resources. Acta Hort 706:141–144

Nou IS, Watanabe M, Isuzugawa K, Isogai A, Hinata K (1993) Isolation of *S*-alleles from a wild population of *Brassica campestris* L. at Balcesme, Turkey and their characterization by *S*-glycoproteins. Sex Plant Reprod 6:71–78

Ockendon DJ (2000) The *S*-allele collection of *Brassica oleracea*. Acta Hort 539:25–30

Oikawa E, Takuno S, Izumita A, Sakamoto K, Hanzawa H, Kitashiba H, Nishio T (2011) Simple and efficient methods for *S* genotyping and *S* screening in genus *Brassica* by dot-blot analysis. Mol Breed 28:1–12

Okamoto S, Sato Y, Sakamoto K, Nishio T (2004) Distribution of similar self-incompatibility (*S*) haplotypes in different genera, *Raphanus* and *Brassica*. Sex Plant Reprod 17:33–39

Okazaki K, Hinata K (1984) Analysis of *S*-alleles and *S*-glycoproteins in F_1 hybrid varieties of Japanese radish (*Raphanus sativus* L.). Japan J Breed 34:237–245 (In Japanese)

Okazaki K, Kusaba M, Ockendon DJ, Nishio T (1999) Characterization of *S* tester lines in *Brassica oleracea*: polymorphism of restriction fragment length of *SLG* homologues and isoelectric points of *S*-locus glycoproteins. Theor Appl Genet 98:1329–1334

Sakamoto K, Kusaba M, Nishio T (1998) Polymorphism of the *S*-locus glycoprotein gene (*SLG*) and the S-locus related gene (*SLR1*) in *Raphanus sativus* L. and self-incompatible ornamental plants in the Brassicaceae. Mol Gen Genet 258:397–403

Sato K, Nishio T, Kimura R, Kusaba M, Suzuki T, Hatakeyama K, Ockendon DJ, Satta Y (2002) Coevolution of the *S*-locus genes *SRK*, *SLG*, and *SP11/SCR* in *Brassica oleracea* and *B. rapa*. Genetics 162:931–940

Sato Y, Fujimoto R, Toriyama K, Nishio T (2003) Commonality of self-recognition specificity of *S* haplotypes between *Brassica oleracea* and *Brassica rapa*. Plant Mol Biol 52:617–626

Sato Y, Okamoto S, Nishio T (2004) Diversification and alteration of recognition specificity of the pollen ligand, SP11/SCR, in self-incompatibility of *Brassica* and *Raphanus*. Plant Cell 16:3230–3241

Sato Y, Sato K, Nishio T (2006) Interspecific pairs of class-II *S* haplotypes having different recognition specificity between *Brassica oleracea* and *Brassica rapa*. Plant Cell Physiol 47:340–345

Schierup MH, Mable BK, Awadalla P, Charlesworth D (2001) Identification and characterization of a polymorphic receptor kinase gene linked to the self-incompatibility locus of *Arabidopsis lyrata*. Genetics 158:387–399

Schopfer CR, Nasrallah ME, Nasrallah JB (1999) The male determinant of self-incompatibility in *Brassica*. Science 286:1697–1700

Shiba H, Kakizaki T, Iwano M, Entani T, Ishimoto K, Shimosato H, Che F-S, Satta Y, Ito A, Takada Y, Watanabe M, Isogai A, Takayama S (2002) The dominance of alleles controlling self-incompatibility in *Brassica* pollen is regulated at the RNA level. Plant Cell 14:491–504

Stein JC, Howlett B, Boyes DC, Nasrallah ME, Nasrallah JB (1991) Molecular cloning of a putative receptor protein kinase gene encoded at the self-incompatibility locus of *Brassica oleracea*. Proc Natl Acad Sci USA 88:8816–8820

Suzuki G, Kai N, Hirose T, Fukui K, Nishio T, Takayama S, Isogai A, Watanabe M, Hinata K (1999) Genomic organization of the *S* locus: Identification and characterization of genes in *SLG/SRK* region of S^9 haplotype of *Brassica campestris* (syn. *rapa*). Genetics 153:391–400

Suzuki T, Kusaba M, Matsushita M, Okazaki K, Nishio T (2000) Characterization of *Brassica S*-haplotypes lacking *S*-locus glycoprotein. FEBS Lett 482:102–108

Takasaki T, Hatakeyama K, Suzuki G, Watanabe M, Isogai A, Hinata K (2000) The *S* receptor kinase determines self-incompatibility in *Brassica* stigma. Nature 403:913–916

Takayama S, Shiba H, Iwano M, Shimosato H, Che F-K, Kai N, Watanabe M, Suzuki G, Hinata K, Isogai A (2000) The pollen determinant of self-incompatibility in *Brassica campestris*. Proc Natl Acad Sci USA 97:1920–1925

Takuno S, Fujimoto R, Sugimura T, Sato K, Okamoto S, Zhang S-L, Nishio T (2007) Effects of recombination on the hitchhiking diversity in *Brassica* self-incompatibility locus complex. Genetics 177:949–958

Takuno S, Oikawa E, Kitashiba H, Nishio T (2010) Assessment of genetic diversity of accessions in Brassicaceae genetic resources by frequency distribution analysis of *S* haplotypes. Theor Appl Genet 120:1129–1138

Tarutani Y, Shiba H, Iwano M, Kakizaki T, Suzuki G, Watanabe M, Isogai A, Takayama S (2010) Trans-acting small RNA determines dominance relationships in *Brassica* self-incompatibility. Nature 466:983–987

Thompson KF, Taylor JP (1966) Non-linear dominace relationships between *S* alleles. Heredity 21:345–362

Tonosaki K, Kudo J, Kitashiba H, Nishio T (2013) Allele-specific hybridization using streptavidin-coated magnetic beads for species identification, *S* genotyping, and SNP analysis in plants. Mol Breed 31:419–428

Warwick SI, Sauder CA (2005) Phylogeny of tribe *Brassiceae* (Brassicaceae) based on chloroplast restriction site polymorphisms and nuclear ribosomal internal transcribed spacer and chloroplast trnL intron sequences. Can J Bot 83:467–483

Watanabe M, Ito A, Takada Y, Ninomiya C, Kakizaki Y, Takahata Y, Hatakeyama K, Hinata K, Suzuki G, Takasaki T, Satta Y, Shiba H, Takayama S, Isogai A (2000) Highly divergent sequences of the pollen self-incompatibility (*S*) gene in class-I *S* haplotypes of *Brassica campestris* (syn. *rapa*) L. FEBS Lett 473:139–144

The Use of Genome Information for Intergeneric Hybridization Breeding

14

Sang Woo Bang, Katsunori Hatakeyama and Yoshihito Takahata

Abstract

Radish is an important vegetable crop and has several agriculturally useful traits to improve the breeding of *Brassica* crops, which include cytoplasmic male sterility, resistance to pod shattering, tolerance to saline soils, and resistance to nematode and diseases. To transfer these traits into *Brassica*, a number of intergeneric hybridization have been performed using intergeneric cross and/or protoplast fusion, and it was shown that some of them were introduced successfully into and expressed in *Brassica* or hybrids with *Brassica*. The various kinds of hybrid progenies such as synthetic amphidiploid line, alloplasmic line, alien gene(s) introgression line, and monosomic alien chromosome addition line would be valuable genetic resources not only to breed more useful cultivars but also to analyze each chromosome and gene concerned. Recently, the development of numerous molecular markers for genetic study has been accelerated in radish, and a high-density linkage map has been constructed. These molecular tools have been used for the identification of QTLs associated with important agronomic traits of radish, genetic analysis of hybridization barrier between radish and *Brassica* crops, and comparative genomics of Brassicaceae. The further accumulation of genome information will facilitate the cloning of candidate genes in QTLs and the marker-assisted breeding in radish, in addition to the intergenomic transfer of useful radish genes into *Brassica* crops through intergeneric hybridization.

S.W. Bang
Faculty of Agriculture, Utsunomiya University, Utsunomiya 321-8505, Japan

K. Hatakeyama · Y. Takahata (✉)
Faculty of Agriculture, Iwate University, Morioka 020-8550, Japan
e-mail: ytakahata@iwate-u.ac.jp

14.1 Introduction

Radish (*Raphanus sativus* L., 2n = 18, RR) belongs to the family Brassicaceae and is an economically important vegetable crop that is grown and consumed around the world,

T. Nishio and H. Kitashiba (eds.), *The Radish Genome*,
Compendium of Plant Genomes, DOI 10.1007/978-3-319-59253-4_14

especially in East Asia. The genus *Raphanus* originates from coastal regions along the Mediterranean and Black Seas, and *R. sativus* is ecologically classified into five varieties: var. *radicula* DC. (var. *sativus* Saz., European small radish), var. *niger* (Mill.) Pers. (var. *major* A. Voss., black radish), var. *raphanistroides* Makino (var. *hortensis* Backer, Asian big radish including north Chinese radish, south Chinese radish, and Japanese radish), var. *mougri* Helm (var. *caudatus* (L.) Hooker & Anderson, rat's tail radish), and var. *olerfer* Netz. (var. *oleiformis* Pers. oil radish) (Banga 1976; Hida 1990; Kumazawa 1965). European small radish with smaller root and early maturity is a popular garden vegetable all over the world. The north Chinese radish, which is characterized by smaller, colored, and starchy root, is used as pickles, fresh vegetables, and condiments. The south Chinese radish with pubescent leaf and low starchy and juicy long root is used for salted vegetables and cooking (Kaneko et al. 2007).

The agricultural traits of the genus *Raphanus* were summarized by Warwick (1993). These include cytoplasmic male sterility (CMS) (Ogura 1968; Yamagishi and Terachi 1996; Tanaka et al. 2012), resistance to pod shattering (Agnihotri et al. 1990), tolerance to saline soils (Warwick 1993), and resistance to blackleg (Salisbury 1987), beet cyst nematode (Lelivelt and Krens 1992; Lelivelt et al. 1993a), and clubroot (Ashizawa et al. 1980). Other *Brassica* crops often lack these traits and are susceptible to these pathogens. Therefore, radish is an important genetic resource for the breeding of *Brassica* crops. A number of intergeneric hybridizations have been performed to transfer these agronomically useful traits into *Brassica* crops, since *Raphanobrassica*, an intergeneric hybrid between *R. sativus* and *B. oleracea*, was reported by Karpechenko (1924) (Kaneko et al. 2009). For nearly a century, cytogenetic research based on their hybridization has been conducted, resulting in the discovery of excellent evidence for partial homologous relations and speciation of these species. Recently, several genetic maps of radish have been constructed using RFLP, AFLP, RAPD, and EST-SSR markers, and these have been applied to QTL identification of beet cyst nematode and clubroot resistance, the shape and pigmentation of root, and flowering time, and comparative genomics of the Brassicaceae (Kaneko et al. 2007; Tsuro et al. 2008; Budahn et al. 2009; Kamei et al. 2010; Shirasawa et al. 2011). Kitashiba et al. (2014) determined the *R. savivus* draft genome sequences by a next-generation sequencer (NGS) along with bacterial artificial chromosome (BAC)-end sequences. They constructed a high-density linkage map by adding new DNA markers and combining two different linkage maps. These genetic markers and high-density linkage maps could be used for the identification of candidate genes in QTLs, and for the marker-assisted breeding in radish, in addition to intergenomic transfer of radish genes into the *Brassica* crops through intergeneric hybridization followed by successive backcrossings.

Intergeneric gene transfer has been carried out for over a century in an attempt to transfer valuable genes into crop species from wild relatives. Recently, intergeneric hybridization has been extensively performed between cultivated species, and between cultivated species and wild relatives, to develop more useful cultivars with improved biotic and abiotic stress tolerance (Harberd and McArthur 1980; Kaneko and Bang 2014; Kaneko et al. 2009; Prakash et al. 1999). The prospects of intergeneric hybridization for genetics and plant breeding in Brassicaceae were evaluated by Matsuzawa et al. (1996) who suggested a system for the development of the five hybrid progenies: synthetic amphidiploid line, alien gene(s) introgression line, alloplasmic line, monosomic alien chromosome addition line (MAL), and monosomic alien chromosome substitution line. These hybrid lines would be valuable genetic resources not only to breed more useful cultivars with novel agronomic traits but also to analyze each chromosome and gene concerned. In these kinds of research projects, it is prerequisite to grow true F_1 hybrids and as many of their progenies as possible, although the reproductive isolation in the intergeneric hybridization hampers the development of these hybrids and progenies (Kaneko and Bang 2014).

In this chapter, we will outline the intergeneric hybridization between radish and the *Brassica* crops, their development, preservation and improvement of hybrid lines, and application of molecular markers in order to expedite the progress of breeding.

14.2 Intergeneric Hybridizations Between Radish and *Brassica* Crops

Intergeneric hybridizations between radish and *Brassica* crops have been made to produce different types of *Raphanobrassica* and *Brassicoraphanus* hybrids for cytological and evolutionary studies, and for investigation of their agronomic potential. The amphidiploid *Raphanobrassica* (RRCC genomes), originating from pollinations of radish (RR) with *Brassica oleracea* (CC) pollen, was first obtained by Karpechenko (1924). Its reciprocal crossing, however, was not successful (Karpechenko 1924). Sarashima et al. (1980) obtained the same intergeneric hybrids using embryo rescue. When McNaughton (1973) compared differences between diploid and tetraploid *B. oleracea* plants as the pollen parent in crossings with radish, he found that the crossings with the tetraploid were more successful. Although the amphidiploid *Raphanobrassica* (RRAA), a hybrid between radish (maternal parent) and *B. rapa* (paternal parent), is reportedly much more difficult to produce, a few have been developed (U et al. 1937; Sarashima and Matsuzawa 1989), and several hybrids (RRAA) were obtained using embryo rescue (Matsuzawa et al. 2000). *Brassicoraphanus* (AARR), i.e., a reciprocal amphidiploid hybrid, has been developed using *B. rapa* as a maternal parent with embryo rescue (Tokumasu 1976; Sarashima and Matsuzawa 1989).

A number of intergeneric hybridizations between radish and *B. napus* have been performed in order to transfer agronomically useful traits from radish into *B. napus*, but only a few hybrid plants have been obtained owing to their poor crossability. The use of *Raphanobrassica* and *Brassicoraphanus* hybrids as the bridge plants in the crossings with *B. napus* has been much more successful than the use of intergeneric hybridization between radish and *B. napus* (Agnihotri et al. 1990; Lelivelt et al. 1993a; Peterka et al. 2004). Several biotechnological methods have been performed to overcome the barrier existing in intergeneric hybridization between *B. napus* and radish: ovary culture (Luo et al. 1989), ovule culture (Dolstra 1982), flower culture (Metz et al. 1995), in vitro organ culture (Luo et al. 2000), and protoplast fusion (Lelivelt and Krens 1992; Pelletier et al. 1983; Sakai and Imamura 1990; Sakai et al. 1996; Wang et al. 2006). Agnihotri et al. (1990) successfully used ovary culture followed by embryo rescue to obtain plantlets by the hybridization between *B. napus* and *Raphanobrassica*. Hybridizations between two different *Raphanobrassica* hybrids (RRAA and RRCC) and *B. napus* (AACC) were performed in order to develop *B. napus*-radish monosomic addition lines (MALs) (Akaba et al. 2009a; Budahn et al. 2008). Ohtsu et al. (2015) obtained intergeneric hybrid seeds between several *B. napus* lines and *R. sativus* cv. 'Shogoin-daikon' without any embryo rescue, and they developed true intergeneric hybrids (AACCRR) subsequent to chromosome doubling, and then generated *B. napus*-radish chromosome addition lines through successive backcrossing to *B. napus*.

A complete series of chromosome addition lines enables markers and genes to be physically allocated to chromosomes without a need for detectable morphological polymorphisms. The series could serve as a bridge plant between genetic maps in radish and physical location on single radish chromosomes by deletion strains. Such a series could be helpful for a more directed and stable introgression of interesting genes into *Brassica* genomes and contribute to the study of the evolutionary relationship of the genus *Raphanus* with other genera in the tribe *Brassiceae*.

14.3 Homoeologous Chromosomes Between Radish and *Brassica* Crop Genomes

The existence of homoeologous chromosomes between radish and *Brassica* genomes has been reported by many researchers. Namai (1976) and Sarashima (1991) observed that the intergeneric hybrids between radish and *B. oleracea* showed a number of bivalent chromosomes at metaphase I, six being the maximum number, and also trivalent and tetravalent chromosomes in very rare occasions. Sarashima (1991) also observed four bivalent chromosomes in F_1 hybrids between *B. rapa* and radish, and up to five bivalent chromosomes in the reciprocal F_1 hybrids. Mizushima (1980) indicated the number of allopairs between R and A genomes is more than three after considering autopairs in an identical genome. Pollen mother cells (PMCs) in the RACC hybrids had varying numbers of univalent and bivalents, but always more than nine bivalents, suggesting that there are homoeologous chromosomes between the R and A genomes (Paulmann and Röbbelen 1988; Agnihotri et al. 1990). A cytological examination of the RCAA plants showed that the majority of PMCs had the expected 38 chromosomes, but there was a higher level of chromosome pairing than the expected one (10 bivalents) between just the A genome chromosomes, suggesting that the R and C chromosomes also make pairing (Lelivelt et al. 1993a). These results indicate that the introgression and gene transfer are possible by homoeologous chromosome pairing between radish and *Brassica* crops. Voss et al. (1999) also observed a translocation based on GISH analysis of BC_3 progenies between radish and *B. napus*, resulting from natural recombination events between the R genome and the A and C genomes, respectively. Recently, a comparison of the genome structure of radish with those of the A, B, and C genomes of *Brassica* based on high-density linkage maps using EST-based SNP markers revealed extensive chromosome homoeology among their genomes, resulting in facilitated transfer of genes from radish to *Brassica* crops (Li et al. 2011).

14.4 Intergeneric Gene Transfer from Radish to the *Brassica* Crops

In practical breeding of *Brassica* crops, desirable traits of radish have been introduced successfully into *Brassica* crops, for example, cytoplasmic male sterility, its fertility restorer, and resistance to diseases and pests such as Turnip mosaic virus (TuMV), clubroot, and beet cyst nematode. Molecular methods including GISH and several polymorphic DNA markers are being widely used in breeding by marker-assisted selection, and in identifying integrated alien chromosomes, chromosome segments, and genes (Table 14.1).

14.4.1 Introgression of the CMS Restorer Gene(s)

One of the CMS systems used by rapeseed (*B. napus*) breeders is the alloplasmic system derived from the radish male sterile cytoplasm such as Ogura CMS (Ogura 1968; Delourme and Budar 1999) and Kosena CMS (Sakai and Imamura 1990; Sakai et al. 1996; Koizuka et al. 2000). Although restored fertility is essential for production of F_1 seeds using the radish CMS, no restorer genes for this system are known to exist in rapeseed. In order to transfer the restorer fertility (*Rf*) genes from radish to the rapeseed recipient, intergeneric crossing or protoplast fusion between them was performed (Paulmann and Röbbelen 1988; Sakai et al. 1996). The Ogura CMS restorer gene (*Rfo*) was transferred into the C genome of *B. napus* from radish (Delourme and Eber 1992; Delourme et al. 1998). In situ hybridization and genetic mapping further confirmed that the radish fragment with the *Rfo* gene was introgressed into the distal region of linkage group 19, corresponding to chromosome 9 of *B. napus* C genome (Hu et al. 2008). Recently, the *Rfo* gene was introgressed from radish to *B. juncea*, and mapped on chromosome 9 of *B. juncea* A genome (Tian et al. 2014). Based on molecular markers tightly linked to the *Rfo* gene, the radish introgressed fragment

Table 14.1 Agronomic traits of radish introduced successfully into and expressed in *Brassica* crops or hybrids with *Brassica*

Traits	Gene(s)	*Brassica* crops	Means to transfer	References
CMS restorer	*Rfo*	*Brassica napus*	Intergeneric cross/protoplast fusion	Delourme and Eber (1992)
				Delourme et al. (1998)
		Brassica juncea	Unknown	INRA (cited by Tian et al. 2014)
				Tian et al. (2014)
	Rfk1	*B. napus*	Protoplast fusion	Sakai et al. (1996)
				Koizuka et al. (2000)
TuMV resistance	Unknown	*Raphanobrassica*	Intergeneric cross	Krämer et al. (2003)
Nematode resistance	QTL $Hs1^{Rph}$	*B. napus-R. satius* MALs	Intergeneric cross	Budahn et al. (2009)
Clubroot ressitance	'c' chromosome	*B. napus-R. satius* MALs	Intergeneric cross	Akaba et al. (2009a, b)
	A major gene and two minor genes on 'c,' 'b,' and 'f' chromosomes, respectively	*B. napus-R. satius*	Intergeneric cross	Ohtsu et al. (2015)

in the *B. napus* Ogura CMS Rf lines has been shortened (Delourme et al. 1998; Giancola et al. 2003; Primard-Brisset et al. 2005). The unimproved *B. napus* Rf line carried a large radish introgression of at least 50 cM with 21 markers, whereas the improved *B. napus* Rf line lost 9 radish markers including ScH03 and PGIint, and showed improved growth vigor and fertility (Primard-Brisset et al. 2005). Although *Rfo* was transferred into *B. juncea* in INRA, the original restorer line exhibited poor vigor and low fertility and was black-seeded. Tian et al. (2014) improved the agronomic performance of the restorer line by marker-assisted breeding.

To establish a Kosena CMS-Rf system for F_1 hybrid seed production in rapeseed, a nuclear-encoded radish Rf gene was introduced into *B. napus* by protoplast fusion (Sakai et al. 1996). Two restorer loci were identified in Kosena CMS radish, and of these, *Rfk1* was reported to be essential to restore fertility of the *B napus* Kosena CMS cybrid (Koizuka et al. 2000). *Rfk1* was cloned by map-based cloning and shown to encode a pentatricopeptide repeat protein (Koizuka et al. 2003). Molecular characterization of the *Rf* genes will not only provide us insight into the interaction between mitochondrial and nuclear genes, but it will also contribute to the improvement of the F_1 seed production system using cytoplasmic male-sterility.

14.4.2 Introgression of the Resistance to Nematode

The beet cyst nematode (BCN), *Heterodera schachtii* Schm., is a widespread and serious parasite in beet and oilseed rape cultivating areas of the world. *H. schachtii* is a bisexual nematode, and its multiplication is dependent on the development of the male and female nematodes (Mousa and Budahn 2012). Radish and white mustard (*Sinapis alba*) are useful sources of the resistance to BCN, and some attempts to transfer these resistances to *B. napus* crops by sexual and somatic hybridization have been carried out (Lelivelt and Krens 1992; Lelivelt et al. 1993a, b; Budahn et al. 2009). Voss et al. (1999) tried to introduce the resistance to nematodes from radish to rapeseed through the radish chromosome addition line of rapeseed by successive back-crossing. They were able to identify two radish chromosomes added to the rapeseed genome using the GISH analysis and confirmed that one of them had a close relation to nematode resistance. Peterka et al. (2004) found that the BCN resistance gene(s) was located on the 'd' chromosome of radish using *B. napus-R. sativus*-monosomic addition lines (MALs), and the resistance mediated by the 'd' chromosome acted in a dominant manner in the genetic background of susceptible rapeseed. The chromosome specified by Voss et al. (1999) might have corresponded to the 'd' chromosome of radish reported by Peterka et al. (2004). Budahn et al. (2009) clarified that the resistance to BCN is controlled by a major QTL $Hs1^{Rph}$ (quantitative trait locus for resistance against *Heterodera schachtii* from *Raphanus sativus*), which was found by the log-likelihood (LOD) plot for BCN resistance reaching the maximum score in the interval between markers E41M59-297 and OPI19-471 at 71.0 cm on the radish 'd' chromosome. Radish chromosome-specific anchor markers including the desired genes, such as $Hs1^{Rph}$, with known map positions were made available for future recombination experiments to incorporate chromosome segments carrying desired genes from radish into rapeseed by means of the intergeneric recombination between rapeseed genome and the 'd' chromosome of radish. The anchor markers allow us to follow the dissection of the 'd' chromosome or other radish chromosomes in such studies and could assist in selection of introgression lines containing QTL $Hs1^{Rph}$ (Budahn et al. 2009).

14.4.3 Introgression of the Resistance to Clubroot

Clubroot disease is caused by an obligate pathogen *Plasmodiophora brassicae* Wor., and it is one of the most serious diseases of cruciferous crops. The pathogen survives as resting spores for long periods in the soil, and it is hard to control the disease using cultural practices or agrochemicals. The breeding of resistant cultivars is an effective approach to eliminate the use of fungicides and minimize crop losses caused by infection with this pathogen. Most radish cultivars and lines are highly resistant to the clubroot pathogen, but some are susceptible to it (Yoshikawa 1993). Kamei et al. (2010) performed a QTL analysis for clubroot resistance (CR) of radish using an F_2 population derived from a crossing of a CR Japanese radish and a clubroot susceptible Chinese radish, and proposed that a large part of radish CR is controlled by a single gene or closely linked genes, although minor effects of other genomic regions cannot be ruled out. The CR locus was named *Crs1*, of which linked markers should be useful for breeding CR cultivars of radish.

Many Japanese radish cultivars possess clubroot resistance, especially cv. 'Shogoin-daikon' is known to be immune against clubroot (Ashizawa et al. 1980). Akaba et al. (2009b) evaluated CR due to each single chromosome of radish cv. 'Shogoin-daikon' using all nine *R. sativus*-monosomic addition lines (MALs) of *B. napus* (*B. napus–R. sativus* MALs, a-i types which are not unified with those of Peterka et al. (2004) and Budahn et al. (2008) mentioned above), and identified that the CR against Ano-01 isolate is largely controlled by the 'c' chromosome of radish. However, there might be other minor and/or complementary gene(s) for resistance to clubroot in radish, because many plants carrying the 'c' chromosome showed slight disease symptoms. Ohtsu et al. (2015) reported that the CR of radish (cv. 'Shogoin-daikon') against field isolate pathogen is controlled by a single major gene and two minor genes using *B. napus–R. sativus* chromosome addition lines, and that the CR major gene is located on the 'c' chromosome, and two minor genes are located on the 'b' and 'f' chromosomes, respectively, using RAPD markers developed by Akaba et al. (2009a). Ohtsu et al. (2015) developed the CR progenies having AAC genomes (2n = 29) without any radish chromosome from the *B. napus–R. sativus* MALs through successive backcrossing to *B. rapa*, suggesting that the CR genes of radish might have been transferred into the A and/or C genomes of *Brassica*. These backcrossing progenies should be useful as the bridge plants to introduce the CR genes into *Brassica* crops. In addition, the 'c,' 'b,' and 'f' chromosome-specific RAPD markers of radish proposed by Akaba et al (2009a) could be a helpful tool for the development of DNA markers linked to the CR loci.

As another disease resistance, resistance to Turnip mosaic virus (TuMV) was found in *R. sativus*, and a high level of resistance to various TuMV pathotypes was also detected in sexually developed *Raphanobrassica* hybrids (Krämer et al. 2003).

14.5 Genetic Analysis of Hybridization Barrier Between Radish and *Brassica* Crops

Although radish is useful as a gene source for disease and pest resistance and cytoplasmic male sterility as mentioned above, the hybridization barrier between radish and *Brassica* crops hampers the production of hybrids. In many cases of intergeneric crossing between *B. rapa* as a maternal parent and radish as a paternal parent, embryo rescues are essential for obtaining the hybrid plants, but successful hybridization has been known to depend on combinations of cultivars or lines (Karpechenko 1924). Although most *B. rapa* lines cannot produce intergeneric hybrids, a 'Shogoin-kabu' line has shown high ability to set seeds of intergeneric hybrids (Kaneko et al. 1993). Tonosaki et al. (2013) conducted genetic analysis of hybrid seed formation ability of *B. rapa* in intergeneric crossings with radish and demonstrated that unsuccessful hybrid seed formation was due to post-zygotic, i.e., an arrest of embryo development at an early stage. QTL analysis using an F_2 population derived from a cross between a 'Shogoin-kabu' having relatively high ability to set seeds on intergeneric crossing with radish and an inbred line of 'Chiifu' with no seed sets revealed that two reproducibly detected QTLs had contrary effects on the number of hybrid seeds, and epistatic interaction between them existed. Genes located in these two QTLs were suggested to be involved in the hybrid formation ability of *B. rapa* in intergeneric crossing between *B. rapa* and radish. Dissection of the molecular mechanism controlling the intergeneric hybrid formation ability will be able to contribute to developing the means to overcome crossing barriers and improve the breeding of new cultivars having novel genetic traits from distantly related species.

References

Agnihotri A, Shivanna KR, Raina SN, Lakshmikumaran M, Prakash S, Jagannathan V (1990) Production of *Brassica napus* × *Raphanobrassica* hybrids by embryo rescue: an attempt to introduce shattering resistance into *B. napus*. Plant Breed 105:292–299

Akaba M, Kaneko Y, Ito Y, Nakata Y, Bang SW, Matsuzawa Y (2009a) Production and characterization of *Brassica napus-Raphanus sativus* monosomic addition lines mediated by the synthetic amphidiploid "*Raphanobrassica*". Breed Sci 59:109–118

Akaba M, Kaneko Y, Hatakeyama K, Ishida M, Bang SW, Matsuzawa Y (2009b) Identification and evaluation of clubroot resistance of radish chromosome using a *Brassica napus-Raphanus sativus* monosomic addition line. Breed Sci 59:203–206

Ashizawa M, Yoshikawa H, Hida K (1980) Studies on the breeding of clubroot-resistance in cole crops. II. Screening of cole crops of clubroot-resistance (2). Bull Veg Ornam Crops Res Stn A7:35–75

Banga O (1976) Radish, *Raphanus sativus* (Cruciferae). In: Simmonds NW (ed) Evolution of crop plants. Longman, London, pp 60–62

Budahn H, Schrader O, Peterka H (2008) Development of a complete set of disomic rape-radish chromosome-addition lines. Euphytica 162:117–128

Budahn H, Peterka H, Mousa MAA, Ding Y, Zhang S, Li J (2009) Molecular mapping in oil radish (*Raphanus sativus* L.) and QTL analysis of resistance against beet cyst nematode (*Heterodera schachtii*). Theor Appl Genet 118:775–782

Delourme R, Budar F (1999) Male sterilisy. In: Gomez-Campo C (ed) Biology of *Brassica* and coenospecies. Elsevier, Amsterdam, pp 185–216
Delourme R, Eber F (1992) Linkage between an isozyme marker and a restorer gene in radish cytoplasmic male sterility of rapeseed (*Brassica napus* L.). Theor Appl Genet 85:222–228
Delourme R, Foisset N, Horvais R, Barret P, Champagne G, Cheung WY, Landry BS (1998) Characterisation of the radish introgression carrying the *Rfo* restorer gene for the *Ogu*-INRA cytoplasmic male sterility in rapeseed (*Brassica napus* L.). Theor Appl Genet 97:129–134
Dolstra O (1982) Synthesis and fertility of × *Brassicoraphanus* and ways of transferring *Raphanus* characters to *Brassica*. Pudoc Wageningen I11. Agric Res Rep 917:1–90
Giancola S, Marhadour S, Desloire S, Clouet V, Falentin-Guyomarc'h H, Laloui W, Falentin C, Pelletier G, Renard M, Bendahmane A, Delourme R, Budar F (2003) Characterization of a radish introgression carrying the fertility restorer gene *Rfo* in rapeseed, using the *Arabidopsis* genome sequence and radish genetic mapping. Theor Appl Genet 107:1442–1451
Harberd DJ, McArthur ED (1980) Meiotic analysis of some species and genus hybrids in the Brassiceae. In: Tsunoda S, Hinata K, Gomez-Campo C (eds) *Brassica* crops and wild allies, biology and breeding. Jpn Sci Soc Press, Tokyo, pp 65–87
Hida K (1990) Root crops: Radish. In: Matsuo T (ed) Collected data of plant genetic resources, vol 2. Koudansha, Tokyo, pp 823–834 (in Japanese)
Hu X, Sullivan-Gilbert M, KubikT Danielson J, Hnatiuk N, Marchione W, Greene T, Thompson S (2008) Mapping of the Ogura fertility restorer gene *Rfo* and development of *Rfo* allele-specific markers in canola (*Brassica napus* L.). Mol Breed 22:663–674
Kamei A, Tsuro M, Kubo N, Hayashi T, Wang N, Fujimura T, Hirai M (2010) QTL mapping of clubroot resistance in radish (*Raphanus sativus* L.). Theor Appl Genet 120:1021–1027
Kaneko Y, Bang SW (2014) Interspecific and intergeneric hybridization and chromosomal engineering of Brassicaceae crops. Breed Sci 64:14–22
Kaneko Y, MatsuzawaY Namai H, Sarashima M (1993) Genetical and breeding evaluation of chromosome addition lines of radish with single kale chromosome. I. Phenotypic expression of some monosomic addition lines for radish and turnip varieties. Bull Coll Agric Utsunomiya Univ 15:27–37
Kaneko Y, Kimizuka-Takagi C, Bang SW, Matsuzawa Y (2007) Radish. In: Kole C (ed) Genome mapping and molecular breeding in plants, vol 5. Vegetables. Springer, New York, pp 141–160
Kaneko Y, Bang SW, Matsuzawa Y (2009) Distant hybridization. In: Gupta SK (ed) Biology and breeding of crucifers. Taylor & Francis Group, New York, pp 207–247
Karpechenko GD (1924) Hybrids of *Raphanus sativus* L. × *Brassica oleracea* L. J Gent 14:375–396
Kitashiba H, Li F, Hirakawa H, Kawanabe T, Zou Z, Hasegawa Y, Tonosaki K, Shirasawa S, Fukushima A, Yokoi S, Takahata Y, Kakizaki T, Ishida M, Okamoto S, Sakamoto K, Shirasawa K, Tabata S, Nishio T (2014) Draft sequences of the Radish (*Raphanus sativus* L.) genome. DNA Res 21:481–490
Koizuka N, Imai R, Iwabuch M, Sakai T, Imamura J (2000) Gemetic analysis of fertility restration and accumulation of ORF125 mitochondrial protein in the kosena radish (*Raphanus sativus* cv. Kosena) and a *Brassica napus* restorer line. Theor Appl Genet 100:949–955
Koizuka N, Imai R, Fujimoto H, Hayakawa T, Kimura Y, Kohno-Murase J, Sakai T, Kawasaki S, Imamura J (2003) Genetic characterization of a pentatricopeptide repeat protein gene, orf687, that restores fertility in the cytoplasmic male-sterile Kosena radish. Plant J 34:407–415
Krämer R, Scholze P, Marthe F, Ryschka U, Klocke E, Schumann G (2003) Improvement of disease resistance in cabbage: 1. Turnip mosaic virus, TuMV. Gesunde Pflanzen 55:193–198
Kumazawa S (1965) Radish. In: Vegetable gardening. Youkendo, Tokyo, pp 295–322 (in Japanese)
Lelivelt CLC, Krens FA (1992) Transfer of resistance to the beet cyst nematode (*Heterodera schachtii* Schm) into the *Brassica napus* L. gene pool through intergeneric somatic hybridization with *Raphanus sativus* L. Theor Appl Genet 83:887–894
Lelivelt CLC, Lange W, Dolstra O (1993a) Intergeneric crosses for the transfer of resistance to the beet cyst nematode from *Raphanus sativus* to *Brassica napus*. Euphytica 68:111–120
Lelivelt CLC, Leunissen EHM, Frederiks HJ, Helsper JPFG, Krens FA (1993b) Transfer of resistance to the beet cyst nematode (*Heterodera schachtii* Schm) from *Sinapis alba* L. (white mustard) to the *Brassica napus* L. gene pool by means of sexual and somatic hybridization. Theor Appl Genet 85:688–696
Li F, Hasegawa Y, Saito M, Shirasawa S, Fukushima A, Ito T, Fujii H, Kishitani S, Kitashiba H, Nishio T (2011) Extensive chromosome homoeology among Brassiceae species were revealed by comparative genetic mapping with high-density EST-based SNP markers in radish (*Raphanus sativus* L.). DNA Res 18:401–411
Luo P, Li XF, Wang ZQ, Lang ZQ (1989) Studies on wide cross between rapeseed and oil radish. In: Proceedings of the International Congress of SABRAO, pp 467–470
Luo P, Lang ZQ, Deng J, Wang ZQ (2000) Application of in vitro organ culture in wide-cross breeding of rapeseed. Euphytica 114:217–221
Matsuzawa Y, Kaneko Y, Bang SW (1996) Prospects of the wide cross for genetics and plant breeding in *Brassiceae*. Bull Coll Agric Utsunomiya Univ 16:5–10

Matsuzawa Y, Funayama T, Kamibayashi M, Konnai M, Bang SW, Kaneko Y (2000) Synthetic *Brassica rapa-Raphanus sativus* amphidiploid lines developed by reciprocal hybridization. Plant Breed 119:357–359

McNaughton LH (1973) Synthesis and sterility of *Raphanobrassica*. Euphytica 22:70–88

Metz PLJ, Nap JP, Stiekema WJ (1995) Hybridization of radish (*Raphanus sativus* L.) and oilseed rape (*Brassica napus* L.) through a flower-culture method. Euphytica 83:159–168

Mizushima U (1980) Genome analysis in *Brassica* and allied genera. In: Tsunoda S, Hinata K, Gomez-Campo C (eds) *Brassica* crops and wild allies, biology and breeding. Jpn Sci Soc Press, Tokyo, pp 89–106

Mousa MAA, Budahn H (2012) Genetic behavior of resistance to the beet cyst nematode (*Heterodera schachtii* Schm) in radish (*Raphanus sativus* L.). Afr J Microbiol Res 6:3755–3760

Namai H (1976) Cytogenetic and breeding studies on transfer of economic characters by means of interspecific and intergeneric crossing in the tribe *Brassiceae* of Cruciferae. Mem Fac Agric Tokyo Univ Educ 22:101–171 (In Japanese with English summary)

Ohtsu F, Tsutsui K, Akaba M, Bang SW (2015) Identification of clubroot resistance genes using a *Brassica napus-Raphanus sativus* chromosome addition line, and introduction of the clubroot resistance. Breed Res 17(Suppl. 2):85 (In Japanese)

Ogura H (1968) Studies on the new male sterility in Japanese radish, with special reference to the utilization of this sterility towards the practical raising of hybrid seeds. Men Fac Agric Kagoshima Univ 6:39–78

Paulmann W, Röbbelen G (1988) Effective transfer of cytoplasmic male sterility from radish (*Raphanus sativus* L.) to rape (*Brassica napus* L.). Plant Breed 100:299–309

Pelletier G, Primard C, Vedel F, Chetrit P, Remy R, Rousselle P, Renard M (1983) Intergeneric cytoplasmic hybridization in Cruciferae by protoplast fusion. Mol Gen Genet 191:224–250

Peterka H, Budahn H, Schrader O, Ahne R, Schutze W (2004) Transfer of resistance against the beet cyst nematode from radish (*Raphanus sativus*) to rape (*Brassica napus*) by monosomic chromosome addition. Theor Appl Genet 109:30–41

Prakash S, Takahata Y, Kirti PB, Chopra VL (1999) Cytogenetics. In: Gomez-Campo C (ed) Biology of Brassica coenospecies. Elsevier, Amsterdam, pp 59–106

Primard-Brisset C, Poupard JP, Horvais R, Eber F, Pelletier G, Renard M, Delourme R (2005) A new recombined double low restorer line for the *Ogu*-INRA cms in rapeseed (*Brassica napus* L.). Theor Appl Genet 111:736–746

Sakai T, Imamura J (1990) Intergeneric transfer of cytoplasmic male sterility between *Raphanus sativus* (cms line) and *Brassica napus* through cytoplast-protoplast fusion. Theor Appl Genet 80:421–427

Sakai T, Liu HJ, Iwabuch M, Kohno-Murase J, Imamura J (1996) Introduction of a gene from fertility restored radish (*Raphanus sativus*) into *Brassica napus* by fusion of X-irradiated protoplasts from a radish restorer line and iodacetoamide-treated protoplasts from a cytoplasmic male-sterile cybrid of *B. napus*. Theor Appl Genet 93:373–379

Salisbury PA (1987) Blackleg resistance in weedy crucifers. Crucif Newsl 12:90

Sarashima M (1991) Breeding of artificial amphidiploid. In: Breeding of *Brassica* crops through interspecific and intergeneric hybridizations (II). Ochiai Books Utsunomiya, pp 5–124 (in Japanese)

Sarashima M, Matsuzawa Y (1989) Intergeneric hybridization between radish (*Raphanus sativus* L.) and two monogenomic species of *Brassica* (*B. campestris* L. and *B. nigra* Koch.). Bull Coll Agric Utsunomiya Univ 14:99–104 (In Japanese with English summary)

Sarashima M, Matsuzawa Y, Kimura T (1980) Intergeneric hybridization between *Brassica oleracea* and *Raphanus sativus* by embryo culture. Crucif Newsl 5:25

Shirasawa K, Oyama M, Hirakawa H, Sato S, Tabata S, Fujioka T, Kimizuka-Takagi C, Sasamoto S, Watanabe A, Kato M, Kishida Y, Kohara M, Takahashi C, Tsuruoka H, Wada T, Sakai T, Isobe S (2011) An EST-SSR likage map of *Raphanus sativus* and coparative genomics of the Brassicaceae. DNA Res 18:221–232

Tanaka Y, Tsuda M, Yasumoto K, Yamagishi H, Terachi T (2012) A complete mitochondrial genome sequence of Ogura-type male sterile cytoplasm and its comparative analysis with that of normal cytoplasm in radish (*Raphanus sativus* L.). BMC Genom 13:352

Tian E, Roslinsky V, Cheng B (2014) Molecular marker-assisted breeding for improved Ogura cms restore line (*RfoRfo*) and mapping of the restore gene (*Rfo*) in *Brassica juncea*. Mol Breed 34:1361–1371

Tokumasu S (1976) The increase of seed fertility of *Brassicoraphanus* through cytological irregularity. Euphytica 25:463–470

Tonosaki K, Michiba K, Bang SW, Kitashiba H, Kaneko Y, Nishio T (2013) Genetic analysis of hybrid seed formation ability of *Brassica rapa* in intergeneric crossings with *Raphanus sativus*. Theor Appl Genet 126:837–846

Tsuro M, Suwabe K, Kubo N, Matsumoto S, Hirai M (2008) Mapping of QTLs controlling root shape and red pigmentation in radish, *Raphanus sativus* L. Breed Sci 58:55–61

U N, Midusima U, Saitô K (1937) On diploid and triploid *Brassica-Raphanus* hybrids. Cytologia 8:319–326

Voss A, Luhs WW, Snowdon RJ, Friedt W (1999) Development and molecular characterization of rapeseed (*Brassica napus* L.) resistant against beet cyst nematodes. In: Proceedings of the 10th international rapeseed congress, Canberra, Australia, 1999

Wang YP, Sonntag K, Rudloff E, Groeneveld I, Gramenz J, Chu CC (2006) Production and characterization of somatic hybrids between *Brassica napus* and *Raphanus sativus*. Plant Cell Tiss Organ Cult 86:279–283

Warwick SI (1993) Guide to the wild germplasm of *Brassica* and allied crops. Part IV. Wild species in the tribe *Brassiceae* (Cruciferae) as sources of agronomic traits. Tech Bull 17E:1–9

Yamagishi H, Terachi T (1996) Molecular and biological studies on male sterile cytoplasm in the Cruciferae. III. Distridution of Ogura-type cytoplasm among Japanese wild radishes and Asian radish cultivars. Theor Appl Genet 93:325–332

Yoshikawa H (1993) Studies on breeding of clubroot resistance in cole crops (in Japanese with English summary). Bull Natl Res Inst Veg Ornam Plants Tea Japan Ser A7:1–165

F_1 Hybrid Breeding Using Genome Information

15

Satoshi Niikura

Abstract

Most radish cultivars on the Japanese and Korean markets today are F_1 hybrids. In the other countries, F_1 cultivars are also increasing. The radish root is a product of field conditions and genotype responses to environment and season. A marketable product depends on a sustained balance of vegetative characters throughout the growth. Breeding for the root size and shape, the disease resistance, and physiological disorder tolerance must maintain the optimum balance of these characters under each cropping environment and consider their heterosis. In vegetable F_1 hybrid breeding, characters relevant to seed production are evaluated and selected during the reproductive stage, after the selection of vegetative characters. Therefore, the identification of selection markers for seed production characters can greatly improve selection efficiency. Even if a cross has improved vegetative characters, it is not grown if its parents cannot stably produce large amounts of high-purity F_1 seeds with a high germination rate. This chapter describes the current state of F_1 hybrid breeding and future needs.

15.1 Introduction

Through a long history of trade and selection, the cultivated radish, *Raphanus sativus* L., has been developed into several distinct forms. European forms include the small round red radish and the black radish. The cold-resistant northern Chinese and Korean varietal groups have good storage quality, and the heat-resistant southern Chinese varietal group has juicy flesh. The northern and southern Chinese forms arrived in Japan about 1300 years ago, and local cultivars are now adapted to most parts of the country. Breeding in the seventeenth and eighteenth centuries (Edo period) resulted in a wide range of shapes, sizes, and colors, and Japanese radish, called *daikon*, are now available all year round (Photos 15.1,

S. Niikura (✉)
Tohoku Seed Co., Utsunomiya, Tochigi 321-3232, Japan
e-mail: breeding@tohokuseed.co.jp

T. Nishio and H. Kitashiba (eds.), *The Radish Genome*,
Compendium of Plant Genomes, DOI 10.1007/978-3-319-59253-4_15

15.2). Annual radish production is about 7 million t worldwide (http://www.prota4u.org/), 1.45 million t in Japan (2014, http://www.e-stat.go.jp/), and 1.22 million t in Korea (USDA 2009). Thus, radish has become an important vegetable, especially in eastern Asia.

F_1 breeding improves crop yields through hybrid vigor. The first F_1 radish cultivar, 'Spring Cross,' was released in 1961 by the Takii Seed Company in Japan. The development of the 'Miyashige' F_1 varietal group expanded the range of uses, and the long protrusion of the root above the soil surface made it easier for farmers to grow radish, ensuring a stable supply of this staple vegetable. Further, F_1 hybrids between varietal groups led to ecotype expansion and made year-round supply possible in Japan. Further, breeding of the 'Miyashige' varietal group incorporated the late bolting of the 'Tokinashi' varietal group to create spring radish, the heat resistance of the 'Minowase' varietal group to create summer radish, and the cold resistance of the 'Nerima' varietal group to create winter radish.

Most radish cultivars on the Japanese market today are F_1 hybrids. Firm-fleshed F_1 radish grown for kimchi was introduced into Korea in the 1980s, and most radish cultivars grown in Korea now are also F_1s. Most cultivars grown in China, India, Vietnam, and Thailand are open-pollinated, but F_1 cultivars in the southern Chinese varietal group were introduced recently. F_1 cultivars of European small radish were introduced into Europe and America about 20 years ago and now are increasing.

The recent remarkable developments in molecular biology have greatly advanced marker-assisted selection of genes for breeding characters such as self-incompatibility and bolting. From the breeder's point of view, I would like to describe information on DNA markers linked to genes that control characters that are costly, time-consuming, or difficult to evaluate. However, effective selection based on genome information depends on strict evaluation of plant materials in various environments. At the moment, however, few characters have this basis. This chapter describes the current state of breeding information and future needs.

15.2 Vegetative Growth

The radish root is a product of interaction between genotypes and environments. A marketable product depends on a sustained balance of vegetative characters throughout the growth. For example, if the root growth rate falls behind the leaf growth rate, the root neck can crack and succumb to soft rot bacteria. Breeding must maintain the optimum balance of these characters under each cropping environment.

15.2.1 Size and Shape

15.2.1.1 Root Size and Shape

Root size and shape reveal very large genetic differences. The small round radish grown in Europe and America is harvested about a month after sowing at a weight of 10 g and a length of a few centimeters. The 'Moriguchi-daikon' landrace is grown in sandy soils of Gifu and Aichi prefectures in Japan and reaches a length of 2 m. The southern Chinese varietal group features an oval root shape (Photo 15.1). The 'Sakurajima-daikon' landrace in Japan is grown in volcanic soils of Kagoshima Prefecture and reaches 20 kg in 4 months (Photos 15.2).

Results of the quantitative evaluation of root shape based on principal component analysis using elliptic Fourier descriptors. Iwata et al. (1998) show that the shape of the upper and lower parts of roots is under strong genetic control and that it is possible to select for size and shape independently (Iwata et al. 2000, 2004). Yet, the genes that control root shape are not known. Japanese wild radish (Chap. 2) does not enlarge (Photo 15.3), whereas an F_2 population derived from an F_1 cross between 'Hamadaikon' and the Japanese landrace 'Miyashige' shows continuous segregation in root size and shape (Photo 15.4). Genes that control heterosis and root shape will be revealed

Photo 15.1 There are large genetic differences in root shape, size, and color among radishes. Examples include the Japanese wild radish (*far left*), the southern Chinese landrace (*2nd from left*); the Japanese landraces 'Horyo' with horn-shaped root, the very long and slender 'Moriguchi,' 'Miura,' 'Miyashige,' and 'Ohkura' (*3rd to 7th from left*); the Korean landrace for kimchi (*8th from left*); the Japanese landraces 'Kozena' for leaf use, and 'Osaka shiju-nichi' (*9th and 10th from left*); 'Shishyu-ji' with dark-green neck, 'Iwakuni-aka' with red skin (*far left* and *2nd on the lower row*); the European black radish (*3rd on the lower row*); the Japanese landrace 'Shogoin' (*4th on the lower row*); and the small red and white European radishes (*center*). The *ruler on the right* is marked in intervals of 10 cm

through the use of a series of genes profiled in Chap. 8 in such segregating populations. The results will also help to explain the mechanisms that control root enlargement in radish.

15.2.1.2 Heterosis

Heterosis in F_1 breeding is important for cultivars bred for size. For example, F_1 heterosis in a cross between the Japanese landrace 'Miyashige' and a French black radish is very large (Photo 15.5). Greater genetic distance between the parents generally promotes greater heterosis, but the result is evident only after the hybrids are grown. Thus, the analysis of genetic relationships to allow the degree of heterosis to be estimated and studies of the molecular mechanisms of heterosis in radish are needed.

15.2.1.3 Inbreeding Depression

F_1 breeding requires parental inbred lines raised through selfing to ensure genetic uniformity. However, radish, an allogamous crop, shows inbreeding depression, which limits root enlargement and makes it difficult to produce parental seeds and to predict their F_1 plant growth. This makes it necessary to ascertain the

Photo 15.2 Japanese landrace 'Sakurajima,' which holds the Guinness World Record for root weight. *The ruler on the right* is marked in intervals of 20 cm (Courtesy of Dr. Tanaka, Kagoshima Prefecture Institute for Agricultural Department)

Photo 15.3 Japanese wild radish 'Hamadaikon.' The marker pen is 14 cm long

degree of inbreeding depression while advancing generations. Thus, we need genome information about the population size that prevents inbreeding depression by measuring population heterozygosity, which is revealed only after crop growth.

15.2.2 Diseases

Difficulties in chemical control of diseases make breeding for resistance to soilborne diseases important. The safe evaluation of the disease resistance demands containment of the pathogen. The isolation of DNA markers or genes for resistance can obviate the need to test for disease resistance by bioassay or ELISA and greatly improve selection efficiency.

15.2.2.1 Fusarium Yellows

Fusarium yellows of radish is caused by *Fusarium oxysporum* f. sp. *raphani*, which proliferates well at >25 °C. Plant vessels turn black and cannot transport water and nutrients, and the plant becomes yellowish and dies (Photo 15.6). Plants are evaluated by a long-established method (Wellmann 1939). Resistance is controlled by multiple genes (Ashizawa et al. 1979). Genetic analysis by Tohoku Seed Company using the resistant inbred line 99–1006, raised from the Japanese 'Risoh' varietal group,

Photo 15.4 F_2 segregating population derived from an F_1 cross between Japanese wild radish 'Hamadaikon' and Japanese landrace 'Miyashige'

Photo 15.5 Japanese landrace 'Miyashige' (*left*), French black radish (*right*), and their F_1 (*center*), which shows heterosis

Photo 15.6 Radish yellows caused by *Fusarium oxysporum* f. sp. *raphani*

suggests that this line has three complementary recessive resistance genes (Fig. 15.1).

15.2.2.2 Verticillium Black Spot Disease

Black spot disease is caused by *Verticillium dahliae*, which proliferates well at 20–24 °C. Plant vessels turn black, but as plants seldom die, the disease can become apparent only after shipment or purchase. Plants are evaluated as for yellow disease, but there are no reports of genetic analysis.

15.2.2.3 Striped Root Surface

Rhizoctonia solani and *Pythium ultimum* var. *ultimum* are the main causative pathogens. Symptoms are promoted by high temperature and humidity. The root skin develops discolored marks (Photo 15.7), which become evident only after harvesting and washing. Genetic differences are evident (Photo 15.8), but no genetic analysis has been done.

15.2.2.4 Lumpiness

High temperatures and dry conditions just after sowing promote abnormal enlargement of part of the root (Photo 15.9). The cause is not yet known. No abnormalities are evident within the root. Genetic differences are evident, but no genetic analysis has been done.

15.2.2.5 Other Diseases

Other diseases targeted for breeding for resistance include infection with turnip mosaic virus (transmitted by an aphid), bacterial black spot, and bacterial soft rot.

15.2.3 Physiological Disorders

Breeding to transfer a character from one varietal group into another can upset the balance of growth and result in physiological disorders, which have a large influence on radish quality. Global warming is likely to exacerbate the problem. Some physiological disorders are visible only within the root and become evident only after harvesting and

Fig. 15.1 Frequency distribution of yellow disease symptoms in an F_2 population derived from the F_1 cross between resistant parent 99-1006 and susceptible parent 99–127. The theoretical segregation ratio is 1 (resistant): 63 (susceptible)

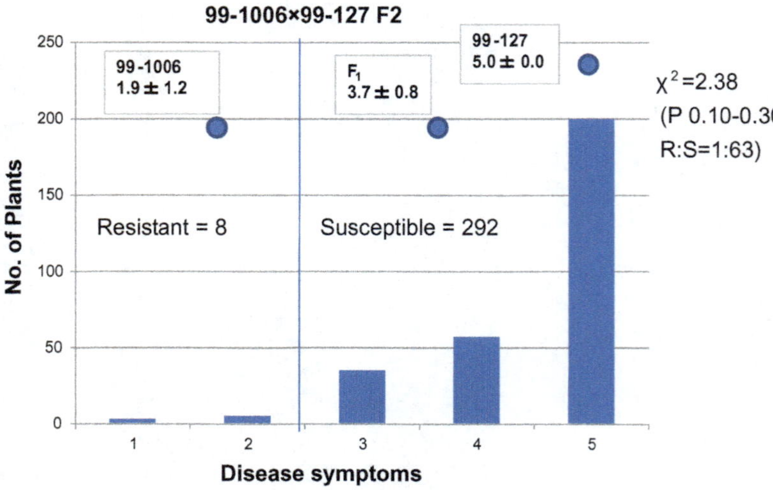

Photo 15.7 Severe symptoms of striped root surface

slicing. Thus, evaluation can take much time and effort. Moreover, reliable selection requires stable expression of the symptoms, which can vary greatly with climate and soil conditions. Further, slicing to reveal symptoms prevents direct selection and seed production. In relation to the absorption and translocation of water and nutrients, aquaporin, and boron transporter genes have been isolated and analyzed (Suga and Maeshima 2004; Takano et al. 2002). If such studies can reveal the underlying mechanisms, we will be able to breed for tolerance to physiological disorders systematically.

15.2.3.1 Internal Browning

Radish needs much more boron than other crops. Boron deficiency caused by high temperatures or drought can reveal large genetic differences in resistance (Fukuoka and Enomoto 2001; Fukuoka et al. 2010; Photos 15.10, 15.11), but no genetic analysis has been done.

15.2.3.2 Pithiness

Aging or an imbalance in growth between leaf and root can promote pithiness. The root water content decreases and cells become sponge-like (Photo 15.12), reducing the commercial value of the crop. The southern Chinese varietal group and European small radish tend to be susceptible. Genetic analysis revealed that pithiness is genetically controlled (Hagiya 1958).

15.2.3.3 Hollow Root

High soil temperatures, reduction of cytokinin biosynthesis, and activation of the loss of xylem parenchyma cells during the early to middle growth stages can cause the formation of a lysigenous intercellular space within the root and

Photo 15.8 Genetic differences in symptoms of striped root surface

Photo 15.9 Severe symptoms of lumpy root

Photo 15.10 Severe symptoms of internal browning

the accumulation of lignin in susceptible cultivars. As the root enlarges, the hollow remains because it cannot be filled with parenchyma cells (Photos 15.13, 15.14).

The Ishikawa Prefecture landrace 'Gensuke-daikon' in Japan is prone to this disorder (Kano 1989). Diallel analysis of the Kagoshima Prefecture landrace 'Sakurajima-daikon' in Japan showed small broad- and narrow-sense heritability, and the predominant influence of the environment (Tanaka et al. 2011).

15.3 Reproductive Growth

In vegetable breeding, characters relevant to seed production are evaluated and selected during the reproductive stage, after the selection of vegetative characters. Thus, radish breeding is very different from the breeding of crops grown for their seeds, such as rice, wheat, corn, and soybean. Therefore, the identification of selection markers for seed production characters can greatly improve selection efficiency. Objectives of F_1 breeding include seed yield, F_1 purity, and germination rate. So even if a cross has improved vegetative characters, it is not grown if its parents cannot stably produce large amounts of high-purity F_1 seeds with a high germination rate. This section discusses the current state of

Photo 15.11 Genetic differences in symptoms of internal browning

Photo 15.12 Severe symptoms of pithiness

breeding at the stages of bolting, flowering, pollination, fertilization, and seed set, and problems to overcome.

15.3.1 Flowering Habit

15.3.1.1 Bolting Characteristics

For crop production, late-bolting cultivars, which need longer treatment of seed vernalization, are sown in the colder months. Genetic differences in bolting time are very large (Photo 15.15). For example, when treated at the cotyledon stage for 21 days at 6 °C, all plants of the autumn-sowing Japanese landraces 'Minowase-daikon,' 'Nerima-daikon,' and 'Sakurajima-daikon' bolted, but the spring-sowing landrace 'Tokinashi-daikon' did not bolt (Cheon et al. 2004). Genetic studies of radish bolting (see Chap. 11) have built on a genetic network study of flowering in *Arabidopsis thaliana* (He and Amashino 2005). However, many bolting behaviors cannot be explained yet. For example, stages of bolting of four cultivars depended on sowing dates and locations in Japan (Table 15.1). The differences may be explained by an interaction between day length and temperature. So far, studies of low-temperature treatment under long-day conditions and devernalization by high temperature have been reported (Sagwansupyakorn et al. 1986; Cheon and Saito 2003), but studies of interactions among

Photo 15.13 Severe symptoms of hollow root

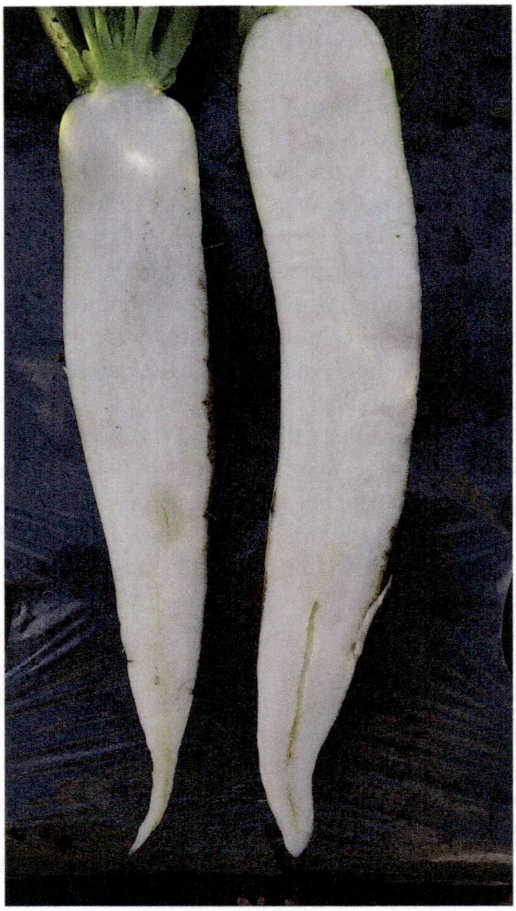

Photo 15.14 Difference in site of the development of hollow root. The hollow usually develops in the lower part of the root (*right*), but can develop in the midsection (*left*)

Photo 15.15 An Example of genetic difference in bolting. Both cultivars were sown on the same date, but that in the rear is already blooming before that in front has not bolted

environmental conditions and between environment and genotype are needed. Our experience of vernalization effect at the seed-setting stage on the same F_1 cultivar seeds produced using their parental lines with same genotypes at three different seed production areas on the same year suggests epigenetic effects (Fig. 15.2). In this experiment, these three lots were seeded and vernalized for 10 days at 5 °C in refrigerator and then cultivated in a common manner. After that, the bolting plants in each seed lot were counted. The seed lot produced in area A, Japan, showed the highest cumulative bolting percentage, 52% at 59 days after sowing. On the other hand, the

Table 15.1 Stages of bolting (from 0 = bud to 5 = full bloom) of four commercial Japanese radish cultivars differed between sowing dates and locations (Tochigi, 36°N, 140°E; Hokkaido, 42°N, 143°E) in 2014

	Region:	Tochigi	Hokkaido	Hokkaido
	Sowing date:	15 Mar	3 Jun	17 Jun
	Assessment date:	16 May	1 Aug	29 Aug
Cultivar	Days:	62	59	73
Natsu-tsukasa		5	2.5	0
Natsu-tsukasa Syun		1.9	1.5	0
Haruno-Ibuki		0.2	0.5	1
Yume-homare		0.1	2.5	1.5

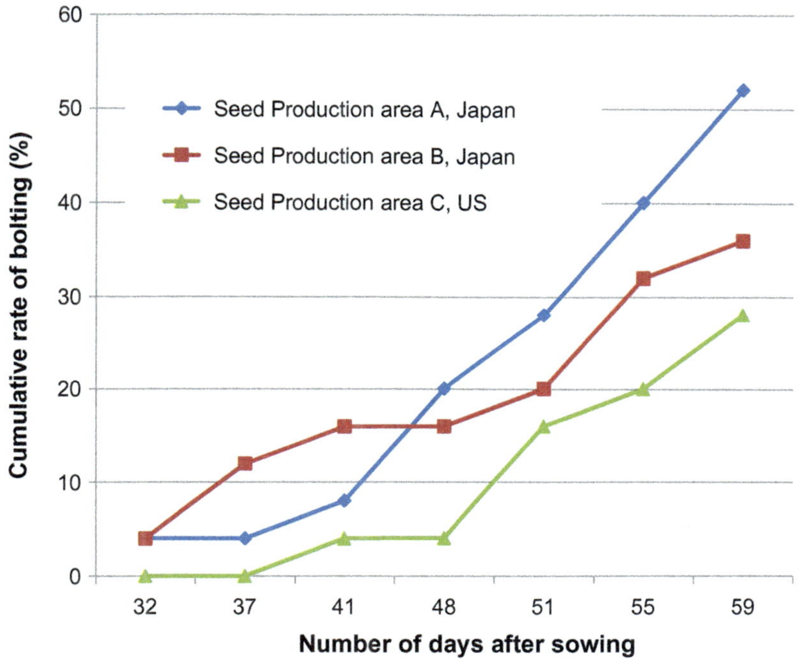

Fig. 15.2 Differences in progression of bolting of an F_1 cultivar among three seed production areas

seed lot produced in area C, US, showed the lowest percentage, 28%. These F_1 seeds have the same genotypes. The differences of bolting could be explained by the circumstances at seed-setting stage in each seed production area, suggesting epigenetic effects. As to the epigenetic effect, there is a report that the expression of vernalization-integral gene, *FLOWERING LOCUS C* (*FLC*) of *Arabidopsis thaliana* is controlled by various epigenetic regulations (Sung and Amasino 2004). When the mechanisms of bolting and their underlying genetics are understood, it will be possible to specify the sowing dates of the parental lines to ensure simultaneous flowering at any location and to breed the stable late-bolting cultivars.

15.3.1.2 Flowering Phenology

When the flowering dates of parental lines in the F_1 seed production field do not coincide (Photo 15.16), seed production and purity decline. Until recently, no handy method to quantify flowering phenology had been available. With the recent development of an image analysis method (Guo et al. 2013; Photos 15.17a–d), this is performed only by taking photographs and processing under the software. Using this method, we can know the exact number of parental flowers through their flowering time easily. And so, it is now possible to plan planting ratios and sowing times of the parental lines, and to perform genetic analysis of flowering phenology.

Photo 15.16 An Example of difference in flowering timing between one parental line of *the middle row* (*green*) and the other parental line of the rest flowering rows in the F_1 seed production field

15.3.2 Pollination and Fertilization

Pollinating insects are necessary in seed production fields. The most important pollinator is the honeybee, but the bees' strong floral selectivity poses problems for F_1 breeding.

15.3.2.1 Floral Morphology

Pollinators must move between the parental lines in F_1 seed production fields. But floral characteristics show large genetic differences (Sasaki and Takahashi 2002; Yoshioka et al. 2005), and honeybees use the size, guide marks, and scent of flowers to distinguish sources (Menzel et al. 1974; Photos 15.18, 15.19a–d), and favor one parental line over the other. For example, honeybees made few trips between a male-sterile radish line and the pollen parent (Kobayashi et al. 2010). The result is low seed yield.

Pollination efficiency also depends on whether floral morphology facilitates cross-pollination and inhibits self-pollination: the physical distance between stamen and stigma within a flower (Photo 15.20) is important for F_1 purity (Kobayashi et al. 2006). In addition, honeybees sometimes rob nectar from the base of the flower between sepal and petal (Photo 15.21), thereby failing to pollinate the flower. If characteristics that honeybees recognize can be identified, they will become selection targets for improving floral morphology.

15.3.2.2 Self-incompatibility (*S*) Haplotype

For successful pollination, the *S* (self-incompatibility) haplotypes of the parents must be different. The *S* haplotype can be identified from the degree of seed set or by observation of pollen tube growth of test cross flowers under a fluorescence microscope, but such evaluation must wait until flowering. The use of a PCR-RFLP method (Brace et al. 1993; Niikura and Matsuura 1998; Nishio et al. 1994), as described in Chap. 13, allows breeders to select the desired *S*-genotype during vegetative growth, contributing greatly to selection efficiency.

15.3.2.3 Strength of Self-incompatibility

The strength of self-incompatibility can be evaluated from the degree of seed set of flowers self-pollinated by hand. Even so, we have found that some parental lines selected by this method

Photo 15.17 Flowering stages of an inbred line: (**a**, **b**) full bloom, (**c**, **d**) end of flowering. *Black* and *white images* show flowers extracted from color images by Guo et al. (2013)

Photo 15.18 Genetic differences in radish flower characteristics

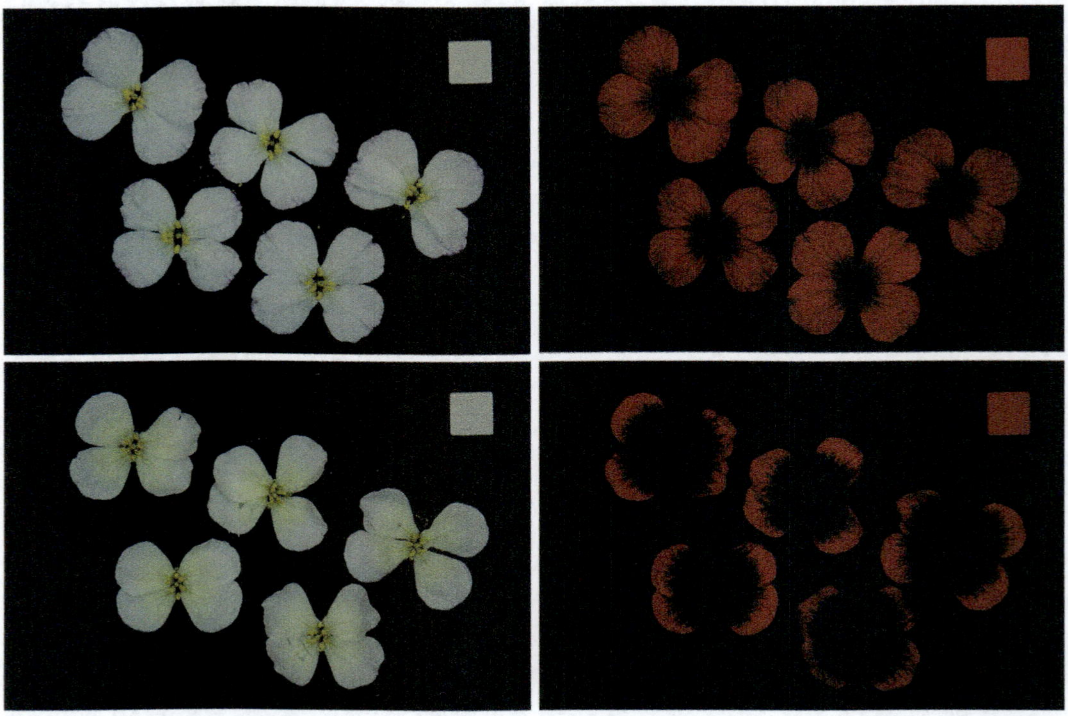

Photo 15.19 Flowers of two radish inbred lines (**a, c**) with different pollinator guide marks under ultraviolet light (**b**, **d**, courtesy of Prof. Ohsawa, Tsukuba Univ, Japan)

produced F_1 seeds with low purity. Genetic analysis of inbred lines selected by an insect pollination self-incompatibility test (Horisaki et al. 2003) suggested genetic factors independent of the *S* haplotype. QTLs with a similar effect were found by the same method in *Brassica rapa* (Hatakeyama et al. 2010), but QTL analysis in radish is needed.

Photo 15.20 An Example of genetic differences in floral characters of radish

Photo 15.21 A Honeybee robs nectar from the base of a radish flower

15.3.2.4 Response to CO_2

Genetic differences among radish cultivars are evident in seed number per pod (Photo 15.22), but seed number is much smaller than those in other *Brassica* crops. This makes the production of parental seeds of the F_1 cultivars by artificial self-pollination inefficient. So, to break self-incompatibility in order to produce parental seeds, CO_2 treatment is used (Nakanishi et al. 1969). Evaluation of genetic differences in response to CO_2 revealed that the reaction is controlled by genetic factors that are independent of the *S* haplotype in the stigma (Niikura and Matsuura 2000). Because the response to CO_2 tends to be stronger when self-incompatibility is lower, these characteristics must be separated genetically, so the underlying genes must be isolated.

15.3.2.5 Male Sterility and Fertility Restorer Genes

Studies of male-sterile genes in Ogura-type and Kozena-type cytoplasm and the fertility restorer gene (Bonhomme et al. 1992; Koizuka et al. 2003) are discussed in Chap. 7. The use of male sterility to breed F_1 cultivars has increased because of high F_1 seed purity (Akamatsu 2005). In the USA, excessive dependence on the T-type male-sterile cytoplasm in corn caused extensive losses due to southern leaf blight. The current situation in radish, which relies on just a few male-sterile cytoplasm, seems to be similar to that in corn. We need to develop new male-sterile cytoplasm by using interspecific crosses among Brassicaceae. In addition, a study of how

Photo 15.22 Genetic differences among radish pods

Ogura-type cytoplasm has little pleiotropic effect on floral characters other than the stamen without limiting the attraction for pollinator insects is needed.

15.3.3 Seed Production

15.3.3.1 Seed Yield

Stable production of a large amount of seeds is necessary in F_1 breeding. Yield components include the number of flowers, the rate of seed set, seed number per pod, 1000-seed weight, and yield per plant. Visit of pollinators from the pollen parent is critical to the rate of seed set.

15.3.3.2 Testa Color

In *Arabidopsis*, testa color is associated with seed dormancy, longevity, and germination (Debeaujon et al. 2000). Testa color varies among radish cultivars (Photo 15.23), so it will be necessary to investigate its relation to germination of radish seed.

15.3.3.3 Germination Ability

Because radish seed is sown directly in the field, high temperatures or heavy rain at sowing can greatly influence germination. In addition, increased agricultural mechanization and labor consolidation will create more demand from farmers for robust seed with reliable germination ability under variable conditions. The study of germination ability under abiotic stress conditions will also be necessary.

15.3.3.4 Seed Longevity

Seed longevity is associated with testa color in *Arabidopsis* (Debeaujon et al. 2000). QTLs controlling seed longevity have been found in rice (Miura et al. 2002). As seed longevity is thus hereditary, its genetic control in radish will need further study.

15.4 Future Prospects for Radish Breeding

Radish breeding will benefit from genome editing. Examples of improvements include the development of self-compatibility for easier parental seed production, incorporating disease resistance, modification of root skin and flesh color, and inactivation of the fertility restorer gene to reduce the time it takes to create.

Good root growth depends on the correct balance of many breeding characters. The

Photo 15.23 Genetic differences in size, color, and shape of radish seeds

acquisition of time-series genomic data necessary for future breeding will depend on a molecular phenology approach (Kudoh 2015) and on practical genomic selection (Meuwissen et al. 2001). Comparison of time-series data with actual vegetative growth will allow breeders to ascertain how to incorporate each character into a cultivar without destroying the existing balance.

History shows that the spread of F_1 cultivars accelerates the decline of landraces and the associated valuable cultural heritage. Yet the 'Sakurajima-daikon', 'Moriguchi-daikon,' 'Akita,' and 'Shinshu-ji' landraces are recently being F_1 cultivars in Japan, mainly because uneven production of landrace crops in the field reduces the crop value and improves the stability of seed production. By thus improving the stability of harvest, the farmers aim at improving their profits. Attention to seed production characters is therefore also necessary. However, because the genetic diversity between the parents of each F_1 cultivar of the landrace is small, the development of DNA markers requires high precision. Genome information will also be necessary for calculating the number of plants needed to maintain the genetic diversity within each landrace.

Molecular biology has made remarkable technical advances. However, seed companies cannot develop it on their own, so the contribution of many researchers is needed. On the other hand, seed companies are good at in raising materials for analysis, and breeders are best placed to conduct breeding. We envisage a consortium between researchers, breeders, and seed companies to develop the technology of breeding.

References

Akamatsu T (2005) Utilization and problem of male sterility in vegetable crops of Cruciferae. Breed Res 7 (Suppl 1, 2):42–43

Ashizawa M, Hida K, Yoshikawa H (1979) Studies on the breeding *Fusarium* resistance in radish. I. Screening of radish varieties for *Fusarium* resistance. Bull Veg Ornam Crops Res Sta 6:39

Bonhomme S, Budar F, Lancelin D, Small I, Defrance MC, Pelletier G (1992) Sequence and transcript analysis of the *Nco2.5* Ogura specific fragment correlated with cytoplasmic male sterility in *Brassica* cybrids. Mol Gen Genet 235:340–348

Brace J, Ockendon DJ, King GJ (1993) Development of a method for the identification of *S*-alleles on *Brassica oleracea* based on digestion of PCR-amplified DNA with restriction endonucleases. Sex Plant Reprod 6:133–138

Cheon CJ, Saito T (2003) An approach to the characterization of devernalization due to high temperature interruption during vernalization in radish plants using "Flower formation index". Environ Control Biol 41:353–359

Cheon CJ, Watanabe Y, Saito T (2004) An approach to the characterization of varietal difference in inflorescence formation due to vernalization in radish plants using "Flower formation index". Environ Control Biol 42:305–314

Debeaujon I, Leon-Kloosterziel KM, Koornneef M (2000) Influence of the testa on seed dormancy, germination and longevity in Arabidopsis. Plant Physiol 122:403–413

Fukuoka N, Enomoto T (2001) The occurrence of internal browning induced by high soil temperature treatment and its physiological function in *Raphanus* root. Plant Sci 161:117–124

Fukuoka N, Ikeshita Y, Enomoto T (2010) Relationship between the occurrence of internal browning and size of xylem parenchymatous cells in roots of Japanese radish. J Jpn Soc Hortic Sci 79(1):27–33

Guo W, Rage UK, Ninomiya S (2013) Illumination invariant segmentation of vegetation for time series wheat images based on decision tree model. Comput Electron Agric 96:58–66

Hagiya K (1958) Studies on the occurrence of 'pithy tissue' in root crops (5). On the occurrence of the 'pithy tissue' in F_1 and F_2 generations of varietal crosses in radish, and in contemplation of breeding the resistant lines to its occurrence. J Jpn Soc Hortic Sci 27:68

Hatakeyama K, Horisaki A, Niikura S, Narusaka Y, Abe H, Yoshiaki H, Ishida M, Fukuoka H, Matsumoto S (2010) Mapping of quantitative trait loci for high level of self-incompatibility in *Brassica rapa* L. Genome 53(4):257–265

He Y, Amashino RM (2005) Role of chromatin modification on flowering-time control. Trends Plant Sci 10:30–35

w?>Horisaki A, Tanaka N, Niikura S (2003) The effectiveness of insect-pollination test to evaluate the level of self-incompatibility and their genetic analysis in radish (*Raphanus sativus* L.). Theor Appl Genet 107:1009–1013

Iwata H, Niikura S, Matsuura S, Takano Y, Ukai Y (1998) Evaluation of variation of root shape of Japanese radish (*Raphanus sativus* L.) based on image analysis using elliptic Fourier descriptors. Euphytica 102:143–149

Iwata H, Niikura S, Matsuura S, Takano Y, Ukai Y (2000) Diallel analysis of root shape of Japanese radish (*Raphanus sativus* L.) based on elliptic Fourier descriptors. Breed Sci 50:73–80

Iwata H, Niikura S, Matsuura S, Takano Y, Ukai Y (2004) Genetic control of root shape at different growth stages in radish (*Raphanus sativus* L.). Breed Sci 54:117–124

Kano Y (1989) Effects of time of high and low temperature treatments on the growth of Japanese radish cv. 'Gensuke' and the occurrence of hollow root. J Jpn Soc Hortic Sci 57:626–632

Kobayashi K, Horisaki H, Niikura S, Ohsawa R (2006) Inter-accession variation in floral morphology in radish (*Raphanus sativus* L.). Euphytica 152:87–97

Kobayashi K, Tsukamoto S, Tanaka A, Niikura S, Ohsawa R (2010) Selective flower visitation behavior by pollinators in a radish F_1 seed production field. Breed Sci 60:203–211

Koizuka N, Imai R, Hayakawa T, Kimura Y, Kohno-Murase J, Sakai T, Kawasaki S, Imamura J (2003) Genetic characterization of a pentatricopeptide repeat protein gene, orf687, that restores fertility in the cytoplasmic male sterile Kosena radish. Plant J 34:407–415

Kudoh H (2015) Molecular phenology in plants: in natura systems biology for the comprehensive understanding of seasonal responses under natural environments. New Phytol. doi:10.1111/nph.13733

Menzel R, Erber J, Masuhr T (1974) Learning and memory in the honey bee. In: Browne LB (ed) Experimental analysis of insect behavior. Springer, New York, pp 195–217

Meuwissen TH, Hayes BJ, Goddard ME (2001) Prediction of total genetic value using genome-wide dense marker maps. Genetics 157:1819–1829

Miura K, Lin SY, Yano M, Nagamine T (2002) Mapping quantitative trait loci controlling seed longevity in rice (*Oryza sativa* L.). Theor Appl Genet 104:981–986

Nakanishi T, Esashi Y, Hinata K (1969) Control of self-incompatibility by CO_2 gas in *Brassica*. Plant Cell Physiol 10:925–927

Niikura S, Matsuura S (1998) Identification of self-incompatibility alleles (*S*) by PCR-RFLP in radish (*Raphanus sativus* L.). Euphytica 102:379–384

Niikura S, Matsuura S (2000) Genetic analysis of the reaction level of self-incompatibility to a 4% CO_2 gas treatment in the radish (*Raphanus sativus* L.). Theor Appl Genetics 101:1189–1193

Nishio T, Sakamoto K, Yamaguchi J (1994) PCR-RFLP of the *S*-locus for the identification of breeding lines in cruciferous vegetables. Plant Cell Rep 13:546–550

Sagwansupyakorn C, Shinohara Y, Saito Y (1986) Effects of light intensity and temperature on devernalization of Japanese radish. J Jpn Soc Hortic Sci 55:56–61

Sasaki K, Takahashi T (2002) A flavonoid from *Brassica rapa* flower as the UV-absorbing nectar guide. Phytochemistry 61:339–343

Suga S, Maeshima M (2004) Water channel activity of radish plasma membrane aquaporins heterologously expressed in yeast and their modification by site-directed mutagenesis. Plant Cell Physiol 45:823–830

Sung S, Amasino RM (2004) Vernalization in *Arabidopsis thaliana* is mediated by the PHD finger protein VIN3. Nature 427:159–164

Takano J, Noguchi K, Yasumori M, Kobayashi M, Gajdos Z, Miwa K, Hayashi H, Yoneyama T, Fujiwara T (2002) Arabidopsis boron transporter for xylem loading. Nature 420:337–340

Tanaka Y, Kuwaduru N, Nagata S (2011) Diallel analysis of root weight and hollow root in Japanese radish '*Sakurajima daikon*'. Hortic Res (Jpn) 10(1):9–13

USDA (2009) http://gain.fas.usda.gov/Recent%20GAIN%20Publications/Cabbage%20Market%20Trends_Seoul_Korea%20-%20Republic%20of_11-5-2010. Accessed 28 Dec 2015

Wellmann FL (1939) A technique for studying host resistance and pathogenicity in tomato Fusarium wilt. Phytopathology 29:945–956

Yoshioka Y, Horisaki A, Kobayashi K, Syafaruddin Niikura S, Ninomiya S, Ohsawa R (2005) Intraspecific variation in the ultraviolet colour proportion of flowers in *Brassica rapa* L. Plant Breed 124:551–556

Index

T. Nishio and H. Kitashiba (eds.), *The Radish Genome*,
Compendium of Plant Genomes, DOI 10.1007/978-3-319-59253-4

Printed by Printforce, the Netherlands